# Lecture Notes in Economics and Mathematical Systems 619

Luca Bertazzi · M. Grazia Speranza
Jo A. E. E. van Nunen (Eds.)

# Innovations in Distribution Logistics

 Springer

Professor Luca Bertazzi
Professor M. Grazia Speranza
Università degli Studi di Brescia
Dipartimento Metodi Quantitativi
Contrada da S. Chiara 50
25122 Brescia
Italy
bertazzi@eco.unibs.it
speranza@eco.unibs.it

Professor Jo. A. E. E. van Nunen
Erasmus University
Rotterdam School of Management
Department of Decision
and Information Sciences
Burg. Oudlaan 50, Roomnr. T09-30
3062 PA Rotterdam
The Netherlands
jnunen@rsm.nl

ISBN 978-3-540-92943-7     e-ISBN 978-3-540-92944-4
DOI 10.1007/978-3-540-92944-4

Lecture Notes in Economics and Mathematical Systems ISSN 0075-8442

Library of Congress Control Number: 2008944011

© 2009 Springer-Verlag Berlin Heidelberg

*Cover design*: SPi Publishing Services, India

Printed on acid-free paper

9 8 7 6 5 4 3 2 1

springer.com

# Preface

The globalization of markets and the availability of new information technologies have stimulated research in distribution logistics in the last few years. Classical transportation, routing and inventory problems are being more and more studied from a supply chain management perspective and the on-line information made available by the new technologies is more and more used in the decision processes.

A scientific approach to the most innovative logistic problems, combined with a high level of attention towards practice, has motivated a series of "International workshops in distribution logistics (IWDL)". This volume includes some of the papers that have been presented at the IWDL2006, the workshop of the IWDL series which was in Brescia (Italy) in October 2006.

In this volume we focus on the distribution network, which represents the most complex and critical part of the logistic system, and cover several areas, in particular distribution network design, supply chain optimization, internal logistics, routing problems, transportation and e-business, and location problems.

Due to the variety of the topics in the domain of distribution logistics covered by this volume, we decided to present the contributions in alphabetical order of the name of the first author. The reader will find through the index the topic he/she is most interested in. This volume is thought to be an important reference for researchers and also for PhD and Master students.

We would like to thank the authors of the papers for their contributions. All the papers submitted for publication in this volume were been subjected to a refereeing process and we are deeply indebted to the referees whose professional help has been fundamental to ensure a high quality level of this volume.

Brescia, Rotterdam                                    *Luca Bertazzi, University of Brescia, Italy*
February 2009                                  *M.Grazia Speranza, University of Brescia, Italy*
                                         *Jo van Nunen, Erasmus University, The Netherlands*

# Contents

# List of Contributors

**Enrico Angelelli** University of Brescia, Italy,
angele@eco.unibs.it

**Luca Bertazzi** University of Brescia, Italy,
bertazzi@eco.unibs.it

**Nicola Bianchessi** University of Brescia, Italy,
bianche@eco.unibs.it

**Thomas Bieding** Vodafone D2 GmbH, Germany,
thomas.bieding@vodafone.com

**Marielle Christiansen** Norwegian University of Science and Technology, Norway,
marielle.christiansen@iot.ntnu.no

**Teodor Gabriel Crainic** University of Quebec in Montreal, Canada,
theo@crt.umontreal.ca

**René de Koster** RSM Erasmus University, Netherlands,
rkoster@rsm.nl

**Laura Di Giacomo** LIX, École Polytechnique, France,
digiacomo@lix.polytechnique.fr

**Ettore Di Lena** Ente Nazionale Energia Elettrica, Italy,
ettore.dilena@enel.it

**Kjetil Fagerholt** Norwegian University of Science and Technology, Norway,
kjetil.fagerholt@iot.ntnu.no

**Ricardo Giesen** Pontificia Universidad Católica de Chile, Chile,
Giesen@ing.puc.cl

**Davide Giglio** University of Genova, Italy,
davide.giglio@unige.it

**Yeming Gong** RSM Erasmus University, The Netherlands,
ygong@rsm.nl

**Simon Görtz** University of Wuppertal, Germany,
simon.goertz@wiwi.uni-wuppertal.de

**Roar Grønhaug** Norwegian University of Science and Technology, Norway,
roar.gronhaug@iot.ntnu.no

**Gianfranco Guastaroba** University
of Brescia, Italy,
guastaro@eco.unibs.it

**Patrick Jaillet** Massachusetts Institute
of Technology, USA,
Jaillet@MIT.edu

**Andreas Klose** University of Aarhus,
Denmark,
aklose@imf.au.dk

**Jarl Eirik Korsvik** Norwegian
University of Science and
Technology, Norway,
korsvik@ntnu.no

**Gilbert Laporte** HEC Montréal,
Canada,
gilbert@crt.umontreal.ca

**Federico Liberatore** University
of Milan, Italy,
fliberatore@crema.unimi.it

**Arne Løkketangen** Molde University
College, Norway,
Arne.Lokketangen@hiMolde.no

**François V. Louveaux** Facultés
Universitaires Notre-Dame de la Paix,
Belgium,
francois.louveaux@fundp.ac.be

**Hani S. Mahmassani** Northwestern
University, USA,
Masmah@northwestern.edu

**Renata Mansini** University of Brescia,
Italy,
rmansini@ing.unibs.it

**Riccardo Minciardi** University of
Genova, Italy,
riccardo.minciardi@unige.it

**Giacomo Patrizi** La Sapienza,
Università di Roma, Italy,
g.patrizi@caspur.it

**Livia Pomaranzi** Chefaro Pharma
Italia, Italy,
livia.pomaranzi@chefaro.it

**Giovanni Righini** University of Milan,
Italy,
righini@dti.unimi.it

**Simona Sacone** University of Genova,
Italy,
simona.sacone@unige.it

**Matteo Salani** École Politechnique
Fédérale de Lausanne, Switzerland,
matteo.salani@epfl.ch

**Frédéric Semet** LAMIH-ROI,
Université de Valenciennes et du
Hainaut-Cambrésis, France, and
Université de Montréal, Canada,
frederic.semet@univ-valenciennes.fr

**Federico Sensi** Enprovia Software
Engineering, Italy,
Federico-Sensi@libero.it

**Silvia Siri** University of Genova, Italy,
silvia.siri@unige.it

**Maria Grazia Speranza** University
of Brescia, Italy,
speranza@eco.unibs.it

**Arnaud Thirion** Facultés Universitaires
Notre-Dame de la Paix, Belgium,
thirionarnaud@hotmail.com

**Barbara Tocchella** University
of Brescia, Italy,
barbara.tocchella@ing.unibs.it

# Management Policies in a Dynamic Multi-Period Routing Problem

Enrico Angelelli[1], Nicola Bianchessi[2], Renata Mansini[3], and M. Grazia Speranza[4]

[1] Dipartimento Metodi Quantitativi, University of Brescia, Italy angele@eco.unibs.it
[2] Dipartimento Metodi Quantitativi, University of Brescia, Italy bianche@eco.unibs.it
[3] Dipartimento di Elettronica per l'Automazione, University of Brescia, Italy
   rmansini@ing.unibs.it
[4] Dipartimento Metodi Quantitativi, University of Brescia, Italy speranza@eco.unibs.it

**Summary.** In this paper we analyze the Dynamic Multi-Period Routing Problem (DMPRP), where a fleet of uncapacitated vehicles has to satisfy customers' pick-up requests. The service of each customer can take place the day the request is issued or the day after. At the beginning of a day a set of requests are already known and have to be served during the day. Additional requests may arrive during the day while the vehicles are traveling. In this context we perform different types of analysis, each one characterized by the comparison of alternative management policies. The first analysis compares a policy which decides, at the time the request is issued, whether to accept or reject it to a policy that accepts all the requests and decides, at a later time, which ones to forward to a back-up service company. The second evaluates the advantages of a collaborative service policy where a fleet of vehicles is managed by a unique decision maker with respect to a policy where the same vehicles are managed independently. Finally, in the last analysis a policy where each new request is taken into account as soon as it is issued is compared to a policy where all the requests issued during a day are analyzed at the end of the day. Extensive computational results evaluating the number of lost requests and the distance traveled provide interesting insights.

**Key words:** Dynamic multi-period routing problems, Postponable requests, Management policies

## 1 Introduction

Dynamic settings are receiving an increasing attention in routing problems thanks also to a wider use of communication devices in vehicles equipment. Nowadays, the use of GPS systems allows a central unit to constantly know the location of vehicles

L. Bertazzi et al. (eds.), *Innovations in Distribution Logistics*, Lecture Notes
in Economics and Mathematical Systems 619, DOI: 10.1007/978-3-540-92944-4,

and to take dynamic decisions on the basis of the overall situation of vehicles and customers. Such situation evolves during the day and previous plans may be modified because new requests are issued by customers or because some unexpected event took place such as a delay due to traffic congestion. It is expected that such dynamic management improves the competitiveness of a company, allowing a better service at a lower cost.

While the literature on static routing problems is wide, the literature on dynamic routing problems is limited, though it has consistently grown in the last years. Comprehensive surveys on dynamic problems can be found in Psaraftis [10, 11] and, more recently, in Ghiani et al. [5]. Among the most relevant contributions in this domain we recall Savelsbergh and Sol [12] and Yang et al. [13] for the management of dynamic fleet of vehicles; Gendreau et al. [4] and Ichoua et al. [6] for real-time vehicle routing and dispatching problems in long-distance courier services; Mitrović-Minić et al. [8] and Mitrović-Minić and Laporte [9] for the dynamic pick-up and delivery problem with time windows and Madsen et al. [7] for a dynamic dial-a-ride system characterized by multiple capacities and multiple objectives. Finally, Angelelli et al. [2, 3] perform a competitive analysis for some policies in a simple dynamic multi-period setting.

The dynamic setting we consider in this paper is the Dynamic Multi-Period Routing Problem (DMPRP) introduced in [1]. The problem is characterized by pick-up requests arriving in real time to the central depot of a courier company. A fleet of uncapacitated vehicles is available for the service. Every morning these vehicles leave the depot and have to return to the depot at the end of the day. Thanks to modern communication technology, the company knows the exact position of its vehicles at any time instant and is able to forecast their positions in the near future. The company can react to on-line requests and possibly modify the previous traveling plans. The distinctive features of this dynamic problem with respect to those analyzed in the literature is that requests can be served within two days from their issuance. This means that when a request is issued, a deadline is associated to it. Thus, every day, the requests can be either *off-line* when they are known in advance (i.e. they have been issued the day before but not serviced yet) or *on-line* when they come over in real-time, while the vehicles are traveling. Moreover, requests are also classified as *postponable* or *unpostponable*. If a request is unpostponable it has to be inserted in the currently traveled routes, on the contrary if it is postponable it can be served either today or tomorrow. The objective of the company is to maximize the number of serviced requests while minimizing the average operational cost per day.

The most common approach used in the literature to solve a dynamic problem is based on a repeated re-optimization of the off-line problem. In [1], the authors introduce different *short term strategies* characterized by a look-ahead period and a short term objective. For each strategy, a corresponding re-optimization problem is defined and iteratively solved by means of a Variable Neighborhood Search (VNS) meta-heuristic. An extensive computational analysis of the impact each short term strategy has on the long term objective of the problem is provided.

In the present paper, we evaluate and compare alternative management policies, by making use of the solution framework presented in [1]. The scope of the paper is to focus on the possible advantages that a company can achieve by applying some management policies concerning, for instance, the treatment of new customers and of on-line information. In this perspective we perform three different types of analysis, each one characterized by two alternative management policies. In the first we compare the management policy on the basis of which an immediate decision is taken, at the time a request is issued, on whether to accept or reject it to the policy which accepts all the requests and, at a later time, forwards some of the requests, if needed, to a back-up service company. Then, we study the advantages of a collaborative service policy where a fleet of vehicles is managed by a unique decision maker with respect to a less flexible policy where the same vehicles are managed independently. Finally, the third analysis compares a policy where each new request is taken into account as soon as it is issued to a policy where all the requests issued during a day are analyzed at the end of the day. Comparison is made by means of extensive computational results evaluating the number of lost requests and the distance traveled.

## 2 The Dynamic Multi-Period Problem

A fleet of uncapacitated vehicles $V = \{v_1, \ldots, v_m\}$ is available to satisfy requests issued by customers. The positions of the vehicles are known to the central depot at any time during the day. Moreover, the vehicles can communicate with the central depot. At the beginning of each day a set of requests are known that have to be served during the day (*unpostponable* requests). These requests are assigned to the vehicles and the vehicles leave the depot and start traveling on the basis of an initial plan. During the day new requests may be issued by customers. Unpostponable requests can be accepted only until a fixed time $L$ in the morning (e.g. noon or 1:00 PM). We define as *postponable* all the requests issued during the day that can be served in the same day or postponed to the day after. The time length of each working day is equal to $\tau$. This is also the maximum time available to each vehicle route, i.e. we will refer to the length of a route by meaning a time length. Decisions are repeated over a time horizon of $T$ days.

All requests are requests of a pick-up service. In fact, it is assumed that delivery requests are not consistent with this dynamic setting, since if a delivery request is issued during the day, then a vehicle cannot be deviated to serve the new customer. Moreover, if a vehicle leaves the depot with the load to be delivered to a customer, the service of that customer cannot later be assigned to a different vehicle. In case the company satisfies both pick-up and delivery requests, the assumption is that the fleet is divided into two parts, a part dedicated to the delivery service and the other part dedicated to the pick-up service. The part dedicated to the delivery service works as traditionally in a static context where the vehicles follow during the day the plan assigned to them at the beginning of the day. The part dedicated to the pick-up service is managed dynamically.

The central depot may elaborate new plans during the day and communicates the changes to the vehicles. The changes in a vehicle plan may concern the inclusion of new customers, the deletion of customers or both. The vehicle can receive the new plans at any time and possibly deviate from its previous route while traveling between two customers. The goal is the minimization of the total service cost over the whole horizon. Such major target has been formalized through two hierarchical objectives. The first one is the maximization of the number of requests directly served by the company, which is equivalent to the minimization of the number of not served requests, i.e. rejected or forwarded to the backup service, depending on the policy. The second one is the minimization of the length of the routes traveled per serviced request.

## 2.1 The Solution Framework

In [1] the authors introduced in a rolling horizon solution framework the concept of a *Short Term Strategy* (STS). A STS includes the definition of a re-optimization problem that is solved by means of a Variable Neighborhood Search (VNS) heuristic. Before the beginning of the day and then at regular intervals (re-optimizations intervals) the re-optimization problem is solved. The first re-optimization problem considers unpostponable requests only and provides for each vehicle a route that starts and ends at the depot. The subsequent re-optimization problems take into account all known requests (postponable and unpostponable) and provide for each vehicle a route that starts at the forecasted position of the vehicle at the end of the re-optimization according to the previously planned routes and ends at the depot.

Time is denoted during the day with $t \in [0, \tau]$. We indicate by $R_P(t)$ and $R_U(t)$ the set of postponable and unpostponable requests at a given time $t$, respectively. We also denote by $R(t) = R_P(t) \cup R_U(t)$ the total set of the requests known at time $t$. Let $\Delta t$ be the length of the re-optimization interval and let $t' = t + \Delta t$. The set $R(t')$ differs from $R(t)$ for the inclusion of all the new requests which have become available during the last re-optimization interval $\Delta t$ and for the elimination of all the requests served in the meantime.

A maximum time $OptTime \leq \Delta t$ is made available to the algorithm that solves each re-optimization problem. The solution found is implemented until the end of the next re-optimization phase. The generated routes are followed by the vehicles from time $t + OptTime$ to time $t' + OptTime$, that is until the routes obtained with the subsequent re-optimization have become available.

## 2.2 The Short Term Strategy

In [1] several *Short Term Strategies* have been analyzed and compared. A *Short Term Strategy* (STS) consists of the following main components:

1. A *look-ahead period*: The period of time over which the re-optimization problem is defined;
2. A *short term objective*: The criterion used to evaluate the quality of a solution in the re-optimization problem;
3. A *re-optimization problem*: The off-line problem which is formulated and solved, after a look-ahead period and a short term objective have been defined;

4. *A re-optimization interval*: The length of the time interval between the solution of two consecutive re-optimization problems.

Let $r_P^1$ and $r_U^1$ represent the number of postponable and unpostponable requests served today, respectively. Moreover, let $r^1$ and $l^1$ denote the total number of served requests and the total length of the routes traveled in the current day, respectively. Let $r^2$ and $l^2$ denote the number of served requests and the total length of the routes traveled the day after, respectively. The class of strategies that in [1] turned out to be the most successful has a *2-day look-ahead* period, and the following objective function:

$$min \quad \alpha l^1 + (1 - \alpha)l^2 + (r_P^1 + r^2)K_2 + r_U^1 K_1 \tag{1}$$

where $\alpha$ is a real number such that $0 \leq \alpha \leq 1$, and $K_1$, $K_2$ are negative constant values such that $K_1 \ll K_2 \ll 0$. The function maximizes the number of unpostponable served requests (term $r_U^1 K_1$) and, as second hierarchical objective, maximizes the total number of postponable requests to be served within the day after (term $(r_P^1 + r^2)K_2$). Actually, the requests that are postponable today will become unpostponable tomorrow and have to be served within tomorrow. Finally, the third hierarchical objective is the minimization of the weighted sum of the lengths of the routes traveled today and tomorrow (term $\alpha l^1 + (1 - \alpha)l^2$). We decided to set $\alpha = 1^-$ so that a decrease in the distance traveled today is to be preferred to any decrease in the distance traveled tomorrow.

Throughout the paper we indicate by *2-day look-ahead($\Delta t$)* the *2-day look-ahead* strategy with the short term objective function (1) and $\alpha$ set to $1^-$. Each of the analyzed management policies, with one exception only, can be implemented by means of a straightforward implementation of the *2-day look-ahead($\Delta t$)* strategy.

# 3 Comparison of Alternative Management Policies

In the following we consider three real situations where different management policies can be implemented. We analyze and compare different policies by making use of a proper implementation of the *2-day look-ahead($\Delta t$)* strategy.

## 3.1 Accept/reject vs. Delay Policy

When a new request is issued, the central unit of a company can immediately check whether it is possible to serve it given the available fleet of vehicles. The decision has to be taken on the basis of the already accepted requests possibly modifying the current routing plan. The company accepts to serve the request if it can be feasibly inserted in the current plan. Otherwise the company rejects the request. We call this policy *accept/reject*. Since we assume that there are requests that can be served in the day of their issuance or postponed to the day after, a more customer oriented version of the *accept/reject* policy is of interest: an *accept/reject* policy where the day in which the service will be accomplished is fixed once for all and is made known to the customer as soon as his/her request is accepted (*fixed day accept/reject* policy).

In the same situation, the company may follow an alternative policy where no request is rejected. Since the fleet may not be sufficient to satisfy all the issued requests, a contract with a back-up service company has to be made in such a way that, in case of need, some of the customers will be served by the back-up company. In this case, the back-up company is informed on the customers to be served at the end of the morning. This policy, called *delay*, allows the company to postpone the decision about the customers to serve directly and the customers to be served by the back-up company. At the same time the company offers a better service, since no customer is ever rejected.

The *delay* policy can be implemented by means of the *2-day look-ahead($\Delta t$)* with any fixed value of the parameter $\Delta t$. On the contrary, to implement the *accept/reject* policy the $\Delta t$ characterizing the strategy should be small enough to allow the definition of a new re-optimization problem as soon as a new request is issued. Moreover, the solution framework itself has to be modified so that if a request is accepted it will be certainly served and will be never excluded by future re-optimizations. Otherwise, if an arriving request is not immediately included in the current solution it will be definitely discarded from the system.

### 3.2 Collaborative vs. Individual Transportation Policy

Traditionally, transportation companies have focused their attention on controlling and reducing their own costs to increase profitability. More recently, companies have started to explore the possibility to share information with other companies and to develop common transportation plans with further reduction of costs. A collaborative transportation policy might open up cost saving opportunities that are impossible to achieve with an internal company policy.

We compare a *collaborative transportation* policy where the route plans of a fleet of vehicles are designed by a unique decision maker who brings together all customer requests to a policy where the same vehicles are managed independently (*individual transportation* policy). The situation refers to the real case where the service provided by a company with a large fleet of vehicles is compared to that provided by different smaller companies which globally own the same number of vehicles but whose route plans are managed independently. In both cases we consider a *delay* policy.

### 3.3 Dynamic vs. Static Policy

A dynamic policy is attractive because it may reduce operational costs and guarantee a better service level. However, it also implies additional costs due to the devices needed for the communication and a different organization. Do the benefits compensate the costs? We compare a dynamic policy to a policy where the vehicles follow the route plans made available at the beginning of the day and based on the requests that arrived the day before. Since for this policy all requests issued during a day are analyzed only at its end we call it *static* policy. In practice, this policy may occur in those companies where vehicles do not have a communication system equipment which makes possible for the central unit to know their location at each time instant and to dynamically change their route plans. The *static* policy can be implemented by setting the re-optimization interval $\Delta t$ equal to the working time $\tau$. Whereas the *dynamic* policy can be implemented by means of the *2-day look-ahead($\Delta t$)* strategy with any value of the parameter $\Delta t$.

## 4 Computational Results

### 4.1 Testing Environment

The computational analysis has been carried out on *random scenarios*, where the requests are uniformly distributed over a service area of $100 \times 100\ km^2$. Each scenario is characterized by a planning horizon of $T = 5$ days and a daily working time of $\tau = 10$ h (from 8:00 AM to 6:00 PM). A scenario is also characterized by a parameter $\lambda$ of a Poisson distribution according to which the requests are dynamically generated. The parameter $\lambda$ is the mean arrival rate of requests per day and is assumed to take one of the following values: $100, 200, 300, 400, 500$. Since a scenario can be characterized by the presence of unpostponable requests, we have assumed that, with a probability equal to $\frac{1}{3}$, the requests issued before 1:00 PM are unpostponable. The service is provided by means of a fleet of three uncapacitated vehicles, each of them traveling at a constant speed of $25\ km\,h^{-1}$. In order to make all results comparable, an additional day is considered to complete the work of the not yet served requests. In fact, if not so, an improper advantage might be obtained by postponing as many requests as possible from day $T$ to day $T + 1$.

Given a scenario, we generate five instances which differ for the number of the daily requests and for their geographical location. In particular, each request coordinates are randomly selected among the customer coordinates in the sets r1 and r2 of the Solomon's instances for the Vehicle Routing Problem with Time Windows. In all the instances, the coordinates of the depot are those of the Solomon's ones. Moreover, requests are listed for each day in increasing order of their release time.

In the following computational experiments some features of this general data setting have been modified according to the type of policies taken into account. For instance, the generated scenarios may not have unpostponable requests. The number of not served requests and the distance traveled are evaluated for each policy in order to discuss the advantages of their application.

All computational experiments have been carried out on a 1.5GHz Intel Pentium IV machine with 512MB of RAM.

### 4.2 Re-Optimization Time Interval Influence

Before analyzing the different policies behavior we have conducted some preliminary experiments testing alternative *2-day look-ahead* strategies differing for the value of the re-optimization interval $\Delta t$ (respectively, set to 30, 150, 300, 600, 900, 1,800, 3,600). We aim at possibly identifying the best re-optimization time.

To implement the short term strategies we have set the parameter *OptTime* representing the re-optimization time given to the VNS meta-heuristic equal to $max\{\frac{1}{12}\Delta t, \ 30 \ s\}$. In [1] the re-optimization time was set to $\frac{1}{12}\Delta t$ and various values of the re-optimization interval $\Delta t$ were tested. However, as the minimum tested value was $\Delta t = 3,600$ s, the value of the re-optimization time turned out to take reasonable values, never smaller than 300 s. In the experiments we discuss here the tested values of the re-optimization interval are much smaller and implied the need to avoid the situation where the re-optimization time could take values smaller than 30 s. We evaluated, on the basis of a set of preliminary experiments, that the VNS does not require more than 30 s to solve the tested instances.

The trend of the number of lost requests as a function of $\Delta t$ is shown in Fig. 1 where the results for each scenario depending on $\lambda$ are provided. It is evident how, independently of the scenario, the minimum number of lost requests is found for $\Delta t = 30$ s. In particular, if we consider those scenarios where the number of lost requests is quite high (i.e. $\lambda \geq 300$) this number reduces on average by 50.66% when moving from $\Delta t = 3,600$ to $\Delta t = 30$ s.

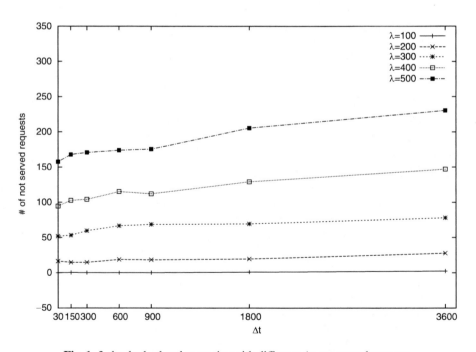

**Fig. 1.** *2-day look-ahead* strategies with different $\Delta t$: not served requests

In Fig. 2 we plot the average distance traveled per served request as function of $\Delta t$ under the different scenarios. Again, independently of $\lambda$, the average distance traveled to serve a request tends to reduce when $\Delta t$ shrinks: the lower the time between two consecutive re-optimizations the more efficient the transportation service.

These preliminary experiments have lead to the final decision to set $\Delta t = 30$ for all the policies but the *static* one.

## 4.3 Accept/reject vs. Delay Policy

In the following we discuss the computational results obtained when comparing *accept/reject* policies to the *delay* policy.

Table 1 is divided into three parts, one for each of the analyzed policies. The first column in each part provides the average number of lost requests when the corresponding policy is applied under the five different scenarios. Each column $gap_d$ measures the average percent increase in the number of not served requests of the analyzed policy with respect to the *delay* policy. Similarly, $gap_{a/r}$ measures the percentage increase of the number of lost requests of the *fixed day accept/reject* policy when compared to its basic version without fixed day. A negative percentage value (as for $\lambda = 200$) means that, on average, the fixed day policy has provided a lower number of lost requests.

Table 2 shows the average total distance traveled by the vehicles. The meaning of each column is the same as for Table 1.

If we consider the scenarios with $\lambda \geq 300$ the *accept/reject* policy looses on average 71.91% more requests than the *delay* policy. The result is even worse if we consider the *fixed day accept/reject* policy.

**Table 1.** *Delay* vs. *accept/reject* policies: number of not served requests

| $\lambda$ | Delay # | Accept/reject # | $gap_d(\%)$ | fixed day accept/reject # | $gap_d(\%)$ | $gap_{a/r}(\%)$ |
|---|---|---|---|---|---|---|
| 100 | 0.00 | 0.00 | – | 0.40 | 100.00 | 100.00 |
| 200 | 16.40 | 21.00 | 28.05 | 19.00 | 15.85 | −9.52 |
| 300 | 52.00 | 88.40 | 70.00 | 101.60 | 95.38 | 14.93 |
| 400 | 94.60 | 162.60 | 71.88 | 175.80 | 85.84 | 8.12 |
| 500 | 157.6 | 274.00 | 73.86 | 276.60 | 75.51 | 0.95 |

When considering traveled distances, one can notice that, for each scenario, all policies tend to completely use the available vehicles: But for only one case ($\lambda = 300$), the average distances traveled by applying the three policies differ from each other for a percentage less than 1%.

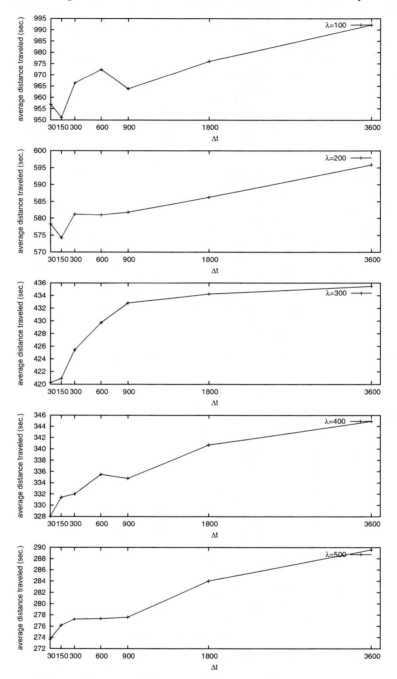

**Fig. 2.** *2-day look-ahead* strategies with different *Δt*: average distance traveled per served request

**Table 2.** *Delay* vs. *accept/reject* policies: distance traveled

| $\lambda$ | Delay $l$ | Accept/reject $l$ | $gap_d(\%)$ | Fixed day accept/reject $l$ | $gap_d(\%)$ | $gap_{a/r}(\%)$ |
|---|---|---|---|---|---|---|
| 100 | 513,482.63 | 511,915.97 | −0.31 | 516,256.46 | 0.54 | 0.85 |
| 200 | 588,000.22 | 588,730.57 | 0.12 | 587,590.58 | −0.07 | −0.19 |
| 300 | 615,194.60 | 618,585.07 | 0.55 | 621,880.01 | 1.09 | 0.53 |
| 400 | 627,632.89 | 632,123.91 | 0.72 | 633,111.51 | 0.87 | 0.16 |
| 500 | 633,372.85 | 636,490.40 | 0.49 | 636,841.83 | 0.55 | 0.06 |

### 4.4 Collaborative vs. Individual Transportation Policy

In this class of experiments we have selected the scenario characterized by $\lambda = 300$ and, since the value of $\lambda$ is fixed, we increased the number of instances to 10.

Since the fleet consists of three uncapacitated vehicles we compared the policy of a company which can decide for the route planning of all the three vehicles to the *individual transportation* policy applied by three different companies each one exploiting one vehicle only. Since, in each instance, the requests are sorted in increasing order of their release time, we have generated three smaller instances. The three small instances consider one request every three requests starting with the first, the second and the third request of the sorted list, respectively. From a practical point of view, this is equivalent to assume that each of the three vehicles in the *individual transportation* policy cover the same $100 \times 100$ $km^2$ geographical area.

Table 3 provides the number of not served requests in each of the 10 instances by the *collaborative transportation* policy (column 2) and by the three *individual transportation* policies applied to the derived instances (columns 3–5). The last line of the table provides the average values. By moving from *individual transportation* transportation systems to a collaborative one the number of lost requests decreases, on average, by an order of magnitude.

Table 4 shows the distance traveled by the three vehicles in the collaborative system (column 2) and that traveled by each of the three vehicles in the individual one (columns 3–5) as well as their sum (column 6). The last column provides the percent gap between values of column 2 and column 6 computed as $\frac{\sum_{i=1}^{i=3} l_i - l}{l}$. As before, the last line provides average values out of the 10 instances. The advantages provided by a *collaborative transportation* policy are evident also in terms of traveled distance. The total distance traveled by the sum of three vehicles managed individually is, on average, higher by 4.77% even if the number of serviced requests is much lower (cfr. column 5 in Table 3).

**Table 3.** Collaborative vs. individual transportation policy: number of not served requests

| Instance | # | $\#_1$ | $\#_2$ | $\#_3$ | $\sum_{i=1}^{i=3} \#_i$ |
|---|---|---|---|---|---|
| 1 | 46 | 184 | 183 | 168 | 535.00 |
| 2 | 54 | 192 | 185 | 198 | 575.00 |
| 3 | 44 | 203 | 195 | 193 | 591.00 |
| 4 | 54 | 183 | 188 | 193 | 564.00 |
| 5 | 62 | 192 | 183 | 198 | 573.00 |
| 6 | 53 | 194 | 194 | 185 | 573.00 |
| 7 | 46 | 191 | 187 | 175 | 553.00 |
| 8 | 44 | 187 | 180 | 181 | 548.00 |
| 9 | 39 | 183 | 183 | 187 | 553.00 |
| 10 | 66 | 205 | 199 | 193 | 597.00 |
| | 50.80 | 191.40 | 187.70 | 187.10 | 566.20 |

**Table 4.** Collaborative vs. individual transportation policy: distance traveled

| Instance | $l$ | $l_1$ | $l_2$ | $l_3$ | $\sum_{i=1}^{i=3} l_i$ | gap (%) |
|---|---|---|---|---|---|---|
| 1 | 611,830.00 | 213,116.62 | 213,115.80 | 213,234.55 | 639,466.97 | 4.52 |
| 2 | 619,250.07 | 214,464.66 | 214,033.89 | 213,555.14 | 642,053.69 | 3.68 |
| 3 | 617,810.59 | 215,376.95 | 213,560.64 | 214,432.88 | 643,370.47 | 4.14 |
| 4 | 607,299.61 | 214,673.09 | 212,627.13 | 213,197.80 | 640,498.02 | 5.47 |
| 5 | 619,782.75 | 214,658.65 | 213,588.85 | 214,084.49 | 642,331.99 | 3.64 |
| 6 | 595,790.27 | 214,671.30 | 213,490.68 | 209,136.81 | 637,298.79 | 6.97 |
| 7 | 609,354.19 | 214,887.06 | 214,401.58 | 214,989.29 | 644,277.93 | 5.73 |
| 8 | 615,016.16 | 214,169.54 | 212,961.75 | 214,310.70 | 641,441.99 | 4.30 |
| 9 | 614,192.44 | 214,833.80 | 214,613.17 | 214,843.07 | 644,290.04 | 4.90 |
| 10 | 613,412.14 | 214,297.79 | 212,961.39 | 213,079.88 | 640,339.06 | 4.39 |
| | 612,373.82 | 214,514.95 | 213,535.49 | 213,486.46 | 641,536.90 | 4.77 |

### 4.5 Dynamic vs. Static Policy

Since on-line unpostponable requests cannot be managed by a static policy, in this class of experiments we have considered postponable requests only. A comment is required in order to better understand the results. Our general assumption is that we want to guarantee the service of a request within the end of the day following its issuance. This means that, on average, 1.5 days are available to serve a request when the *dynamic* policy is applied and only 1 day when the *static* policy is chosen.

In Table 5 for each scenario (value of $\lambda$) the average number of not served requests for the *static* policy (column 3) and for the *dynamic* one (column 2) are shown. If we analyze the values for $\lambda \geq 300$ the number of not served requests increases, on average, by 7 times when the *static* policy is considered instead of the *dynamic* one.

**Table 5.** *Dynamic* vs. *static* policy: number of not served requests

| $\lambda$ | $\#_{dynamic}$ | $\#_{static}$ |
|-----|-------|--------|
| 100 | 0.00  | 1.40   |
| 200 | 3.20  | 62.80  |
| 300 | 19.40 | 193.60 |
| 400 | 38.20 | 329.40 |
| 500 | 79.60 | 501.80 |

In Table 6 we report the average distance traveled to serve a request when the *dynamic* policy (column 2) and the *static* policy (column 3) are applied, respectively. By analyzing the results it is evident how, independently of the value of $\lambda$, the *dynamic* policy leads to a more efficient transportation system: the gap between the values in columns 2 and 3 is, on average, 9.49%.

**Table 6.** *Dynamic* vs. *static* policy: average distance traveled per served request

| $\lambda$ | $l_{dynamic}$ | $l_{static}$ | gap(%) |
|-----|--------|--------|-------|
| 100 | 799.14 | 898.10 | 12.38 |
| 200 | 494.62 | 536.41 | 8.45  |
| 300 | 366.59 | 388.61 | 6.01  |
| 400 | 283.00 | 301.36 | 6.49  |
| 500 | 232.44 | 251.30 | 8.12  |

## Conclusions

In this paper we analyzed, from a management point of view, a dynamic environment for a carrier. The vehicles are equipped with communication devices that make it possible to a central control unit to evaluate in real-time new service requests and re-route the vehicles whenever beneficial. We tested different scenarios where each scenario is characterized by a different intensity of traffic. The first result we have obtained is that a reduction of the interval between two consecutive re-optimizations of the service from 1 h down to 30 s reduces the number of lost customers and the distance traveled. Thus, in all the subsequent experiments we have fixed the re-optimization interval to 30 s.

We then studied a number of different management policies a carrier may decide to follow to carry out the service to its customers. We analyzed the policies by evaluating two performance criteria: the number of lost customers and the distance traveled by the vehicles.

One of the management issues a carrier has to face is whether to give an immediate accept/reject answer to a service request on the basis of the previously accepted requests and the fleet of available vehicles or to accept all customers and, in case of need, to make use of a back-up company at a later time. The results of the experiments have shown that the latter policy is much more effective as the number of customer served by the vehicles of the carrier increases on average by more than 70% and the distance traveled increases only slightly.

It is well known that a large carrier can take advantage of its size to increase the average load of a vehicle and reduce the number of empty trips with respect to a smaller carrier. Small carriers are frequent in Europe and in Italy in particular. Small carriers may merge or at least implement a collaborative strategy to improve their overall performance. The comparison of the behavior of three carriers that own one vehicle each to the behavior of a hypothetical carrier that owns the three vehicles shows that a collaborative strategy dramatically increases the number of served customers and at the same time reduces the traveled distance.

Finally, we compared a dynamic environment with a static one. The dynamic environment requires investment costs in communication devices and a more complex and dynamic organization. Is it worthwhile? We have shown that the dynamic environment reduces the number of lost customers by almost an order of magnitude while reducing at the same time the traveled distance.

While in most cases a model and a solution algorithm for a routing problem are designed and tested with an operational point of view we have taken in this paper a managerial point of view and have provided, thanks to the availability of a software for the optimization of a dynamic routing environment, to quantify the advantages and disadvantages of different management policies.

# References

[1] E. Angelelli, N. Bianchessi, R. Mansini, and M.G. Speranza. Short term strategies for a dynamic multi-period routing problem. Quaderno di ricerca n. 278, Dipartimento Metodi Quantitativi, Università degli Studi di Brescia, 2007, to appear on Transportation Research Part C.

[2] E. Angelelli, M.W.P. Savelsbergh, and M.G. Speranza. Competitive analysis for dynamic multi-period uncapacitated routing problems. *Networks*, 49:308–317, 2007.

[3] E. Angelelli, M.W.P. Savelsbergh, and M.G. Speranza. Competitive analysis of a dispatch policy for a dynamic multi-period routing problem. *Operations Research Letters*, 35:713–721, 2007.

# Service Design Models for Rail Intermodal Transportation

Teodor Gabriel Crainic

NSERC Industrial Research Chair in Logistics Management, Department of Management
and Technology, School of Business and Management
and CIRRELT, University of Quebec in Montreal, Canada theo@crt.umontreal.ca

**Summary.** Intermodal transportation forms the backbone of the world trade and exhibits significant growth resulting in modifications to the structure of maritime and land -based transportation systems, as well as in the increase of the volume and value of intermodal traffic moved by each individual mode. Railroads play an important role within the intermodal chain. Their own interests and environment-conscious public policy have railroads aiming to increase their market share. To address the challenge of efficiently competing with trucking in offering customers timely, flexible, and "low"-cost transportation services, railroads propose new types of services and enhanced performances. From an Operations Research point of view, this requires that models be revisited and appropriate methods be devised. The paper discusses some of these issues and developments focusing on tactical planning issues and identifies challenging and promising research directions.

**Key words:** Intermodal transportation, Freight rail carriers, Tactical planning, Full-asset-utilization policies, Intermodal shuttle networks, Design-balanced service network design

## 1 Introduction

Intermodal transportation forms the backbone of the world trade and exhibits significant growth. The value of multimodal shipments in the U.S., including parcel, postal service, courier, truck-and-rail, truck-and-water, and rail-and-water, increased from about 662 billion US dollars to about 1.1 trillion in a period of nine years (1993 to 2003 [31]). In the same period, the total annual world container traffic grew from some 113.2 millions of TEUs (20 feet equivalent container units) to almost 255 millions, reaching an estimated 304 millions of TEUs by 2005.

Intermodal transportation involves, sometimes integrates, at least two modes and services of transportation to improve the efficiency of the door-to-door distribution

L. Bertazzi et al. (eds.), *Innovations in Distribution Logistics,* Lecture Notes
in Economics and Mathematical Systems 619, DOI: 10.1007/978-3-540-92944-4,

[4] M. Gendreau, F. Guertin, J-Y. Potvin, and E. Taillard. Parallel tabu search for real-time vehicle routing and dispatching. *Transportation Science*, 33:381–390, 1999.

[5] G. Ghiani, F. Guerriero, G. Laporte, and R. Musmanno. Real-time vehicle routing: Solution concepts, algorithms and parallel computing strategies. *European Journal of Operational Research*, 151:1–11, 2003.

[6] S. Ichoua, M. Gendreau, and J-Y Potvin. Diversion issues in real-time vehicle dispatching. *Transportation Science*, 34:426–438, 2000.

[7] O.B.G. Madsen, H.F. Ravn, and J.M. Rygaard. A heuristic algorithm for a dial-ride problem with time windows, multiple capacities and multiple objectives. *Annals of Operations Research*, 60:193–208, 1995.

[8] S. Mitrović-Minić, R. Krishnamurti, and G. Laporte. The double-horizon heuristic for the dynamic pickup and delivery problem with time windows. *Transportation Research Part B*, 38:669–685, 2004.

[9] S. Mitrović-Minić and G. Laporte. Waiting strategies for the dynamic pickup and delivery problem with time windows. *Transportation Research Part B*, 38:635–655, 2004.

[10] H.N. Psaraftis. Dynamic vehicle routing: Status and prospects. *Annals of Operations Research*, 61:143–164, 1995.

[11] H.N. Psaraftis. Dynamic vehicle routing problems. In B.L. Golden and A.A. Assad, editors, *Vehicle Routing: Methods and Studies*, pages 223–248. Elsevier Science, Amsterdam, 1998.

[12] M.W.P. Savelsbergh and M. Sol. Drive: Dynamic routing of independent vehicles. *Operations Research*, 46:474–490, 1998.

[13] J. Yang, P. Jaillet, and H. Mahmassani. Real-time multivehicle truckload pickup and delivery problems. *Transportation Science*, 38:135–148, 2004.

# Optimizing the Storage Area Dimension
# in a Production System

Luca Bertazzi

Department of Quantitative Methods, University of Brescia, Italy bertazzi@eco.unibs.it

**Summary.** We study the problem of determining the optimal dimension of a work-in-process storage area in a two-line production system with delays and breakdowns. We propose a stochastic model and prove theoretical results that allow us to implement an exact algorithm for the solution of the model. We optimally solve a real instance and carry out a sensitivity analysis to evaluate if the optimal solution is stable when the initial data are perturbed.

**Key words:** Production, Storage area, Stochastic models

## 1 Introduction

The problem of determining the right storage area dimension plays an important role in production systems. In fact, companies need to reduce costs, but at the same time to operate in markets that change every day and stimulate them to have a diversified and customer-oriented production. To operate in this way, a correct dimensioning of the storage areas and an efficient planning and scheduling are necessary. Serial production lines have been studied to find models for determining the right buffer dimension. The typical aim is to find rules for the storage space allocation (see [1, 2, 3, 4]). What we want to do is different: To develop a stochastic model that minimizes the sum of storage costs and of delay and break costs that occur during the process. The system we study is a two-line assembly production system, where a work-in-process storage area precedes the assembly stage. We solve a real instance provided by a primary company located in Brescia (Italy) and which is part of a multinational group which produces industrial vehicles. The production is customer-oriented: The daily production program is based on the demand of the customers. Each customer places an order to one of the dealers of the company; then, the dealer transfers the order to the commercial direction located in Turin (Italy), which sends it to the production service direction located in Brescia. Finally, in the production plant located

L. Bertazzi et al. (eds.), *Innovations in Distribution Logistics,* Lecture Notes
in Economics and Mathematical Systems 619, DOI: 10.1007/978-3-540-92944-4,
© 2009 Springer-Verlag Berlin Heidelberg

in Brescia the order is transformed in an executive program. During this step, a storage space has to be allocated in order to guarantee a regular production at minimum cost.

The paper is organized as follow. In Sect. 2 the problem is described and formulated. In Sect. 3 an exact algorithm is proposed and some particular cases are analyzed. Finally, in Sect. 4 the real instance is optimally solved and a sensitivity analysis is carried out.

## 2 Problem Description and Formulation

The production plant located in Brescia is a two-line production system that produces specific components of industrial vehicles (see Fig. 1). Line 1 produces chassis, while line 2 produces cabs. Each item produced by line 1 is matched with a specific item produced by line 2. Line 1 is affordable, while line 2 has two types of problems: First, the items can take longer time than their scheduled time to finish their processing cycle, whenever the items are not conform to the assigned quality standards. Second, breakdowns may happen. Since any delay in line 2 can stop the production process, a work-in-process storage area is made available in line 2.

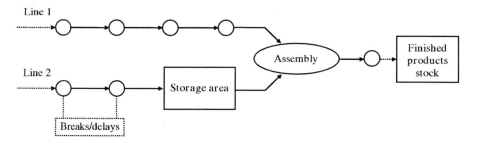

**Fig. 1.** The production system

Our aim is to determine the right dimension of this storage area. It is necessary to balance the opposite effects of different cost components. In fact, there is not only the inventory cost, but also the penalty cost due to a wrong dimension of the storage area. Our aim is to find the dimension of the storage area that minimizes the expected total cost $E(C)$, given by the sum of the investment cost $I$ and the expected operative cost $E(O)$. The investment cost is deterministic, while the expected operative cost, which depends on the break probability, can be expressed as follows. Let $R$ be the break event and $\bar{R}$ be the complementary no-break event, $p(R)$ be the probability that a break occurs during the production, $p(\bar{R}) = 1 - p(R)$ be the probability that no-break occurs during the production, $E(c|R)$ be the expected cost if there is a break in one of the machines and $E(c|\bar{R})$ be the expected cost in the opposite case. Then,

$$E(O) = E(c|\bar{R})p(\bar{R}) + E(c|R)p(R),$$

where the expected break cost $E(c|R)$ is the sum of the cost for production loss ($PL$), of the expected cost for delay $E(D|R)$ and the expected cost at the restart of the production $E(S|R)$. Therefore, the optimization model can be written in compact form as follows:

$$\min_{x} I + E(c|\bar{R})p(\bar{R}) + [PL + E(D|R) + E(S|R)]p(R), \qquad (1)$$

where each cost component depends on the dimension $x$ of the storage area expressed in terms of number of hours of stock. Our aim is to find the value of $x$ that minimizes the total expected cost $E(C)$. Let us now formulate each cost component.

## 2.1 Investment Cost

Every node in a production process implies an investment cost for establishment and for development. When the node is a storage area, the investment cost is related to the dimension of the area. If $C_0$ denotes the investment cost of one unit of storage area per year, then the total investment cost is:

$$I = C_0 x.$$

## 2.2 Expected Operative Cost

The expected operative cost $E(O)$ is composed of two different cost components, the first depending on the delays and the second on the breaks occurred during the production.

### Expected No-Break Cost

The first component of the expected operative cost is the so called expected no-break cost (or failed sequence cost). This cost is generated when an item takes longer time than its scheduled time to finish its processing cycle. This can happen when the item is not conform to the assigned quality standards and, therefore, it has to be corrected to attain the desired level. These corrections break the planned production sequence. Therefore, two situations can happen: Either the late-comer item has a delay not greater than the dimension of the storage area (in hours) or it has a delay greater than it. In the former case, the item becomes available in time to be matched with its complementary and therefore the corresponding no-break cost is equal to zero. In the latter case, the item cannot be matched with its complementary in the production line and therefore it must be driven out of the production line. Additional working time is necessary to match it with its complementary and therefore a no-break cost is charged. Let $k = 1, 2 \ldots, \beta$ be the number of late-comer items per production hour, $h$ be the number of working hours per day, $\delta$ be the number of working days per year, $p(k|\bar{R})$ be the probability to have $k$ late-comer items per hour, given that the no-break event occurs, and $C_1$ be the unit cost to match one late-comer item with its complementary out of the production line. Then, the expected no-break cost can be expressed as:

$$E(c|\bar{R}) = \sum_{k=1}^{\beta} E(c|k, \bar{R}) p(k|\bar{R}),$$

where $E(c|k, \bar{R})$ is the expected cost to have $k$ late-comer items per hour. Let $r = 1, 2, \ldots, \alpha$ be the number of hours of delay. Then, if $E(c|r, k, \bar{R})$ is the expected cost to have $r$ hours of delay and $k$ late-comer items, then $E(c|k, \bar{R})$ can be expressed as:

$$E(c|k, \bar{R}) = \sum_{r=1}^{\alpha} E(c|r, k, \bar{R}) p(r|k, \bar{R}),$$

which is equal to

$$E(c|k, \bar{R}) = C_1 kh\delta \sum_{r=x+1}^{\alpha} p(r|k, \bar{R}),$$

as $E(c|r, k, \bar{R})$ is equal to 0 for $1 \leq r \leq x$ and to $C_1 kh\delta$ for $x < r \leq \alpha$, because it does not depend on the number of hours of delay. If we denote by $p(rit > x|k, \bar{R}) = \sum_{r=x+1}^{\alpha} p(r|k, \bar{R})$, the expected no-break cost is:

$$E(c|\bar{R}) = C_1 h\delta \sum_{k=1}^{\beta} k p(rit > x|k, \bar{R}) p(k|\bar{R}). \tag{2}$$

**Expected Break Cost**

There is a break cost when an $R$ event happens. This type of events are often caused by breakdowns. In this case, the main problem is not due to costs, but to the fact that the production process is stopped. This type of events would be really catastrophic if there are not sufficient stocks that make possible to go ahead with the production until the failed process restarts. The expected break cost is the sum of the following three components:

1. *Cost for production loss*
   The cost for production loss represents the most evident effect of a breakdown. If there is no storage area or it is not adequate, there is a clear loss of added value. If $C_2$ denotes the added value of one hour of production, then the cost per production loss is:

   $$PL = C_2 \delta \max(\alpha - x, 0).$$

2. *Expected cost for delay*
   There is a cost for delay when, in the case of a break, an inadequate storage area makes not possible to restore the right sequence in the production. Let $n$, with $n = 1, 2, \ldots, x$, be the index of the hours of items in the storage area. It represents

the numbers of hours the corresponding items remain in the storage area before matching their complementary items.

Consider first the case with $x \leq \alpha$. For each $n, k$ of the corresponding items, with $k = 1, 2, \ldots, \beta$ and probability $p(k|R)$, can be in the storage area just to replace late-comer items. If the delay of the corresponding late-comer items is greater than $n$, these items cannot be replaced in time and therefore must be matched with their complementary items out of the production line. This happens with probability $p(rit > n|k, R) = \sum_{r=n+1}^{\alpha} p(r|k, \bar{R})$. Therefore, if $x \leq \alpha$, the expected cost for delay is:

$$E(D|R) = C_1 \delta \sum_{n=1}^{x} \sum_{k=1}^{\beta} k p(rit > n|k, R) p(k|R) \qquad x \leq \alpha.$$

Consider now the case with $x > \alpha$. In this case, the items corresponding to $n = 1, 2 \ldots, x - \alpha$ are already in the right sequence. Instead, for each $n = x - \alpha + 1, \ldots, x$, $k$ of the corresponding items, with $k = 1, 2, \ldots, \beta$ and probability $p(k|R)$ respectively, can be in the storage area just to replace late-comer items. If the delay of the corresponding late-comer items is greater than $n$, these items cannot be replaced in time and therefore must be matched with their complementary items out of the production line. This happens with probability $p(rit > n|k, R) = \sum_{r=n+1}^{\alpha} p(r|k, \bar{R})$. Therefore, if $x > \alpha$, the expected cost for delay is:

$$E(D|R) = C_1 \delta \sum_{n=x-\alpha+1}^{x} \sum_{k=1}^{\beta} k p(rit > n|k, R) p(k|R) \qquad x > \alpha.$$

3. *Expected cost at the restart*

   At the restart of the production, another type of cost has to be charged. In fact, after the stop of the production, the number of hours of stock in the storage area is too small to restore the right sequence in the production. This cost has to be charged until the storage area is full. Since $\gamma$ hours of production are needed to recover one hour of stock, the expected cost at the restart is equal to the expected cost for delay multiplied by $\gamma$, that is:

$$E(S|R) = \gamma E(D|R).$$

## 3 Determining the Optimal Stock Dimension

In this section we propose an exact algorithm to solve the stochastic model described above. Let us first prove the following results.

**Lemma 1.** $p(rit > x|k, \bar{R})$ *is a decreasing function in $x$ for $1 \leq x < \alpha$ and $p(rit > x|k, \bar{R}) = 0$ for $x \geq \alpha$.*

*Proof.* We know that $\sum_{r=1}^{\alpha} p(r|k,\bar{R}) = 1$. Therefore, since $p(rit > 1|k,\bar{R}) = 1 - p(1|k,\bar{R})$ and $p(rit > x|k,\bar{R}) = p(rit > x - 1|k,\bar{R}) - p(x|k,\bar{R})$ for $1 < x < \alpha$, then $p(rit > x|k,\bar{R})$ is a decreasing function in $x$ for $1 \leq x < \alpha$, and $p(rit > x|k,\bar{R}) = 0$ for $x \geq \alpha$.

The following corollary holds.

**Corollary 1.** $E(c|\bar{R})$ *is a decreasing function in $x$ for $1 \leq x < \alpha$ and $E(c|\bar{R}) = 0$ for* $x \geq \alpha$.

**Lemma 2.** $\sum_{n=x-\alpha+1}^{x} p(rit > n|k,R)$ *is a decreasing function in $x$ for $\alpha < x < 2\alpha - 1$ and is equal to 0 for $x \geq 2\alpha - 1$.*

*Proof.* For $x = \alpha + 1$, $\sum_{n=x-\alpha+1}^{x} p(rit > n|k,R) = p(rit > 2|k,R) + p(rit > 3|k,R) + \ldots + p(rit > \alpha|k,R) + p(rit > \alpha + 1|k,R)$. For $\alpha + 1 < x < 2\alpha$, $\sum_{n=x-\alpha+1}^{x} p(rit > n|k,R) = \sum_{n=(x-1)-\alpha+1}^{x} p(rit > n|k,R) - p(rit > x - 1|k,R)$. Since $p(rit > s|k,R) = 0$ for $s \geq \alpha$, then $\sum_{n=x-\alpha+1}^{x} p(rit > n|k,R)$ is a decreasing function in $x$ for $\alpha < x < 2\alpha - 1$ and it is equal to 0 for $x \geq 2\alpha - 1$.

The following corollary holds.

**Corollary 2.** $E(D|R) = 0$ *and $E(S|R) = 0$ for $x \geq 2\alpha - 1$.*

The previous results allow us to prove the following theorem.

**Theorem 1.** *Any dimension of the storage area $x > 2\alpha$ cannot be optimal.*

*Proof.* For $x \geq 2\alpha$, the expected no-break cost $E(c|\bar{R}) = 0$ thanks to Corollary 1, the cost for production loss $PL = 0$ by definition and the expected cost for delay $E(D|R) = 0$ and the expected cost at the restart $E(S|R) = 0$ thanks to Corollary 2. Therefore, since the total cost corresponding to $x \geq 2\alpha$ is $C_0 x$, any solution $x > 2\alpha$ is dominated by $x = 2\alpha$.

Given Theorem 1, the optimal dimension of the storage area $x^*$ can be obtained by complete enumeration of the integer values of $x$ in the interval $[1, 2\alpha]$.

### Analysis of Particular Cases

To complete our analysis, we now study some particular cases, interesting from the practical point of view.

## a) Standard Production Systems

The model studied in the previous section considers a production system with a customer-oriented production. Therefore, each item processed in the line 1 has to be matched to the corresponding item processed in the line 2. Let us now consider the case in which the production is standardized. In this case, the model becomes simpler, because the investment cost $I$ and the cost for loss production $PL$ only are included in the objective function.

**Theorem 2.** *In any standard production system, any dimension of the storage area* $x > \alpha$ *cannot be optimal. Moreover, if* $C_2 > C_0$, *then* $x^* = \alpha$; *otherwise,* $x^* = 1$.

*Proof.* For $x \geq \alpha$, the investment cost $I = C_0 x$ and the cost for production loss $PL = 0$ by definition. Therefore, since the total cost is $C_0 x$, then any solution $x > \alpha$ is dominated by $x = \alpha$. Consider now any solution $1 \leq x \leq \alpha$. The total cost is $C_0 x + C_2(\alpha - x)$. Therefore, if $C_2 > C_0$, then the optimal solution is $x^* = \alpha$, otherwise it is $x^* = 1$.

## b) No-Break Systems

Consider now the case in which there are no process stops caused by a break, but only delays due to products not conform to the assigned quality standard. In this case, the investment cost $I$ and the expected no-break cost $E(c|\bar{R})$ only are included in the objective function.

**Theorem 3.** *In any no-break system, any dimension of the storage area* $x > \alpha$ *cannot be optimal.*

*Proof.* For $x \geq \alpha$, the investment cost $I = C_0 x$ and the expected no-break cost $E(c|\bar{R}) = 0$. Therefore, since the total cost is $C_0 x$, any solution $x > \alpha$ is dominated by $x = \alpha$.

# 4 Computational Results

In this section we apply the exact algorithm developed in the previous section to solve the real instance. Let us first list the data of this instance:

- Maximum number of hours of delay $\alpha$: 6;
- Maximum number of late-comer items per production hour $\beta$: 6;
- Investment cost of one unit of storage area per year $C_0$: 8,300 Euro;
- Unit cost to match one late-comer item with its complementary out of the production line $C_1$: 217 Euro;
- Added value of one hour of production $C_2$: 116,203 Euro;

- Number of working hours per day $h$: 16;
- Number of working days per year $\delta$: 225;
- Number of hours to obtain 1 h of stock $\gamma$: 5;
- Probability that a break occurs in the process $p(R)$: 3.1%;
- Distribution of late-comer items:

| Late-comer items | Cases | Frequency(%) |
|------------------|-------|--------------|
| 1 | 3 | 9.4 |
| 2 | 4 | 12.5 |
| 3 | 5 | 15.6 |
| 4 | 6 | 18.8 |
| 5 | 8 | 25.0 |
| 6 | 6 | 18.8 |
|   | 32 | 100 |

This distribution has been obtained on the basis of historical data. We assume that both the probability $p(k|R)$ of having $k$ late-comer items given that there is a break and the probability $p(k|\bar{R})$ of having $k$ late-comer items given there is no a break are equal to the frequency of having $k$ late-comer items.

- Distribution of hours of delay:

| Hours of delay | Cases | Frequency(%) |
|----------------|-------|--------------|
| 1 | 1 | 6.3 |
| 2 | 2 | 12.5 |
| 3 | 2 | 12.5 |
| 4 | 3 | 18.8 |
| 5 | 5 | 31.3 |
| 6 | 3 | 18.8 |
|   | 16 | 100 |

This distribution has been obtained on the basis of historical data. We assume that the probability $p(r|k,\bar{R})$ of having a delay equal to $r$ given that there are $k$ late-comer items and no break is equal to the frequency of having a delay equal to $r$.

Table 1 shows the application of the exact algorithm to the real instance. It is organized as follows. Column 1 gives the dimension of the storage area, column 2 the corresponding investment cost, column 3 the corresponding expected no-break cost, column 4 the corresponding cost for production loss, column 5 the corresponding expected cost for delay, column 6 the corresponding expected cost at the restart and, finally, column 7 the corresponding expected total cost.

The optimal solution corresponds to a dimension of the storage area of 10 h, with a total expected cost of 89,704.66.

**Table 1.** Iterations in the solution algorithm

| $x$ | $I$ | $E(c\|\bar{R})$ | $PL$ | $E(D\|R)$ | $E(S\|R)$ | $E(C)$ |
|---|---|---|---|---|---|---|
| 1 | 8,300.00 | 2,883,726.56 | 130,728,375.00 | 180,232.91 | 901,164.55 | 6,888,733.98 |
| 2 | 16,600.00 | 2,499,229.69 | 104,582,700.00 | 336,434.77 | 1,682,173.83 | 5,742,994.13 |
| 3 | 24,900.00 | 2,114,732.81 | 78,437,025.00 | 468,605.57 | 2,343,027.83 | 4,592,784.51 |
| 4 | 33,200.00 | 1,537,987.50 | 52,291,350.00 | 564,729.78 | 2,823,648.93 | 3,249,581.48 |
| 5 | 41,500.00 | 576,745.31 | 26,145,675.00 | 600,776.37 | 3,003,881.84 | 1,522,626.54 |
| 6 | 49,800.00 | 0.00 | 0.00 | 600,776.37 | 3,003,881.84 | 161,544.40 |
| 7 | 58,100.00 | 0.00 | 0.00 | 420,543.46 | 2,102,717.29 | 136,321.08 |
| 8 | 66,400.00 | 0.00 | 0.00 | 264,341.60 | 1,321,708.01 | 115,567.54 |
| 9 | 74,700.00 | 0.00 | 0.00 | 132,170.80 | 660,854.00 | 99,283.77 |
| 10 | 83,000.00 | 0.00 | 0.00 | 36,046.58 | 180,232.91 | 89,704.66 |
| 11 | 91,300.00 | 0.00 | 0.00 | 0.00 | 0.00 | 91,300.00 |
| 12 | 99,600.00 | 0.00 | 0.00 | 0.00 | 0.00 | 99,600.00 |

**Sensitivity Analysis**

Our aim is now to evaluate if the optimal dimension of the storage area $x^* = 10$ h is stable when the initial data are perturbed. The sensitivity analysis we carry out is based on the variation of one parameter at a time in a given interval. The results are shown in Tables 2 and 3. Table 2 is organized as follows. Columns 1–2 show the results obtained when the investment cost of one unit of storage area per year $C_0$ is modified from 0 to 20,000. Columns 3–4 show the results obtained when the unit cost to match one late-comer item with its complementary out of the production line $C_1$ is modified from 0 to 400. Columns 5–6 show the results obtained when the added value of 1 h of production $C_2$ is modified from 0 to 400,000. Finally, the columns 7–8 show the results obtained when the probability that a break occurs $p(R)$ is modified from 0 to 10%.

The results show that the optimal solution is equal to 10 for values of $C_0$ between 7,000 and 17,000, while it increases to 11 for values lower than 7,000 and reduces to 9 for values grater than 17,000. The optimal solution is equal to 10 for $C_1$ between 120 and 260, while it reduces up to 6 for values lower than 120 and increases to 11 for values grater than 260. The optimal solution does not vary when the added value of 1 h of production $C_2$ is modified. Finally, the optimal solution is equal to 10 for values of $p(R)$ between 1.5 and 3.5%, while it reduces up to 6 for values lower than 1.5% and increases to 11 for values grater than 3.5%.

Table 3 shows the results obtained by varying the distribution probabilities of both late-comer items and of the hours of delay. For the late-comer items, one thousand distribution probabilities have been generated in the following way. Given a maximum number $a$ of late-comer items, the number of cases for each number

**Table 2.** Sensitivity analysis: varying costs and probability of break

| $C_0$ | $x^*$ | $C_1$ | $x^*$ | $C_2$ | $x^*$ | $p(R)(\%)$ | $x^*$ |
|---|---|---|---|---|---|---|---|
| 0 | 11 | 0 | 6 | 0 | 10 | 0.0 | 6 |
| 1,000 | 11 | 20 | 6 | 20,000 | 10 | 0.5 | 6 |
| 2,000 | 11 | 40 | 6 | 40,000 | 10 | 1.0 | 8 |
| 3,000 | 11 | 60 | 7 | 60,000 | 10 | 1.5 | 10 |
| 4,000 | 11 | 80 | 9 | 80,000 | 10 | 2.0 | 10 |
| 5,000 | 11 | 100 | 9 | 100,000 | 10 | 2.5 | 10 |
| 6,000 | 11 | 120 | 10 | 120,000 | 10 | 3.0 | 10 |
| 7,000 | 10 | 140 | 10 | 140,000 | 10 | 3.5 | 10 |
| 8,000 | 10 | 160 | 10 | 160,000 | 10 | 4.0 | 11 |
| 9,000 | 10 | 180 | 10 | 180,000 | 10 | 4.5 | 11 |
| 10,000 | 10 | 200 | 10 | 200,000 | 10 | 5.0 | 11 |
| 11,000 | 10 | 220 | 10 | 220,000 | 10 | 5.5 | 11 |
| 12,000 | 10 | 240 | 10 | 240,000 | 10 | 6.0 | 11 |
| 13,000 | 10 | 260 | 10 | 260,000 | 10 | 6.5 | 11 |
| 14,000 | 10 | 280 | 11 | 280,000 | 10 | 7.0 | 11 |
| 15,000 | 10 | 300 | 11 | 300,000 | 10 | 7.5 | 11 |
| 16,000 | 10 | 320 | 11 | 320,000 | 10 | 8.0 | 11 |
| 17,000 | 10 | 340 | 11 | 340,000 | 10 | 8.5 | 11 |
| 18,000 | 9 | 360 | 11 | 360,000 | 10 | 9.0 | 11 |
| 19,000 | 9 | 380 | 11 | 380,000 | 10 | 9.5 | 11 |
| 20,000 | 9 | 400 | 11 | 400,000 | 10 | 10.0 | 11 |

$k = 1, 2, \ldots, 6$ of late-comer items has been randomly generated as an integer number between 0 and $a$ on the basis of a uniform distribution. For the hours of delay, one thousand distribution probabilities have been generated in the following way. Given a maximum number $b$ of late-comer items, the number of cases for each number $r = 1, 2, \ldots, 6$ of hours of delay has been randomly generated as an integer number between 0 and $b$ on the basis of a uniform distribution. Table 3 is organized as follows. Column 1 gives the optimal dimension $x^*$ of the storage area. Columns 2, 3 and 4 show, for each value of $x^*$, the percentage of times this value has been the optimal one in the one thousand instances generated with $a = 4$ and $b = 3$, $a = 8$ and $b = 5$ and, finally, $a = 16$ and $b = 10$, respectively.

The results show that the dimension of 10 has been the optimal one in more than 50% of the instances for each of the combinations of $a$ and $b$.

The conclusion of the sensitivity analysis is that the optimal dimension of 10 is stable when the initial data are perturbed.

**Table 3.** Sensitivity analysis: varying distribution probabilities

| $x^*$ | (4,3)(%) | (8,5)(%) | (16,10)(%) |
|---|---|---|---|
| 1 | 0.00 | 0.00 | 0.00 |
| 2 | 0.00 | 0.00 | 0.00 |
| 3 | 0.00 | 0.00 | 0.00 |
| 4 | 0.00 | 0.00 | 0.00 |
| 5 | 0.00 | 0.00 | 0.00 |
| 6 | 0.10 | 0.00 | 0.00 |
| 7 | 0.50 | 0.60 | 0.30 |
| 8 | 5.10 | 5.20 | 3.90 |
| 9 | 24.00 | 25.10 | 25.00 |
| 10 | 51.20 | 50.20 | 53.50 |
| 11 | 19.10 | 18.90 | 17.30 |
| 12 | 0.00 | 0.00 | 0.00 |

## Conclusions

We studied the problem of determining the optimal dimension of a work-in-process storage area in a two-line production system with delays and breakdowns. The stochastic model and the algorithm we proposed allowed us to exactly solve the real instance with a stable solution and can be easily extended to solve more general two-line production systems.

## Acknowledgements

The author wishes to acknowledge the contribution of Marco Bertella at an early stage of this paper and the useful suggestions of Maria Grazia Speranza that allowed to improve the paper.

## References

[1] Conway, R., Maxwell, W., McClain, J.O., Thomas, L.J. (1988), The Role of Work-In-Process Inventory in Serial Production Lines, *Operations Research* 2, 229–241.
[2] Hillier, F.S., So, K.C., Boling, R.W. (1993), Toward Characterizing the Optimal Allocation of Storage Space in Production Line Systems with Variable Processing Times, *Management Science* 39, 126–133.

[3] Papadopoulos, H.T., Vidalis, M.I. (1998), Optimal Buffer Storage Allocation in Balanced Reliable Production Lines, *International Transactions in Operational Research* 5, 325–339.

[4] So, K.C. (1997), Optimal buffer allocation strategy for minimizing work-in-process inventory in uncapacitated production lines, *IIE Transactions* 29, 81-88.

# On-line Routing per Mobile Phone: A Case on Subsequent Deliveries of Newspapers

Thomas Bieding[1], Simon Görtz[2], and Andreas Klose[3]

[1]  Vodafone D2 GmbH, Germany thomas.bieding@vodafone.com
[2]  Faculty of Economics and Social Sciences, University of Wuppertal, Germany
    simon.goertz@wiwi.uni-wuppertal.de
[3]  Department of Mathematical Sciences, University of Aarhus, Denmark
    aklose@imf.au.dk

**Summary.** On-line routing is concerned with building vehicle routes in an on-going fashion in such a way that customer requests arriving dynamically in time are efficiently and effectively served. An indispensable prerequisite for applying on-line routing methods is mobile communication technology. Additionally it is of utmost importance that the employed communication system is suitable integrated with the firm's enterprise application system and business processes. On basis of a case study, we describe in this paper a system that is cheap and easy to implement due to the use of simple mobile phones. Additionally, we address the question how on-line routing methods can be integrated in this system.

**Key words:** On-line and dynamic vehicle routing, Mobile communication, Mobile business and solutions

## 1 Introduction

On-line routing is concerned with building vehicle routes in an on-going fashion in such a way that customer requests arriving dynamically in time are efficiently and effectively served. Although dynamic routing problems and quantitative methods for on-line routing have been discussed in the scientific literature since a seminal paper of Psaraftis [21], the technology required for implementing on-line dispatching of vehicles was till recently not available at a cost allowing a widespread use. An indispensable prerequisite for applying on-line routing methods is mobile communication technology. Drivers must be timely informed about the next stop to approach; the dispatching centre must be informed about the driver's location and the status of the delivery. The availability of a mobile communication system alone is, however, not sufficient. It is important that the communication system is properly interfaced with

L. Bertazzi et al. (eds.), *Innovations in Distribution Logistics,* Lecture Notes
in Economics and Mathematical Systems 619, DOI: 10.1007/978-3-540-92944-4,
© 2009 Springer-Verlag Berlin Heidelberg

the firm's IT system, enterprise applications and data bases. A suitable hardware and software solution for interconnecting enterprise applications and the mobile communication system should also allow to easily integrate on-line routing algorithms.

In the information management area, the more general topic of integrating mobile co-workers in a firm's IT system is also discussed under the buzzword of *mobile business* and *mobile logistics* (see, e.g., [32]). This paper describes a mobile business solution that heavily relies on the use of mobile phones and the wireless application protocol (WAP). An advantage of such a system is its low cost and its ease of use. Furthermore, integrating algorithms for on-line routing should be relatively straightforward. A case concerning the subsequent delivery of newspapers will serve as a base for illustrating the potential benefits of such a system. The next section details this case and Sect. 3 subsumes the case under the broader field of mobile business. Moreover, alternative mobile communication technologies are briefly described. Section 4 discusses then in some detail the above mentioned web/WAP-based mobile communication system, its benefits and potential drawbacks. Section 5 attends to the different decision problems arising when planning vehicle routes for subsequent newspaper delivery. A short overview of the literature on some methods for dynamic vehicle routing is given in Sect. 6. The discussion of these methods will reveal that there seems to be a gap between deterministic reoptimisation heuristics, on the one hand, and relatively simple dispatching rules based on queueing theory and stochastic arguments, on the other hand. Section 7 lists some further application examples. Finally, a summary and outlook is given in Sect. 8.

## 2 A Practical Case: Subsequent Deliveries of Newspapers

Distributing and selling magazines and newspapers is a difficult business. The competitive pressure is substantial, and publishing houses should achieve high print runs in order to earn some money. Subscribers that regularly receive a certain title are thus very important, and publishers usually invest a relatively large amount of money in trying to attract more subscribers. It is therefore important to keep these kind of customers.

The newspaper delivery is usually performed by third-party carriers. In case that a subscriber did not receive his accustomed, e.g., daily newspaper, he will complain about this at a call centre. The operator will then apologise for the inconvenience and assure subsequent delivery of the issue as fast as possible. Customer complaints arrive dynamically and have to be treated on additional delivery routes. For the case of a medium-sized regional newspaper publisher, the whole process of subsequent delivery can be described as follows.

1. A customer complains the non-delivered newspaper issue at the call centre.
2. The operator collects the case's data and feeds them into a software system. The customer chooses between a voucher and a subsequent delivery of the missing issue.

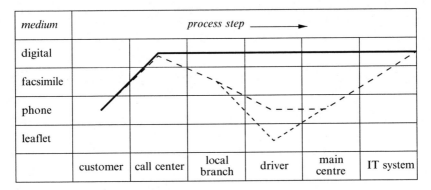

| medium | process step ———▶ | | | | | |
|---|---|---|---|---|---|---|
| digital | | | | | | |
| facsimile | | | | | | |
| phone | | | | | | |
| leaflet | | | | | | |
| | customer | call center | local branch | driver | main centre | IT system |

**Fig. 1.** Information transfer characterised by "media break"

3. In case the customer chooses a subsequent delivery, a corresponding delivery order is triggered. The order is transmitted by facsimile or phone to a local branch of the publisher. A printout of the order is made and handed to a driver that performs the delivery. Drivers that already started their delivery tours may then receive additional orders by trunking.

4. The driver notifies the publisher's dispatching centre about the completed delivery, and an operator marks the order as completed in the firm's software system.

The following figures underpin the dimension of the subsequent deliveries.

- The publisher distributes three local newspapers as well as one supra-regional newspaper in Germany.
- Within the ordinary delivery process, local branches and driver sites are equipped with about 60 additional copies per driver for the purposes of potential subsequent deliveries.
- Each driver executes, on average, 25 subsequent deliveries on each of the 280 delivery days per year. On a peak day, about more than 45 deliveries are performed by a driver.
- Delivery drivers serve predefined delivery areas. A delivery tour starts in the morning and ends at midday. A driver either works as a freelancer or is employed by a carrier; his compensation depends on the time spent on delivery and the distance driven. The average distance of a delivery tour is about 20 km.
- According to the firm's actual figures, 26 drivers are in total daily employed for subsequent delivery. This amounts to $26 \cdot 280 \cdot 20\,km = 145,600\,km$ spent on subsequent delivery per year.

Apparently, a relatively large number of different media (computer, facsimile, phone, paper, trunking) is used for the purposes of transferring information. The dashed line in Fig. 1 shows the status quo of information transmission characterised by a number of "media breaks". Obviously, this kind of communication has a number of drawbacks:

- Permanent communication with the drivers is not possible.

- Drivers cannot promptly notify the dispatching centre (main centre) about a completed delivery.
- The dispatching centre has no ongoing information about the status of a delivery and the driver's route.
- Due to the large number of different media used, the communication process is slow, costly and fault prone.

The solid line in Fig. 1 depicts an ideal state where all required information is centralised in a digital form from the point when a customer complaint arrives. Technologies as well as technical, managerial and organisational methods for achieving this degree of integration of mobile co-workers is a major concern of *mobile business*. The subject of mobile business and possible mobile communication solutions are briefly described in the next section. The mobile communication system should then allow to integrate the driver into the firm's IT system, to streamline and centralise the information process and, finally, to integrate methods for performing the routing in some "intelligent way".

## 3 Mobile Business and Mobile Communication Solutions

The integration of mobile technology into any kind of business process is called mobile business. Mobile business hence deals with the exchange of goods, services and information using mobile devices [35].

The use of mobile technologies emerges new business processes. These are characterised by mobility, reachability, localisation and personalisation. The user of a mobile device such as a mobile phone or a personal digital assistant (PDA) can access networks, products and services while on the move. This is important in context of the use of time critical information like news services or stock tickers. Mobile users can also be reached by other people and service providers regardless of place and time. This feature is especially vital to business processes that regularly need to reach their mobile workforce and pass them new tasks and instructions [23]. A person, who carries a mobile device which is on-line, can be localised. The ability to determine a location of a specific user is very important for location-sensitive services. Logistic companies may be interested in identifying the current position of a vehicle or determining a mobile co-worker who is the closest to the place of the latest customer's request. In addition, a mobile user can also be identified; there is usually a one-to-one correspondence between a mobile device and the person who is carrying it.

Mobile business is based on network technologies, service technologies, mobile middleware and mobile devices as well as mobile localisation and personalisation technologies [34]. In the following, these mobile technologies are briefly described.

Wireless network technologies enable data transfer between devices which are not connected physically. They differ in transmission rate (bandwidth) and their suitability for mobile business application. Regional analogue systems for speech service were the first generation of mobile systems. The first standard based on digital signal transmission was the GSM (Global System for Mobile communication). This second generation (2G) is the starting point of the development of mobile business. In the meanwhile, GSM has spread in most countries. Its low transfer rate of 9.6 Kbps allows only simple applications like SMS, fax or voice mail [23]. The introduction of a faster network (2.5G) has allowed new services. The GPRS (General Packet Radio Service) can attain speeds up to 115 Kbps (theoretically up to 171 Kbps) and enables an 'always-on' connection between a mobile device and the network. The GPRS standard is a packet-switched protocol. Like the wired internet, data are split into parcels called packets, to which a unique address is appended. A mobile GPRS device uses network capacity only when packets are actually being sent or received although it is, in effect, permanently connected to the network. EDGE (Enhanced Data for GSM Evolution) is a faster version of GSM wireless service that enables data to be delivered at rates up to 384 Kbps. It is considered to be a 2.5G technology, although originally designed to evolve GSM to 3G [1]. The transfer rate is comparable to the speed that is offered by low-end ADSL services and could support even streaming short video clips and other multi media applications. Nevertheless, EDGE is an intermediate step to UMTS (Universal Mobile Telephone System). UMTS is counted to the third generation of mobile communication systems. This technology enables transfer rates up to 2 Mbps to stationary or mobile users but still rates up to 382 Kbps to fast moving users like drivers. UMTS promises a wide range of mobile business opportunities, including real-time applications, videoconferencing and massive volumes of transferred data. An alternative to UMTS is WLAN (wireless local area network) technology based on the IEEE 802.11 standard. The bit rates for the widespread 802.11g have gone up to 54 Mbps likely to be shared among 20–30 users. 802.11b enables a transfer rate of 11 Mbps. WLAN enables communication between devices equipped with a wireless network interface in the basic service set. This is the set of all stations that can communicate with each other. As long as a user is moving around within the basic service set he is still connected to the network.

Service technologies like SMS and WAP provide content and applications to mobile devices. SMS (Short Message Service) is a service available on most mobile devices as mobile phones, laptops, PDAs that permits the sending and receiving of short text messages. The length of the message is limited by the constraint of the signalling protocol to 140 bytes. This translates up to 160 characters. The WAP is a global standard for mobile communication and wireless connection to the internet. The restrictions of mobile phones and PDAs make it impossible to provide all services of computer based web browsers to mobile devices. Therefore, WAP browsers are designed to provide at least the basic services. Mobile internet sites are called WAP sites. They are written in, or dynamically converted to WML (wireless markup language). The mobile sites are accessed and interpreted via WAP browser and displayed on the mobile device. Although WAP is now the protocol used for almost all world's mobile internet sites, some authors see the end of this technology [34].

Mobile middleware provides a way to connect and exchange information between mobile and stationary devices. It includes transmission technologies as well as special storage technologies. For example, Bluetooth is a radio standard and communication protocol designed for communication of mobile devices with each other, when they are in range up to 10 m. The used short-range radio frequency is secure and globally unlicensed. Money cards and GMS SIM cards (subscriber identity module) are so-called smart cards. These are pocket-sized cards with embedded integrated circuits. Most smart cards contain memory and microprocessor components and are capable of providing security services. These components allow to identify users.

Mobile localisation and personalisation technologies allow to provide geographic and personalised applications and services. The most important localisation technology is the Global Positioning System (GPS). It is a satellite navigation system based on 24 satellites. The transmitted signals allow GPS receivers to nearly determine the exact location and speed. GPS is free for everybody's use. GPS-technology can easily be integrated in mobile devices. Personalisation technologies collect and provide information to tailor a mobile device application, which is personalised according to the interest of an individual or on user attributes such as role, field of work, or task lists.

Mobile business divides in two categories, mobile B2C (or m-commerce) and mobile solutions. Mobile solutions are used to re-engineer internal business processes [32]. There are two types of application areas: mobile Business-to-Machine and mobile Business-to-Employee (B2E). The first type comprises the controlling and monitoring of machines and plants; the latter the integration of mobile co-workers to the firm's IT infrastructure. Horizontal mobile B2E solutions provide mobile access to standard applications such as groupware solutions. Vertical solutions are employed to improve specific business processes. The objective is a better integration of the mobile co-worker into the business process by means of data acquisition and transmission without media breaks. The opportunities and complexity of mobile technology is sufficient to create a mobile solution. The success of a mobile solution strongly depends on the software solution implemented to the mobile devices and middleware. A complete mobile business solution integrates mobile technology, applications and data base systems into the backend. This encloses the rules of data access and other security aspects, the management of data synchronisation between mobile devices and the firm's backend, and the business logic [32]. The later is the core module of a mobile solution. Business logic deals with real life business objects. It handles the storage of objects, prescribes the interaction of different business objects and defines the routes and the methods by which business objects are accessed, updated and synchronised. Business logic comprises business rules and workflows. Hence, a mobile business solution is generally very complex and hard to achieve by means of standard tools and procedures. In many cases, it is necessary to develop and to implement customised solutions. The next section concretises a mobile solution developed for the case of subsequent newspaper delivery and some other "mobile re-engineering projects".

# 4  A Web/WAP Based Mobile Communication System

This section describes a simply structured mobile communication system used in the case of subsequent newspaper delivery (see also [8]). Due to its ease of use and low cost, the system seems well-suited for small and medium-sized companies.

The system establishes an indirect communication channel between the dispatching centre and the mobile co-worker (driver) via a web and WAP server and mobile phones. Figure 2 shows the system's general architecture. The structure shown in this figure is not exclusively related to the case of subsequent newspaper delivery but also applies to similar mobile business processes. The following parties are involved in the communication process: At a firm's main centre, a central data base/ERP system is kept. The data base feeds a web and WAP server with required information and also extracts data from there. In case of the newspaper delivery example, the main centre is the firm's main branch that is also responsible for dispatching vehicles. Decentralized co-workers may also be connected to the system via the web and WAP server. In contrast to mobile co-workers their location is fixed and they are connected using (non-mobile) remote access technology. In case of the newspaper example, there are no decentralized co-workers. There is, however, the call centre acting as a branch of the main centre and directly connected to its data base by means of remote intranet technology. Finally, there are the mobile co-workers sharing information with the other parties via the web and WAP server. In case of our example, the mobile co-workers are the drivers performing the subsequent delivery of newspapers. Which information is shared how and when between which actors will be explained later below, based on the case of subsequent newspaper delivery. Beforehand, it is shown how the mobile co-workers are integrated to the system.

Mobile co-workers are equipped with mobile devices. The device can be a notebook, PDA, or simply a cellular phone. The last device has a number of advantages in case that its display capabilities are sufficient for the application in question. The device is relatively cheap, most people are very familiar with it, and special agreements and contracts with the cellular radio provider allow to keep transmission costs very low. To this end, GPRS is used for transmitting data. In case of this technique, transmission costs depend on the volumes of transmitted data. Hence, a permanent connection between the mobile co-worker and the central web server can be established, whereas the conventional GSM technology uses a time-based billing system.

For accessing and viewing web pages, the mobile phone must be WAP enabled. A web server as well as a WAP server are then used for processing and answering requests sent by the mobile devices. These two kind of servers just provide different views on the same data extracted from the same data base. The servers are operated by an application service provider. Although the servers as well as other software tools (PERL, PHP) required for extracting information from the central data bases and building dynamic web pages are very cheap or even free software, operating and using such a system requires some expertise that a smaller firm possibly cannot acquire.

Sufficient security of a mobile co-worker's access to the web/WAP server can be ensured in different ways. The mobile devices' MAC addresses are registered at the

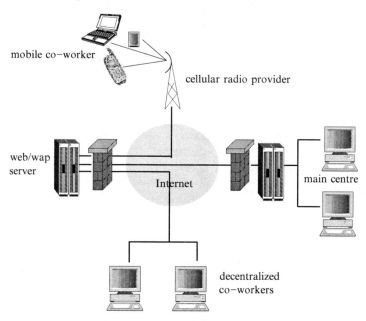

**Fig. 2.** Components of the mobile communication system

server, and only registered devices may access the system. Additionally, authenticated access via a user name and password can be established. For the purposes of establishing secure data transmission, cellular radio providers furthermore offer VPN ("virtual private network") tunnelling techniques. Access to the web/WAP server is then only granted via the provider's VPN tunnel, and only calls approved by the provider's radius server are put through. Moreover, the cellular phone's identification number is then additionally transmitted for security reasons.

Web pages are dynamically built using PHP ("hypertext preprocessor"). The required information (customer's address, product demanded, time until delivery should be completed, additional short information regarding the customer and his location) is extracted from the main centre's data base. The operators at the call centre who process the customer requests feed the data base with a customer's data by means of the same familiar software system that the firm had used for these purposes since a long time. Different techniques can then be applied for transmitting the required data from the data base to the web server. The data base may be directly accessed using ODBC ("open data base connectivity"); the data may also be directly send via the hypertext transfer protocol (HTTP) and then extracted and processed by means of a PERL script. Other possible methods are to use the file transfer protocol (FTP) by first sending the data to a FTP server, or to send the data as a simple e-mail attachment. Privacy and data security is ensured by encryption and a secure socket layer connection.

The following example further illustrates the (real-time) communication between a driver and the central data base via the web/WAP server. Figure 3 shows the

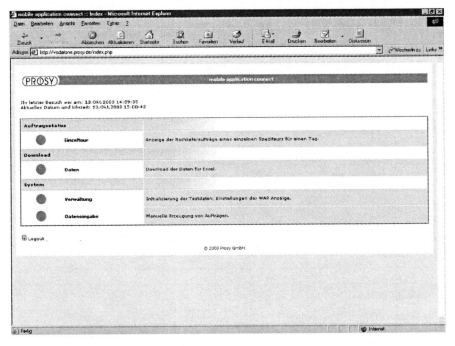

**Fig. 3.** Dispatching centre's view on the system's main page*

*The main page's menu shows the following entries:

Order status
- Single delivery tour:
  → Display customer requests on a given tour assigned to certain carrier

Download
- Data:
  → Download data for spreadsheet access

System
- Administration:
  → Initialize test data; configure WAP display
- Input data:
  → Manually enter customer orders

dispatching centre's view on the system's main page. The main page offers different options for further processing: Data may be downloaded again for further analysis using spreadsheet computation (e.g., for tracking arrival times, response times, actual tour duration, etc.) If necessary, additional customer requests may be entered manually (e.g., in case of missing or misentered data), and some functions for administering the system are provided. Most important is the information about customer requests/complaints of the current day. Figure 4 shows a fictive example of such a list. Customer requests are already assigned here to certain carriers that have to serve the request. The displayed information comprises: the carrier's identification number; the customer's id, name, address, and his demand; the time the request was posed; if and at which time the message was read by the driver, and the status of delivery. The

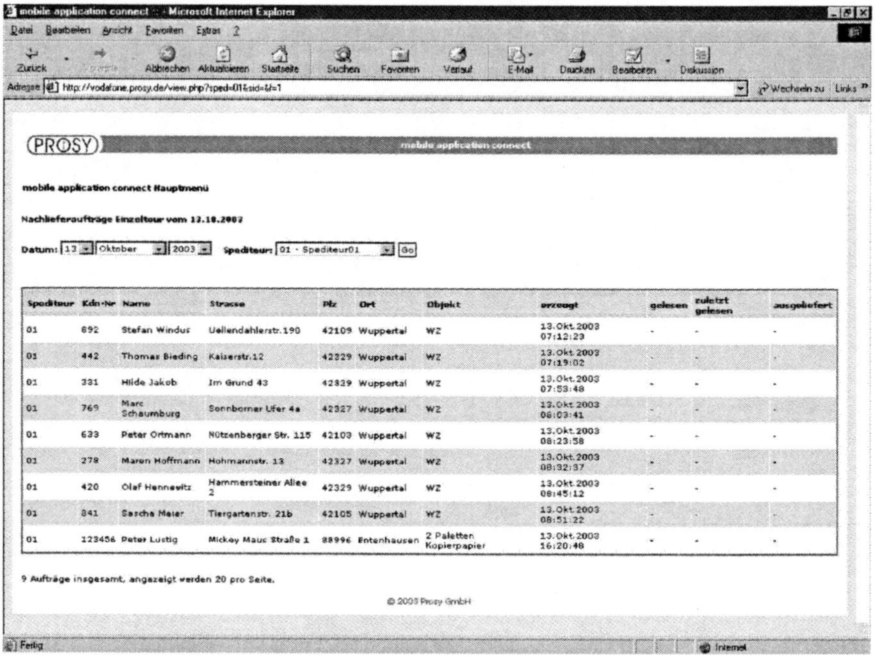

**Fig. 4.** Dispatching centre's view on the list of customer requests*

<hr>

* The columns of the table displaying the customer requests show the following information: Carrier's identification number (1); customer number (2); customer's name (3) and address by street (4), zip code (5), and city (6); product demanded (7); date and time when the request was created (8); if the request was already read by the driver (9); date and time the request was read the last time (10); status of delivery completion (11).

<hr>

driver has access to relevant parts of this information in a different form by means of a WAP enabled cellular phone. The information is displayed in two different forms. First, a simple list of all customer requests to be served by the driver is shown as in Fig. 5a. Each entry of this list just contains the customer's address and the time the request was generated. Using the phone's function keys, the driver can click on an entry in order to see all relevant information concerning the specific customer request (see Fig. 5b). After performing the delivery, the driver marks a request as completed in a similar simple way by just using the function keys of his mobile phone (Fig. 6). This information is promptly transferred back to the web/WAP server as well as the firm's central data base, where now the customer request is marked as served (see Fig. 7). This way sufficient bidirectional and on-going communication between the dispatching centre and the drivers is established.

It seems obvious that this kind of integrated mobile communication system has a number of advantages compared to the previously employed system, even without the incorporation of on-line routing algorithms. Especially, the following points can be raised:

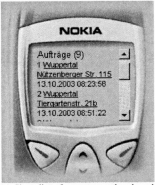

(a) Short list of customer orders by address and creation date and time

(b) Detailed information on customer order: name, address, product demanded, creation date and time, time when delivery has to be completed, etc.

**Fig. 5.** Driver's view on the list of customer requests and a single customer order

(a)*"Should the customer's order be really marked as completed?"*

(b) *"Order marked as completed. Continue"*

**Fig. 6.** After delivery, customer requests are marked as completed

- Drivers are not only informed about new orders; the required bidirectional communication is possible and customer orders are not only fulfilled but also digitally processed in an integrated way by the drivers.
- The aforementioned media breaks arising when transmitting information are reduced to a minimum.
- The driver's actions are integrated in the firm's IT system. The process of subsequent delivery is thereby accelerated and substantial time savings could be realised.
- Drivers get prompt information and are able to better plan next deliveries to be made.
- The system is very cheap and easy to use; drivers are very familiar with a cellular phone.
- Finally, the system is a prerequisite for implementing on-line routing algorithms.

read

delivered

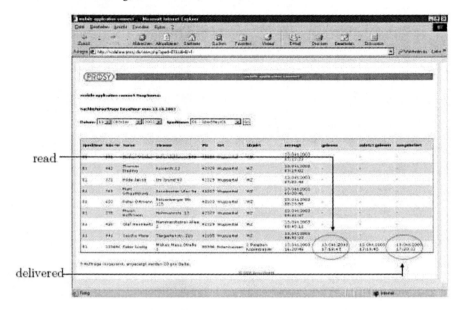

**Fig. 7.** Drivers actions are recorded and written back to the data base*

* See Fig. 4 for a translation of the table's columns.

## 5 Decision Problems to be Addressed

Which kind of subproblems arise when planning dynamic vehicle routes for subsequent newspaper delivery depends on the overall solution approach adopted. Moreover, the firm's willingness and ability to implement organisational changes has to be taken into account. Currently, the firm predefines the drivers' delivery areas, which generally favours a cluster first-route second approach. Especially, the following subproblems may then be distinguished:

1. Design of delivery areas,
2. Location planning,
3. Determining operating times,
4. Operational on-line routing.

Constraints to be taken into account are especially a minimum and maximum operating time of a route as well as a maximum allowed response time. A vehicle's capacity does in this case not play any role. Obviously, the total time spent by the drivers must not exceed a given maximum. However, a too short route duration is also impractical; the earnings granted to a driver are then too low and no driver will like to operate such a route. The response time is the time that passed from the arrival of a customer request until its attendance. Short response times are part of a high customer service, but also very costly. A certain maximum response time must not be exceeded. Below this limit, the response time can be viewed as a soft constraint; attaining short response times is then more an objective than a constraint.

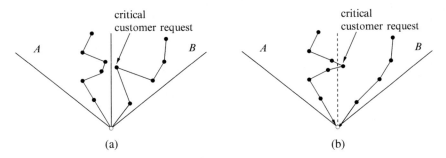

**Fig. 8.** Prefixed delivery areas without (**a**) and with (**b**) "variable borders"

Objectives are more difficult to postulate than constraints. The problem is multi-objective in nature. Customer service and delivery costs have to be balanced, both the service dimension as well as the cost dimension are however difficult to measure. Moreover, the uncertainty of customer requests adds to the complexity of finding appropriate objectives, since in this case not only expected costs but also robustness of solutions is an issue. In order to obtain some operational objectives it is required to use some surrogate performance measures, which usually are (average) response time, travel distance and time, and the number of delivery tours.

In the sequel, the above mentioned subproblems are further described.

### 5.1  Design of Delivery Areas

When using a cluster first-route second approach, the first step consists in dividing the whole delivery area into a number of sub-areas that are then assigned to one or several drivers. Although customer requests arrive dynamically and are "stochastic", the subscribers and their locations itself are of course known. Different alternative methods for obtaining reasonable clusters of subscribers might then be applied for defining delivery areas. Objectives to be taken into account are the minimisation of the number of required drivers as well as the minimisation of the (expected) travel times and distances.

Since customer requests arrive at random, it seems reasonable not to fully pre-fix the delivery areas, but to allow exchanges of customer nodes at the "borders" of delivery areas as illustrated in Fig. 8. In Fig. 8a, the fixed areas result in an un-favourable dynamically emerging route for subregion *B* and some distances can be saved if "border customers" are exchanged between subregions as in Fig. 8b.

In both cases, the clustering will not be part of an on-line algorithm, but computed beforehand by means of a suitable method. Because, subscriber locations are known and also historical customer data can be made available, one might also take a two-stage stochastic programming approach for performing the clustering. The use of two-stage stochastic programs has already been proposed in the literature on stochastic vehicle routing problems [13, 27]. In our case, the first stage decisions would refer to the clustering, while second stage decisions would be the routing or, alternatively, in order to take the dynamically arriving customer demands better into account, recourse actions as, e.g., inserting a customer node in a route.

## 5.2 Location Planning

After completing a customer request, the driver has either to wait until he receives a new customer order or to finish his route and return to his home base. As already mentioned, reliable historical customer data is available in case of the subsequent newspaper delivery. Obviously, dynamic and stochastic routing problems have generally much in common, provided that dependable historical data is available.

Due to the probabilistic nature of a demand case, one might thus rise the question at which location a driver should wait for a next customer request. Possible alternative waiting points are, e.g.,

- A central point as, e.g., the median of the driver's delivery area,
- The closest customer/subscriber location,
- The customer's location with largest probability of a positive demand,
- A suitable customer location selected by considering a combination of the two criteria above.

Larsen et al.[17] investigate further possible rules on which such "repositioning decisions" might be based for the case of the "a priori travelling saleman problem with time windows". The performance (average response time, travel distance and time, number of delivery tours) of different rules for driver repositioning should then be tested within simulation experiments using empirically validated assumptions regarding the customer arrival process.

## 5.3 Determining Operating Times

Neither the exact start and end times of a route are fixed nor is it required to start a route immediately after a customer request has arrived. Rather, it seems advisable to delay the start of a driver's route in order to gather more information. There is obviously a trade-off between the customer service degree and the delivery costs if the routes' start times are postponed. In case of the delivery area of Wuppertal, e.g., 80 % of all customer complaints are known until 10 AM; and 90 % are known until 11 AM. Moreover, routes computed on the base of full information are, on average, 30 % shorter than the routes that emerged dynamically. When delaying the start time of routes, a larger percentage of customer requests gets known in advance; this should allow to perform the routing in such a way that the drivers's waiting and travel times as well as the distances driven are reduced compared to a situation where less information is available. On the other hand, delaying start times results of course in longer response times. Possible strategies for determining starting times are, e.g., to begin a route if a given percentage of the expected number of customer requests is known in advance. Simulation experiments should then be carried out in order to measure the performance (average response time, travel distance and time, number of delivery tours) of different rules for determining start times.

## 5.4 Operational On-Line Routing

The on-line routing consists in inserting dynamically arriving customer requests into one of the emerging routes in some "intelligent" or even "optimal" way. Section 6 describes some methods applicable to this purpose. The dynamic routing of customer requests can, e.g., be based on dynamic versions of standard (parallel) insertion procedures for the vehicle routing problem with time windows [20, 26], possibly followed by some post-optimisation of the customer sequences by means of, e.g., a Lin-Kernighan type local search heuristic [16, 18]. A reoptimisation is then triggered whenever a new customer request arrives. Again it might be useful not to directly communicate the resulting next stop to the driver, but to delay this decision for some time or as long as possible in order to gather more information.

For the purposes of the on-line routing, the mobile communication system has to be enhanced by a GPS as well as a GIS ("geographical information system") component. The on-line routing requires information about the current position of the drivers. To this end, the drivers can, e.g, be equipped with GPS receivers. Although GPS technology works quite accurately, it is fault prone to some extend, and sometimes a GPS signal cannot be received due to some interferences. Recently, GPS devices are getting cheaper; but equipping every driver with such a device can still be a too large investment for a small firm. Nowadays, there are also cellular phones available that are equipped with an additional GPS module. The GPS device should, however, work as an *automatic vehicle locating system* [19] and be able to automatically send positioning information to the web/WAP server if a corresponding request was sent from the server to this device. A GPS system is, however, not a conditio sine qua non. The mobile communication system described above allows to locate the driver exactly, directly after fulfilment of a customer request. The drivers' position between two customer nodes may then be simply estimated using appropriate measures of travel times and distances.

In addition, the customer locations must be geo-coded and travel distances and times need to be computed. Because the newspaper subscribers are known in advance, this task can be completed beforehand using, e.g, commercial geo-coding and route planning servers (see, e.g., [30]). Since in urban areas travel times can vary substantially, it can probably be necessary to use some on-line traffic information system (see, e.g., [11]) in order to obtain more accurate travel time information. On the other extreme, it might also be possible to work with a rather cheap solution that just uses "adjusted" Euclidean distances and travel time estimates in dependence of distances and time of day. Dillmann [9] and Dillmann et al. [10] report on good experiences with such a simplified approach.

Integrating the aforementioned mobile communication system and on-line routing algorithms is relatively straightforward. The algorithms can run on one or several servers in the background. The required input data is pushed forward from the web server whenever a new customer request arrives, and an algorithm's results are written back to the web server.

# 6 Overview of Some Methods for On-Line Routing

This section briefly reviews some selected methods for on-line routing. Especially, relatively simple dispatching rules based on stochastic arguments are contrasted with "reoptimisation" procedures that do not include some kind of lookahead capability. Comprehensive surveys on methods for dynamic routing problems can be found in [14, 15, 21, 22]. Stochastic approaches to dynamic routing problems were proposed by Bertsimas and van Ryzin [4, 5], whilst a tabu search procedure of Gendreau et al. [12] as well as a column generation method of Savelsbergh and Sol [25] can be mentioned as representatives of the second group of methods.

## 6.1 Reoptimisation Approaches

Savelsbergh and Sol [25] propose a column generation approach to a dynamic version of the pickup-and-delivery problem. Column generation is used here to generate a number of feasible routes. The routes as well as the master problem are then reoptimised at fixed points in time during the working day, taking all known unfulfilled transportation requests and partial routes operated so far into account. The algorithm proposed by Savelsbergh and Sol uses heuristics to solve the pricing subproblems as long as this way columns of negative reduced cost can be found. An advantage of the column generation procedure is that a large number of alternative high quality routes is available, in which new transportation requests can then be inserted for the purposes of reoptimising the problem. Furthermore, in order to make feasible integral solutions more quickly available, primal heuristics are often invoked. The method is used for solving a dynamic pickup-and-delivery problem of a large carrier. The loads are relatively large and have to be transported over longer distances. It can thus be assumed that the number of requests that dynamically arrive per time unit is relatively small, which makes such a complex and computational demanding method as a column generation procedure feasible.

Gendreau et al. [12] propose a tabu search heuristic for (repeatedly) solving a dynamic vehicle routing problem with time windows. The procedure is based on the *adaptive memory programming* framework suggested by Rochat and Taillard [24] for the vehicle routing problem and by Taillard et al. [29] as a general unifying metaheuristic solution approach. The method first constructs $I > 1$ initial solutions using a randomised insertion procedure. Generated solutions are improved by means of a tabu search method, and the best solutions encountered so far are stored in a pool. New solutions are then constructed by recombining routes taken from the pooled solutions, completing the resulting partial solution to a route plan and reapplying the tabu search method. A parallel version of this method is implemented by simultaneously applying the tabu search to different initial solutions on different processors. In order to take the dynamics into account, the method runs in the background working on improving solutions as long as no new event (new customer order, completion of an order) occurs. If a new customer request arrives, the request is inserted into every solution of the solution pool; then a reoptimisation is triggered. In case that a customer order is completed, the information about the next stop of the best solution obtained so far is transmitted to the vehicle. An obvious advantage of this procedure is that, as in the case of the column generation method above, a number of diverse (and probably good) solutions is available, into which new customer orders can be inserted. Moreover, in this case, a feasible solution that can be used to dynamically dispatch the vehicles is always available. As a disadvantage of this method, the large requirements on computing power and computing equipment can be mentioned. Furthermore, as all kind of reoptimisation methods, the procedure does not include a lookahead capability that tries to take the stochastic nature of the problem into account.

## 6.2 Stochastic Approaches

Bertsimas and van Ryzin [4, 5] apply relatively simple rules for dispatching vehicles in case of a stochastic and dynamic vehicle routing problem. Bertsimas and Simchi-Levi [3] give a further brief overview of these methods. Swihart and Papastavrou [28] present a similar analysis of a single vehicle pickup and delivery problem. In [4], Bertsimas and van Ryzin consider the *Travelling Repairman Problem*, and in [5] they extend the analysis to the case of multiple and capacitated vehicles. The problem is modelled as a queueing system. To this end, it is assumed that customer requests arrive independently according to a Poisson process, that customer locations are uniformly distributed in a square region of the Euclidean plane, and that the objective is to minimise the customers' total system time, which is the time elapsed between arrival of a demand and the time the server completes the service of the customer. Bertsimas and van Ryzin propose and analyse then a number of different dispatching rules and heuristics. For the case of a single uncapacitated vehicle they give the following heuristics and results:

1. In case of low traffic when the arrival rate approaches zero, the so-called stochastic median strategy (SQM) is optimal. The SQM strategy locates the server at the median of the region and serves the demand in a first come, first served (FCFS) order while returning to the median after each service.
2. A FCFS strategy that services the demands in the order in which they arrive is approximately 36 % above the optimum in case of light traffic.
3. The partitioning strategy extends the above simple rule by first dividing the service region into a number of equally sized subregions. The FCFS rule is then applied to each subregion. This heuristic gives a constant factor approximation in case of high traffic when the interarrival times approach the service times.
4. The travelling salesman (TS) strategy first waits until $n$ customer requests have arrived and afterwards builds an optimal travelling salesman tour over these customers. Then this process is repeated. In case of high traffic and a suitable large value of $n$, the policy gives a better constant factor approximation than the partitioning strategy above.
5. An even better approximation is obtained in case of high traffic by means of spacefilling curves (SC). A spacefilling curve $g$ is a continuous mapping of the

interval $C = [0, 1)$ onto the unit square $S = C \times C$ such that $\|g(a) - g(a')\| \leq 2\sqrt{|a - a'|}$ for all $a, a' \in C$. The SC strategy maps the customer coordinates on the interval $C$, and then repeatedly visits dynamically arriving customers by sweeping the interval $C$ in increasing order.

6. The simple "nearest neighbour" policy just services the closest available demand after every service completion. The method performed well within a number of simulation experiments performed by Bertsimas and Ryzin (note that the objective is to minimise the customers' system time).

In [5], Bertsimas and van Ryzin analyse dispatching rules for the case of $m > 1$ vehicles as well for the case of a capacitated vehicle that can serve at most $q$ customers. The heuristics below are proposed for the $m$ vehicle case.

1. The $m$ stochastic queue median policy extends the SQM strategy for a single vehicle by determining the $m$-median of the service region. Each vehicle returns then to the closest median after completing a single customer service. The policy is optimal in case of light traffic.
2. The randomised assignment strategy randomly assigns customer demands to each vehicle and then services each subprocess using a single server strategy.
3. The travelling salesman strategy waits as above until $n$ customers have arrived. Then an optimal travelling saleman tour is built and the next available vehicle is used to serve the route. Routes are served according to the FCFS principle. The procedure's performance factor depends on the number $m$ of vehicles.
4. The modified travelling salesman strategy divides the service region into $k$ subregions and applies the above TS strategy with parameter $n/k$ to each subregion. The procedure gives a constant factor approximation for the high traffic case.
5. The deterministic strategy works similar as the strategy above. The strategy collects however demands to routes for a given amount of time. Moreover, after a given amount of time has elapsed, the next subregion is served. The procedure has the same worst-case performance as the above TS strategy.
6. The independent partitioning policy divides the service region into $m$ subregions and applies any of the single server strategies for each vehicle.

Finally, Bertsimas and van Ryzin consider the following methods for the case of a capacitated vehicle:

1. The $q$RP-strategy divides the service area into a given number of subregions. If $q$ customer requests arrived in a certain subregion, the first available vehicle is used to serve these customers.
2. The $q$TP-strategy waits until $n \geq q$ customer requests have arrived. An optimal travelling salesman tour over the $n$ customers is then partitioned into feasible vehicle routes.

An advantage of the above procedures is their simplicity and that they take account of the problem's stochastic nature. Compared to reoptimisation procedures, the above rules seem, however, rather rigid and inflexible. The results of Bertsimas and van Ryzin depend furthermore on a number of limiting assumptions. It is questionable if the above approaches also perform reasonably good in case of non-Euclidean distances, non-uniformly distributed customers, non-Poisson arrival processes, and an objective that does not only consist in minimising the customers' system time. Nevertheless, the above strategies may give some guidelines for adding lookahead capabilities to reoptimisation procedures, e.g., by means of locating one or more suitable waiting points in the service region. As already mentioned in Sect. 5.1, a further possible approach for including stochastic considerations might be to first build subregions or clusters of customers by means of a two-stage stochastic programming procedure with recourse, and then to apply a suitable dynamic reoptimisation heuristic to each cluster/subregion. A further way of exploiting stochastic information might also be the construction of a priori routes that are then used as guidelines for dynamically dispatching vehicles. Bertsimas [2] finds closed-form expressions for the expected length of the routes to be performed by a capacitated vehicle if a giant tour is already given and customer demands are stochastic. He also provides lower and upper bounds on an optimal a priori giant tour. Thomas and White [31] discuss "anticipatory route selection" as a further interesting idea for tackling the problem. A single uncapacitated vehicle performs pickups and deliveries, earning a reward for pickups and adding to the total cost when travelling. The probability that a customer requests a pickup depends on the time $t$. The problem is to determine a policy for selecting the next node to visit such that the expected total cost of a trip from an origin to the destination is minimised. The problem is formulated as a Markov decision process and some structural results concerning the cost function and optimal policy are presented. Moreover, for a few small examples the optimal anticipatory policy is computed by means of stochastic dynamic programming. The required computational effort is however far too large, which underpins the need for heuristic procedures for solving the stochastic dynamic program.

# 7 Further Application Areas

The depicted case of subsequent newspaper delivery is a neat example of a specific class of applications characterised by a relatively high degree of dynamics. On the one hand, it is not known when new customer complaints arrive; on the other hand, customer requests should be fulfilled quickly in order to achieve low response times. An on-line connection is thus needed for ensuring the required bidirectional exchange of information between dispatcher, mobile co-worker, and the data base backend. The WAP standard employed for this purpose shows furthermore the advantage that any mobile device supporting this protocol can be used without installing additional software on this device. A simple bookmark suffices to grant easy access to the server. A permanent connection between the server and the mobile devices must however be established.

A number of other mobile business cases show a much less degree of dynamics. The information required by the mobile co-workers is then known before start of their work. Hence, a connection between the mobile co-worker and the data base has just to be established on demand, e.g., for downloading data, updating status information on the data base server, updating data stored on the mobile device, uploading data to the data base after work finish. These kind of applications require specific pieces of software to be installed on the mobile devices itself.

A further application example of the web/WAP-based system of Sect. 4 is the recording and supervision of delivery times for press wholesalers and newspaper publishers. In this business, accurate information about delivery times and route durations is of vital importance for future planning. Bieding [6, 7] describes a JAVA based system that exploits the mobile communication system of Sect. 4 for this purpose. The system extends an older solution based on RFID technology and has successfully been used in practice for about 200 carriers. A specific software allowing to download the required data from the mobile devices during the delivery is installed on the cellular phones. Using this device, the driver confirms arrival and departure and only after this confirmation a status update is sent to the data base. An associated web application allows then to receive detailed information about the delivery times and offers several customised reports.

Private mail companies exploit a similar system for controlling their mail delivery [33]. Detailed information about mail still to be delivered is accessed by couriers via a cellular phone and the described web/WAP-based system. The couriers work in an off-line status; only in case of a status change, information is sent to the data base server. Status changes occur, e.g., in case of a successful delivery, in case that mail is returned to the sender, in case of a second delivery attempt, etc. The associated web application serves here more as an information tool. A further event management system based on a set of rules for automatically monitoring the work process is however additionally installed. For example, if the courier did not download data before a given deadline, an e-mail is automatically sent to the dispatcher. Also, when some mail is not delivered after a given time span, an automatically generated e-mail informs the supervisor about this. The event management system helps to significantly reduce a dispatcher's and supervisor's workload; only in case of a failure, they are informed in detail, allowing them to react immediately. On-line routing methods can also play an important role in this kind of application, e.g., when pick-up orders should be sent to the "nearest" courier.

Mobile business applications are also widely used in the field of private nursing. The employees get detailed working plans sent to their mobile device. In most cases PDAs or also smart phones are used. Not only arrival and departure at the customer places are confirmed, also every single outpatient care activity will be confirmed and corresponding information sent to the company's data base. This way, the complete invoicing process is highly automated on a digital level, as recommended by the German health insurance companies. An additional on-line routing could help to identify and reduce delays.

Also for service technicians, mobile business applications provide several advantages. Service orders can easily be forwarded to the responsible field force employee with all specific information required to successfully complete orders. Assigning orders to technicians can be based, e.g., on criteria like distance and a technician's available capacity.

The advantages of such mobile business applications are obvious. Digital interactive information transmission without any media break by means of integrated applications and systems significantly reduces the amount of communication and enables real-time availability of important information. In fact, this provides essential transparency for monitoring, managing and optimising important processes outside the companies' buildings. In combination with location-based services and on-line routing additional advantages and cost reduction could be realised.

## 8 Summary and Outlook

This paper introduced a web/WAP server based mobile communication system that enables a firm to easily integrate mobile co-workers into its IT system. The use of the system was illustrated for the case of subsequent newspaper deliveries of a medium-sized publishing house in Germany. It was also shown that this system allows a straightforward integration of on-line vehicle routing algorithms, and some possible applicable methods were reviewed.

The mobile communication system described here exemplifies that todays communication and information technology allows to quickly build mobile communication systems that are based on a relatively simple system architecture as well as cheap and easy-to-handle system components. Such systems are of course not only useful for supporting on-line routing; they generally allow to implement mobile business processes in a firm's IT system, to streamline the information flows and to avoid thereby a number of dysfunctional "media breaks". This way the necessary prerequisites for supporting mobile logistic processes by means of optimisation and quantitative methods in general are established. The development, investigation and implementation of suitable optimisation methods for controlling logistic processes on-line is then surely a fruitful research area. Regarding techniques for supporting dynamic routing, there seems to exist a gap between stochastic approaches, on the one hand, and reoptimisation methods, on the other hand. While the first group of methods stresses the stochastic aspects of dynamic routing, the second family of methods, although powerful, miss lookahead capabilities. Integrating both methodological approaches in some way might thus be helpful for obtaining further improvements.

## References

[1] AlShaali S, Varshney U (2005) On the usability of mobile commerce. *International Journal of Mobile Communication* 3:29-37.
[2] Bertsimas DJ (1992) A vehicle routing problem with stochastic demand. *Operations Research* 40:574–585.
[3] Bertsimas DJ, Simchi-Levi D (1996) A new generation of vehicle routing research: Robust algorithms, addressing uncertainty. *Operations Research* 44:286–304.
[4] Bertsimas DJ, van Ryzin G (1991) A stochastic and dynamic vehicle routing problem in the Euclidean plane. *Operations Research* 39:601–615.

[5] Bertsimas DJ, van Ryzin G (1993) Stochastic and dynamic vehicle routing in the Euclidean plane with multiple capacitated vehicles. *Operations Research* 41:60–76.

[6] Bieding TJ (2001) Wann wird der Kunde wirklich beliefert. *Logistik Heute* 6.

[7] Bieding TJ (2004) Planning and controlling the daily delivery to the same customers – problems and adequate solutions. In: Fleischmann B, Klose A (eds) *Distribution Logistics - Advanced Solutions to practical Problems.* Lecture Notes in Economics and Mathematical Systems 544. Springer, Berlin Heidelberg New York.

[8] Bieding TJ, Ehrle C (2004) Nachlieferungen per Handy übertragen. *dnv - Der neue Vertrieb* 19.

[9] Dillmann R (2002) Strategic vehicle problems in practice – a pure software problem or a problem requiring scientific advice. In: Klose A, Van Wassenhove LN, Speranza MG (eds) *Quantitative Approaches to Distribution Logistics and Supply Chain Management.* Lecture Notes in Economics and Mathematical Systems 519. Springer, Berlin Heidelberg New York.

[10] Dillmann R, Becker B, Beckefeld V (1996) Practical aspects of route planning for magazine and newspaper wholesalers. *European Journal of Operational Research* 90:1–12.

[11] Fleischmann B, Gnutzmann S, Sandvoß E (2004) Dynamic vehicle routing based on online traffic information. *Transportation Science* 38:420–433.

[12] Gendreau M, Guertin F, Potvin J-Y, Taillard ED (1999) Parallel tabu search for real-time vehicle routing and dispatching. *Transportation Science* 33:381–390.

[13] Gendreau M, Laporte G, Séguin R (1996) Stochastic vehicle routing. *European Journal of Operational Research* 88:3–12.

[14] Gendreau M, Potvin J-Y (1998) Dynamic vehicle routing and dispatching. In: Crainic TG, Laporte G (eds) *Fleet Management and Logistics.* Kluwer Academic Publishers, London Dordrecht Boston.

[15] Ghiani G, Guerriero F, Laporte G, Musmanno R (2003) Real-time vehicle routing: Solution concepts, algorithms and parallel computing strategies. *European Journal of Operational Research* 151:1–11.

[16] Helsgaun K (2000) An effective implementation of the Lin-Kernighan traveling salesman heuristic. *European Journal of Operational Research* 126:106–130.

[17] Larsen A, Madsen OBG, Solomon MM (2004) The a priori dynamic traveling salesman problem with time windows. *Transportation Science* 38:459-472.

[18] Lin S, Kernighan BW (1973) An effective heuristic for the traveling-salesman problem. *Operations Research* 21:498–516.

[19] Mintsis G, Basbas S, Papaioannou P, Taxiltaris C, Tziavos IN (2004) Applications of GPS technology in the land transportation system. *European Journal of Operational Research* 152:399–409.

[20] Potvin J-Y, Rousseau J-M (1993) A parallel route building algorithm for the vehicle routing and scheduling problem with time windows. *European Journal of Operational Research* 66:331–340.

[21] Psaraftis HN (1988) Dynamic vehicle routing problems. In: Golden BL, Assad AA (eds) *Vehicle Routing: Methods and Studies*. North-Holland, Amsterdam.

[22] Psaraftis HN (1995) Dynamic vehicle routing: Status and prospects. *Annals of Operations Research* 61:143–164.

[23] Okhrin I, Richter K (2005) Mobile Business – Framework, Business Applications and Practical Implementation in Logistics Companies. *Arbeitsberichte Mobile Internet Business* Nr. 1, Frankfurt.

[24] Rochat Y, Taillard ED (1995) Probabilistic diversification and intensification in local search for vehicle routing. *Journal of Heuristics* 1:147–167.

[25] Savelsbergh M, Sol M (1998) Drive: Dynamic routing of independent vehicles. *Operations Research* 46:474–490.

[26] Solomon MM (1987) Algorithms for the vehicle routing and scheduling problem with time window constraints. *Operations Research* 35:254–265.

[27] Stewart WR, Golden BL (1983) Stochastic vehicle routing: A comprehensive approach. *European Journal of Operational Research* 14:371–385.

[28] Swihart MR, Papastavrou JD (1999) A stochastic and dynamic model for the single-vehicle pick-up and delivery problem. *European Journal of Operational Research* 114:447–464.

[29] Taillard ED, Gambardella LM, Gendreau M, Potvin J-Y (1998) Adaptive memory programming: A unified view of meta-heuristics. Technical Report IDSIA-19-98, IDSIA Lugano.

[30] Tarantilis CD, Diakoulaki D, Kiranoudis CT (2004) Combination of geographical information system and efficient routing algorithms for real life distribution operations. *European Journal of Operational Research* 152:437–453.

[31] Thomas BW, White CC (2004) Anticipatory route selection. *Transportation Science* 38:473487.

[32] Wichmann T (2004) Prozesse optimieren mit Mobile Solutions. Berlecon Research Report 03/2004. http://www.berlecon.de/research/reports.php.

[33] Wilhelm H (2006) Differenzierungsmerkmal Sendungsverfolgung. Presentation, 4. Lizenznehmerforum der Bundesnetzagentur, 22. November 2006, Bonn.http://www.bundesnetzagentur.de/media/archive/8380.pdf.

[34] Wirtz BW (2001) *Electronic Business*. Gabler, Wiesbaden.

[35] Zobel J (2001) *Mobile Business and M-Commerce*. Hanser, München.

process. The growth in intermodal traffic thus resulted in significant modifications to the structure of maritime and land-based transportation systems as well as in major increase of the volumes and value of intermodal traffic moved by each individual mode. Thus, for example, in 2003, for the first time ever, intermodal freight surpassed coal as a source of revenue for major, Class I, U.S. railroads, representing 23% of the carriers' gross revenue [31]. The growth of intermodal rail traffic in the U.S., which reached 11 million trailers (26% of total) and containers (76%) in 2004, is the direct result of the rapid growth in the use of containers for international trade, imports accounting for the majority of the intermodal activity [31].

Governmental policy may also contribute to re-structuring intermodal transportation and shifting parts of the land part of the journey from trucking toward rail and water (interior and coastal navigation). This is, for example, the main focus of the European Union as stated in its 2001 White Paper on transportation [20]. The reason for this is to reduce road congestion and promote environmentally friendlier modes of transportation. The instruments favored to implement such policies vary from road taxes to penalize truck-based transportation to the support of new rail services for intermodal traffic.

The performance of intermodal transportation depends directly on the performance of the key individual elements of the chain, navigation companies, railroads, motor carriers, ports, etc., as well as on the quality of their interactions regarding operations, information, and decisions. The Intelligent Transportation Systems and Internet-fueled electronic business technologies provide the framework to address the latter challenges. Regarding the former, carriers and terminals, on their own or in collaboration, strive to continuously improve their performance. Railroads are no exception. Indeed, for intermodal as for general traffic, railroads face significant challenges to efficiently compete with trucking in offering customers timely, flexible, and "low"-cost, long-haul transportation services.

Railroads are rising to the challenge by proposing new types of services and enhanced performances. Thus, North-American railroads have created intermodal subdivisions that operate so-called "land-bridges" providing efficient container transportation by long, double-stack trains between the East and the West coasts and between these ports and the industrial core of the continent (so-called "mini" land-bridges). Most North-American railroads are now enforcing some form of scheduled service. In Europe, where congestion has long forced the scheduling of trains, the separation of the infrastructure ownership from service providing increases the competition and favors the emergence of new carriers and services. Moreover, the expansion of the Community to the east provides the opportunity to introduce new services that avoid the over-congested parts of the European network. New container and trailer-dedicated shuttle-train networks are thus being created within the European Community.

The planning and management processes of these new railroad-based intermodal systems and operations are generally no different from those of "traditional" systems in terms of issues and goals, profitability, efficiency, and customer satisfaction. The "new" operating policies introduce, however, elements and requirements into the planning processes which, from an Operations Research point of view, require that models be revisited and appropriate methods be devised.

This paper aims to discuss some of these issues and developments. It focuses on the tactical planning of rail intermodal services in North America and Europe and is based on a number of observations and on-going projects. Its goal is to be informative, point to challenges, and identify opportunities for research aimed at both methodological developments and actual applications.

## 2 Intermodal and Rail-Based Transportation

Many transportation systems are multimodal, their infrastructure supporting various transportation modes, such as truck, rail, air, and ocean/river navigation, carriers operating and offering transportation services on these modes. Then, broadly defined, intermodal transportation refers to the transportation of people or freight from their origin to their destination by a sequence of at least two transportation modes. Transfers from one mode to the other are performed at intermodal terminals, which may be a sea port or an in-land terminal, e.g., rail yards, river ports, airports, etc. Although both people and freight can be transported using an intermodal chain, in this paper, we focus on the latter.

The fundamental idea of intermodal transportation is to consolidate loads for efficient long-haul transportation performed by large ocean vessels and, on land, mostly by rail and truck. Local pick-up and delivery is usually performed by truck. Most of the freight intermodal transportation is performed by using containers. Intermodal transportation is not restricted, however, to containers and intercontinental exchanges. For instance, the transportation of express and regular mail is intermodal, involving air and land long-haul transportation by rail or truck, as well as local pick up and delivery operations by truck [16]. Moving trailers on rail is also identified as intermodal. In this paper, we focus on container and trailer-based transportation by railroads.

Intermodal transportation systems and railroads may be described as being based on consolidation. A consolidation transportation system is structured as a hub-and-spoke network, where shipments for a number of origin-destination points may be transferred via intermediate consolidation facilities, or hubs, such as airports, seaport container terminals, rail yards, truck break-bulk terminals, and intermodal platforms. An example of such a network with three hubs and seven regional terminals is illustrated in Fig. 1 [7]. In hub-and-spoke networks, low-volume demands are first moved from their origins to a hub where traffic is sorted (classified) and grouped (consolidated). The aggregated traffic is then moved in between hubs by efficient, "high" frequency and capacity, services. Loads are then transferred to their destination points from the hubs by lower frequency services often utilizing smaller vehicles. When the level of demand is sufficiently high, direct services may be run between a hub and a regional terminal.

Railroads operate most of their services according to a double-consolidation policy based on a series of activities taking place at rail hubs, the so-called classification

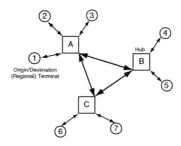

**Fig. 1.** A hub-and-spoke network [7]

or marshaling yards. The first consolidation activity concerns the sorting and grouping of railcars into blocks. A block is thus made up of cars of possibly different origins and destinations, which travel as a single unit between the origin and destination of the block. Consequently, the only operation that could be performed on a block at a yard which is not its destination is to transfer it from one service to another. The second consolidation activity taking place at yards, known as train make up, concerns the grouping of blocks into trains. Although a hub-and-spoke network structure results in a more efficient utilization of resources and lower costs for shippers, it also incurs a higher amount of delays and a lower reliability due to longer routes and the additional operations performed at terminals. Carriers thus face a number of issues and challenges in providing services that are simultaneously profitable and efficient for the firm and high quality and cost effective for customers. Operations Research has contributed a rich set of models and methods to assist addressing these issues and challenges at all levels of planning and management, classically identified as strategic (long term), tactical (medium term), and operational (short term). A more in-depth treatment of these topics may be found, for example, in the reviews of Cordeau, Toth, and Vigo [11], Crainic [12, 13], Crainic and Laporte [17], and Crainic and Kim [16]. In this paper, we focus on tactical planning issues.

## 3 New Rail Intermodal Services

Rail transportations systems evolved according to the geographic, demographic, economic, and sociological characteristics of the countries and continents they belong to. North American and European railroads were no exception. Yet, recently, a number of similar trends emerge. Traditional North American railroad operating policies were based on long-term contracts for the transportation of high volumes of mostly bulk commodities. Cost per ton/mile (or km) was the main performance measure, with somewhat little attention being paid to delivery performance. Consequently, rail services in North America, and mostly everywhere else in the world, were organized around loose schedules, indicative cut-off times for customers, "go-when-full" operating policies, and significant marshaling activities in yards. This resulted in rather long and unreliable trip times that generated both inefficient asset utilization and loss of market share. This was not appropriate for the requirements of intermodal transportation and the North American rail industry responded through [14]:

1. A significant re-structuring of the industry through a series of mergers, acquisitions, and alliances which, although far from being over, has already drastically reduced the number of companies resulting in a restricted number of major players.
2. The creation of separate divisions to address the needs of intermodal traffic, operating dedicated fleets of cars and engines, and marshaling facilities (even when located within regular yards). Double-stack convoys have created the land-bridges that ensure an efficient container movement across North America.
3. An evolution toward planned and scheduled modes of operation and the introduction of booking systems and full-asset-utilization operating policies.

Booking systems bring intermodal rail freight services closer to the usual mode of operation of passenger services by any regular mode of transportation, train, bus, or air. In this context, each class of customers or origin-destination market has a certain space allocated on the train and customers are required to call in advance and reserve the space they require. The process may be phone or Internet based but is generally automatic, even though some negotiations may occur when the train requested by the customer is no longer available. This new approach to operating intermodal rail services brings advantages for the carrier, in terms of operating costs and asset utilization, and the customers (once they get used to the new operating mode) in terms of increased reliability, regular and predictable service and, eventually, better price. A full-asset-utilization operation policy generally corresponds to operating regular and cyclically-scheduled services with fixed composition. In other words, given a specific frequency (daily or every given number of days), each service occurrence operates a train of the same capacity (length, number of cars, tonnage) and composition, that is, the same blocks make up all the occurrences of the service, each block displaying a fixed definition: origin, destination, number of total cars, and number of cars for each origin-destination included in its composition.

A full-asset-utilization operation policy generally corresponds to operating regular and cyclically-scheduled services with fixed composition. In other words, given a specific frequency (daily or every given number of days), each service occurrence operates a train of the same capacity (length, number of cars, tonnage) and composition, that is, the same blocks make up all the occurrences of the service, each block displaying a fixed definition: origin, destination, number of total cars, and number of cars for each origin-destination included in its composition.

Assets, engines, rail cars and even crews, assigned to a system based on full-asset-utilization operation policies can then "turn" continuously following circular routes and schedules (which include maintenance activities for vehicles and rest periods for crews) in the time-space service network, as schematically illustrated in Fig. 2 for a system with three yards and six time periods [3]. The solid lines in the service-network (left) part of Fig. 2 represent services. There is one service from node 1 to node 3 (black arcs) and one service from node 3 to node 2 (gray arcs), both

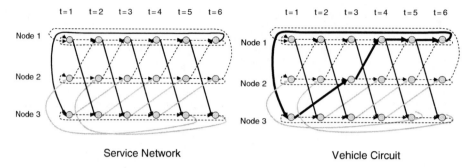

**Fig. 2.** Full-asset-utilization-based service network and vehicle circuit [3]

with daily frequency. Dotted arcs indicate repositioning moves (between different nodes) and holding arcs (between different time representations of same node). One feasible vehicle circuit in the time-space service network is illustrated in the right part of Fig. 2. The vehicle operates the service from node 3 to node 2, starting in time period 1 and arriving in time period 3. Then from period 3 to period 4 the vehicle is repositioned to node 1, where it is held for two time periods. In period 6 the vehicle operates the service from node 1 to node 3, arriving at time 1 where the same pattern of movements starts all over again. The planning of systems operating according to such policies requires the development of new models and methods, as described in the next section.

Most Western Europe railroads have for a long time now operated their freight trains according to strict schedules, similarly to their passenger trains. This facilitated both the interaction of passenger and freight trains and the quality of service offered to customers. Particular characteristics of infrastructure (e.g., low overpasses and infrastructure for electric traction) and territory (short inter-station distances) make for shorter trains than in North America and forbid double-stack trains. Booking systems are, however, being implemented and full-asset-utilization and revenue management operating policies are being contemplated. Moreover, intermodal shuttle-service networks are being implemented in several regions of the European Union to address the requirements of the European Commission policy and the congested state of the infrastructure (e.g., [1, 27]).

Indeed, European railroads face a number of particular challenges. First, the rail infrastructure, as almost the entire transportation infrastructure in Europe, is very congested. Second, the liberalization of the rail industry in Europe has led to the separation of the traditional national rail companies into infrastructure owners and service operators. The former manage the infrastructure and associated network capacity, while the latter operate trains according to the capacity acquired from the infrastructure managers. This liberalization favors the emergence of new rail operators providing specialized services, in particular intermodal rail shuttle services between cities with high traffic demand.

The limited capacity of most of the infrastructure, at least in the western part of the network, together with the increasing number of operators, forces the allocation of capacity according to pre-defined routes and times, which makes planning decisions and the efficient utilization of resources more difficult. The European Union, the member states, and the corresponding rail authorities are implementing steps, however, toward interoperability and an interconnected trans-European rail network for freight trains, the so-called freight freeways [19]. As a result, one assists at the emergence of large service networks across the European continent operated by single operators or by alliances of operators, similar to those seen in the airline industry. The resulting service networks will be complex to plan and operate and appropriate models and methods must be developed. Pedersen and Crainic [27] detail the case and propose a first service network design model.

To alleviate the congestion in the "central" part of the network while working toward the goal of increasing the market share of rail and navigation, new intermodal services are being studied using the networks of countries that have recently joined the Union. Andersen and Christiansen [1] describe such a project. The Longchain Polcorridor study [24] aims to develop a new intermodal transport corridor between Northern and South-Eastern Europe taking advantage of previously unused railway capacity in Poland, the Czech Republic and Austria, and thus create a fast and reliable transport solution than can compete with the more traditional route through Germany. The authors propose a formulation to determine an optimal service level and design that accounts for both operating costs and a number of service quality criteria. An extensive network of inland waterways, sea transport, trucking services, and other railway lines will be used as distribution networks at the extremities of the new network. This requires external synchronization of schedules with partner carriers. Internal synchronization is also required to account for power-equipment switching at particular borders due to different technical standards between participating national railroads. Andersen, Crainic, and Christiansen [3] propose formulations for this case.

## 4 Impact on Planning Models

A study of the trends observed in North America and Europe, illustrated by the cases mentioned in the previous section, indicates a number of converging issues. One may sum up these issues by noticing that the operations and asset management of intermodal railroads are more and more similar to those of long-haul passenger transportation, airlines and fast rail, in particular. Services are thus precisely scheduled and service space is booked in advance. Moreover, schedules are repetitive (cyclic) and synchronized, both internally among the railroad's own services and externally with those of partner carriers. This implies tighter consolidation, classification, transfer, and make-up operations at terminals, as well as scheduling assets for maximum but efficient utilization.

Traditionally, planning was performed through a series of tasks, planning models being used one after another to address particular issues: design of the service network and schedules, power (locomotive) assignment and management, empty railcar repositioning and fleet management, and so on. This approach was not particular to railroads or freight transportation; it was typical of the traditional management structure and planning processes of most industrial firms facing complex issues. From an Operations Research point of view, it reflected the limitations of our capabilities in addressing large-scale combinatorial formulations with complex additional constraints. Managerial structures evolve, however, and our capabilities are continuously being enhanced, both in terms of computer power and methodology sophistication. The trend toward integrated models addressing in a comprehensive formulation several issues previously treated separately, initially observed within the airline industry (e.g., [5]) is now influencing the development of planning methods for railroad operations, most particularly within the field of intermodal transport.

To briefly illustrate these issues and the corresponding challenges, we turn to service network design in the context of full-asset-utilization operating policies. We conclude the section by identifying a number of other "new" planning issues offering exiting research perspectives.

## 4.1 Service Network Design

Recall that service network design is concerned with the planning of operations related to the selection, routing, and scheduling of services, the consolidation and make-up activities at terminals, and the routing of freight of each particular demand through the physical and service network of the company (see, for example, the surveys of Crainic for service network design [12] and long-haul land transportation [13], Crainic and Kim [16] and Macharis and Bontekoning [25] for intermodal transportation, Christiansen *et al.* [10, 9] for maritime navigation, and Cordeau, Toth, and Vigo [11] for railroads. These activities are a part of tactical planning at a system-wide level. The two main types of decisions considered in service network design address the determination of the service network and the routing of demand. In the railroad context, the former refers to selecting the train routes and attributes, such as the frequency or the schedule of each service. The latter is concerned with the itineraries that specify how to move the flow of each demand, including the services and terminals used, the operations performed at these terminals, etc. The objective is generally concerned with the minimization of a global measure of the performance of the system that includes the operating costs of providing services, performing yard operations, and moving freight, as well as service-quality measures usually based on delays to equipment and loads. The term "generalized cost" is often used in these cases.

The basic service network design mathematical models take the form of deterministic, fixed cost, capacitated, multicommodity network design (CMND) formulations [26, 12, 16]. Let $S$ represent the service network, defined on a graph representing the physical infrastructure of the system (yards, stations, and the rail links connecting them), which specifies the transportation services that could be offered. Each potential service $s \in S$ is characterized by a number of attributes such as its route, capacity measured in number of vehicles, length, total weight, or a combination thereof, service class indicating the speed and priority, as well as, eventually power type, preferred traffic or restrictions, etc. When schedules are to be determined, a time-space network $G = \{\mathcal{N}, \mathcal{A}\}$ is introduced (see Fig. 2). Nodes representing the terminals (yards and stations) of the system are repeated at all periods (e.g., days) of the considered planning horizon (e.g., a week) yielding the set $\mathcal{N}$. Nodes in $\mathcal{N}$ representing the same terminal at two consecutive time periods are connected by holding arcs, and departure times from origin, as well as arrival at and departure times from intermediary stops are associated to each service. Set $\mathcal{A}$ is then the union of the holding and service arcs. The service network is used to move commodities $p \in \mathcal{P}$ defined by their origins, destinations, the period of availability at origin and, eventually the due date at destination, the type of product or vehicle to be used, priority class, and so on. The demand for product $p$ is denoted $d_p$. Traffic moves according to itineraries defined within the model as service paths $l \in \mathcal{L}^p$ for commodity $p$, each specifying the intermediary terminals where operations (e.g., consolidation or transfer) are to be performed and the sequence of services between each pair of consecutive terminals where work is performed.

*Flow routing* decisions are then represented by decision variables $h_l^p$ indicating the volume of product $p$ moved using its itinerary $l \in \mathcal{L}^p$, and *service selection* decision variables $y_s$, $s \in \mathcal{S}$, define whether the particular service is operated (i.e., it will leave at the associated departure time) or not. Let $y = \{y_s\}$ and $h = \{h_l^p\}$ be the decision-variable vectors. Let also $f_s$ denote the "fixed" cost of operating service $s$ and $c_l^p$ stand for the unit transportation cost along itinerary $l \in \mathcal{L}^p$ of commodity $p$. The core service network design model minimizes the total generalized system cost, while satisfying the demand for transportation and the service standards:

$$\text{Minimize} \quad \sum_{s \in \mathcal{S}} f_s y_s + \sum_{p \in \mathcal{P}} \sum_{l \in \mathcal{L}^p} c_l^p h_l^p + \phi(y, h) \tag{1}$$

$$\text{subject to} \quad \sum_{l \in \mathcal{L}^p} h_l^p = d_p, \quad p \in \mathcal{P}, \tag{2}$$

$$y_s \in \{0, 1\}, \quad s \in \mathcal{S}, \tag{3}$$

$$h_l^p \geq 0, \quad l \in \mathcal{L}^p, p \in \mathcal{P}, \tag{4}$$

$$(y, h) \in \chi, \tag{5}$$

where $\phi(y, h)$ indicates additional restrictions, e.g., service capacity, expressed as utilization targets, which may be allowed to be violated at the expense of additional penalty costs. Relations 5 stand for the classical linking constraints (i.e., no flow may use an unselected service), as well as for additional constraints reflecting particular characteristics, requirements, restrictions, and policies (e.g., particular routing or load-to-service assignment rules) particular to the carrier considered in application.

The class of models represented by the previous formulation does not account for the utilization of assets. Yet, to adequately plan operations according to a full-asset-utilization operating policy requires the asset circulation issue to be integrated into the service network design model. Adding constraints enforcing the conservation of the flow of vehicles at terminals is the first step in reaching this goal. These constraints take the form

$$\sum_{s \in \mathcal{S}} y_{si^+} - \sum_{s \in \mathcal{S}} y_{si^-} = 0, \quad i \in \mathcal{N}, \tag{6}$$

where $si^+$ indicates the services that arrive and stop or terminate at node (yard) $i \in N$, while $si^-$ stands, symmetrically, for the services that initiate their journeys or stop and depart from node $i$. The resulting models, denoted *design-balanced capacitated multicommodity network design* (DBCMND) by Pedersen, Crainic, and Madsen [28], account for coherent movements in and out of terminals (particularly when"empty" movements are allowed) and yield cyclic and repetitive schedules for the fleet of vehicles associated to services.

This generalization of the CMND model has not been studied much. A few applications may be found in planning maritime liner [9] or ferry [22] routes and express postal services, e.g., [6, 21], where vehicles, ships and airplanes, have high acquisition and utilization costs and their management is central to the efficient operation of the system. For land-based carriers, while empty-vehicle considerations were usually part of the most comprehensive service network design models, the emphasis was not on the management of the fleet. Thus, for example, Powell [29] considered the balance of loaded and empty truck balance at terminals in a static model for designing Less-Than-Truckload motor-carrier services that did not consider vehicle schedules. For rail, Crainic, Ferland, and Rousseau [15] (see also [18]) addressed the issue by adding a product to represent the demand for empty-car repositioning movements. The model was static and no asset schedule or route considerations were explicitly included. More recently, Smilowitz, Atamtürk, and Daganzo [30] developed a time-dependent formulation similar to the one presented above for truck operations within an express postal network and proposed a particularly tailored procedure where, first, the linear programming relaxation of the problem is solved (approximately, for large problem instances) using column generation and, second, a feasible solution is obtained by applying repetitively a sequence of rounding and cut-generation procedures.

The DBCMND is a difficult problem with an added "complexity layer" compared to the CMND and much work is required to study it and develop efficient exact and heuristic solution methods. A few efforts are under way. Pedersen and Crainic [27] studied the design of a network of shuttle intermodal trains in Europe using a DBCMND model that included detailed yard operations (excluding car classification). The resulting formulation was sufficiently small, however, to allow commercial software to be used. Pedersen, Crainic, and Madsen [28] introduced arc and cycle-based DBCMND formulations and proposed a two-stage, tabu search-based meta-heuristic that is shown to be efficient for problem instances up to 700 service arcs and 400 commodities. Andersen, Crainic, and Christiansen [3] extended the arc-based formulation to account for coordination of multiple fleets and synchronization of schedules among subsystems. The analysis was performed within the scope of the design of new north-south intermodal services in Central Europe and emphasized the benefit of increased inter-system integration and coordination. The same authors also studied various DBCMND formulations, where product flows were represented either by arc or path variables, while design decisions were represented either by arcs or cycle variables [2]. Notice that the latter correspond to circuits of vehicles, the service selection (design) decisions becoming thus implicit in the selection of strategies for fleet management. Commercial software was used for experimentation and results showed a very good computational behavior for cycle-based formulations in terms of computational efficiency and quality of the final solution (when the optimal solution could not be reached within the time available). Recently, Andersen *et al.* [4] proposed a first branch-and-price algorithm for the generic cycle-based DBCMND formulation.

The contributions briefly reviewed above are very encouraging, but significant work is still required on models, algorithms, and applications. Regarding models, we need to better understand the DBCMND class of formulations and their properties. Initial results seem to indicate that cycle-based formulations outperform arc-based ones, but more in-depth studies are required to fully characterize the various formulations and explain their respective behaviors. Work is also required in developing tighter lower and upper bounds on the optimal solution to these formulations. Lagrangean relaxation and decomposition approaches have provided interesting results for other classes of network design problems and are worthy of investigation in developing good lower bounds for DBCMND problems. Using the solutions of lower bound methods to compute "good" upper bounds proved difficult for CMND problems and we expect it to present an even greater challenge for DBCMND problems for which identifying feasible solutions is proving to be far from trivial [28].

Turning to applications, stating a DBCMND formulation is generally only the first step to a complete model. Actual railroad planning applications bring a rich set of additional constraints that add both to the realism and the complexity of the formulation. Pedersen and Crainic [27] thus discussed the need for a more general definition of "period" within time-dependent formulations to capture adequately the time intervals when services overlap at terminals and inter-service transfers may be performed. The authors also emphasized the need for a more detailed representation of terminal operations than it is usually the case in service network design models to capture their delay and capacity impacts on the general performance of the system. This aspect is also emphasized by Andersen, Crainic, and Christiansen [3] who detailed the operations in terminals connecting the system studied to adjacent maritime and land systems, and presented a first quantification of the benefits of synchronizing services both internally, among services using possibly different vehicle fleets, and externally with services belonging to neighboring systems. The authors also examined and quantified the impact of a number of fleet-management considerations, such as limits on the length of vehicle routes and bounds on the number of departures for particular services. One need to follow up on these early efforts by focusing on two main areas. One the one hand, the study of integrating fleet management and service network design must be continued to identify relevant issues and the most appropriate modeling approaches, and to analyze their impact on the resulting transportation plan and, ultimately, on the performance of the rail system. On the other hand, the models must be enriched to account for other important planning issues, such as congestion related to yard (and, eventually, line) operations and the management of more than one class of assets.

Last but not least, significant algorithmic work is required. The methods proposed so far addressed simple model formulations and have often been tested on problem instances that do not cover the full dimensions of actual large-scale applications. Both exact and meta-heuristic solution methods must be developed for the models described above. Given the dimensions and complexity (e.g., in the number of interacting and possibly conflicting components) of these formulations, we expect parallel optimization approaches to play an important role in addressing real-life applications.

### 4.2 Additional Issues

Many other issues related to the planning and operations of intermodal and, more generally, consolidation-based freight transportation offer rich research challenges and opportunities.

Consider, for example, that, although bookings tend to "smooth" out demand, the variability inherent to the system is not altogether eliminated since regular operations tend to be disrupted by a number of phenomena. Thus, for example, ocean liner ships do not always arrive at container port terminals according to schedule and custom and security verifications may significantly delay the release of containers. When this occurs, rail intermodal operations out of the corresponding ports are severely strained: there might be several days without arrivals, followed by a large turnout of arriving containers. Optimization approaches [14] may be used to adjust service over a medium-term horizon in such a way that a full-asset-utilization policy is still enforced, but a certain amount of flexibility is added to services to better fit service and demand. Such approaches may become even more effective when appropriate information sharing and container-release time mechanisms are implemented.

Among the other relevant challenging research issues, let's not forget the explicit consideration of stochastic elements in tactical planning models. Preliminary results [23] indicate that the plans thus obtained are different and "better" from a robustness point of view than those of traditional deterministic models, but much more work is needed in this field. Terminal planning issues also require attention. While the literature dedicated to container port terminals is rather rich, there is almost nothing dedicated to rail yards within the intermodal context (the work by Bostel and Dejax [8] is the only exception we are aware of and it is directed toward an innovative but as yet not implemented rail transportation system). On a more operational level, work is required relative to detailed fleet management procedures to mitigate the impact of incidents and accidents on service and to guide the process of getting back to normal operations following such disruptions.

## 5 Conclusions

We have discussed a number of service and operating strategies railroads propose to improve the performance of their operations, increase their market share of intermodal traffic, and efficiently compete with trucking in offering customers timely, flexible, and "low"-cost transportation services. This evolution, including the advance bookings and full-asset-utilization policies increasingly implemented by existing and planned railroad intermodal systems, challenges current models and methods for the design of services and the management of operations. Focusing on tactical planning issues, we have briefly examined these impacts and have identified research challenges and opportunities.

## Acknowledgments

Partial funding for this project has been provided by the Natural Sciences and Engineering Research Council of Canada (NSERC) through its Discovery Grants and Chairs and Faculty Support programs

## References

[1] Andersen, J. and Christiansen, M. Optimal Design of New Rail Freight Services. *Journal of the Operational Research Society*, 2007. forthcoming.

[2] Andersen, J., Crainic, T.G., and Christiansen, M. Service Network Design with Asset Management: Formulations and Comparative Analyzes. Publication CIRRELT-2007-40, CIRRELT - Centre de recherche sur les transports, Université de Montréal, Montréal, QC, Canada, 2007.

[3] Andersen, J., Crainic, T.G., and Christiansen, M. Service Network Design with Management and Coordination of Multiple Fleets. *European Journal of Operations Research*, 2007. forthcoming.

[4] Andersen, J., Grønhaug, R., Christiansen, M., and Crainic, T.G. Branch-and-Price for Service Network Design with Asset Management Constraints. Publication-2007-55, CIRRELT - Centre de recherche sur les transports, Université de Montréal, Montréal, QC, Canada, 2007.

[5] Barnhart, C., Johnson, E.L., Nemhauser, G.L., and Vance, P.H. Crew Scheduling. In Hall, R.W., editor, *Handbook of Transportation Science*, pages 517–560. Kluwer Academic Publishers, Norwell, MA, second edition, 2003.

[6] Barnhart, C. and Schneur, R.R. Network Design for Express Freight Service. *Operations Research*, 44(6):852–863, 1996.

[7] Bektaş, T. and Crainic, T.G. A Brief Overview of Intermodal Transportation. In Taylor, G.D., editor, *Logistics Engineering Handbook*. Taylor and Francis Group, 2007. forthcoming.

[8] Bostel, N. and Dejax, P. Models and Algorithms for Container Allocation Problems on Trains in a Rapid Transshipment Shunting Yard. *Transportation Science*, 32(4):370–379, 1998.

[9] Christiansen, M., Fagerholt, K., Nygreen, B., and Ronen, D. Maritime Transportation. In Barnhart, C. and Laporte, G., editors, *Transportation*, volume 14 of *Handbooks in Operations Research and Management Science*, pages 189–284. North-Holland, Amsterdam, 2007.

[10] Christiansen, M., Fagerholt, K., and Ronen, D. Ship Routing and Scheduling: Status and Perspectives. *Transportation Science*, 38(1):1–18, 2004.

[11] Cordeau, J.-F., Toth, P., and Vigo, D. A Survey of Optimization Models for Train Routing and Scheduling. *Transportation Science*, 32(4):380–404, 1998.

[12] Crainic, T.G. Network Design in Freight Transportation. *European Journal of Operational Research*, 122(2):272–288, 2000.

[13] Crainic, T.G. Long-Haul Freight Transportation. In Hall, R.W., editor, *Handbook of Transportation Science*, pages 451–516. Kluwer Academic Publishers, Norwell, MA, second edition, 2003.

[14] Crainic, T.G., Bilegan, I.-C., and Gendreau, M. Fleet Management for Advanced Intermodal Services - Final report, Report for Transport Canada. Publication CRT-2006-13, Centre de recherche sur les transports, Université de Montréal, Montréal, QC, Canada, 2007.

[15] Crainic, T.G., Ferland, J.-A., and Rousseau, J.-M. A Tactical Planning Model for Rail Freight Transportation. *Transportation Science*, 18(2):165–184, 1984.

[16] Crainic, T.G. and Kim, K.H. Intermodal Transportation. In Barnhart, C. and Laporte, G., editors, *Transportation*, volume 14 of *Handbooks in Operations Research and Management Science*, pages 467–537. North-Holland, Amsterdam, 2007.

[17] Crainic, T.G. and Laporte, G. Planning Models for Freight Transportation. *European Journal of Operational Research*, 97(3):409–438, 1997.

[18] Crainic, T.G. and Rousseau, J.-M. Multicommodity, Multimode Freight Transportation: A General Modeling and Algorithmic Framework for the Service Network Design Problem. *Transportation Research Part B: Methodological*, 20:225–242, 1986.

[19] European Commission. A Strategy for Revitalizing the Community's Railways. Whipe paper, Office for official publications of the European Communities, Luxembourg, 1996.

[20] European Commission. European Transport Policy for 2010: Time to Decide. Whipe paper, Office for official publications of the European Communities, Luxembourg, 2001. Available on line at http://www.europa.eu.int/comm/energy_transport/en/lb_en.html.

[21] Kim, D., Barnhart, C., Ware, K., and Reinhardt, G. Multimodal Express Package Delivery: A Service Network Design Application. *Transportation Science*, 33(4):391–407, 1999.

[22] Lai, M.F. and Lo, H.K. Ferry Service Network Design: Optimal Fleet Size, Routing and Scheduling. *Transportation Research Part A: Policy and Practice*, 38:305–328, 2004.

[23] Lium, A.-G., Crainic, T.G., and Wallace, S. A Study of Demand Stochasticity in Service Network Design. *Transportation Science*, 2007. forthcoming.

[24] Longchain Polcorridor Project Description. EUREKA, http://www.eureka.be/.

[25] Macharis, C. and Bontekoning, Y.M. Opportunities for OR in Intermodal Freight Transport Research: A Review. *European Journal of Operational Research*, 153(2):400–416, 2004.

[26] Magnanti, T.L. and Wong, R.T. Network Design and Transportation Planning: Models and Algorithms. *Transportation Science*, 18(1):1–55, 1984.

[27] Pedersen, M.B. and Crainic, T.G. Optimization of Intermodal Freight Service Schedules on Train Canals. Publication CÏRRELT-2007-51, CIRRELT - Centre de recherche sur les transports, Université de Montréal, Montréal, QC, Canada, 2007.

[28] Pedersen, M.B., Crainic, T.G., and Madsen, O.B.G. Models and Tabu Search Meta-heuristics for Service Network Design with Asset-Balance Requirements. *Transportation Science*, 2007. forthcoming.

[29] Powell, W.B. A Local Improvement Heuristic for the Design of Less-than-Truckload Motor Carrier Networks. *Transportation Science*, 20(4):246–357, 1986.

[30] Smilowitz, K.R., Atamtürk, A, and Daganzo, C.F. Deferred Item and Vehicle Routing within Integrated Networks. *Transportation Research Part E: Logistics and Transportation*, 39:305–323, 2003.

[31] U.S. Department of Transportation. Freight in America. Technical report, Research and Innovative Technology Administration, Bureau of Transportation Statistics, Washington D.C., 2006. Available on line at http://www.bts.gov/publications/freight_in_america/.

# C.A.s.S.a.n.D.r.A.: Computerized AnalysiS for Supply ChAiN DistRibution Activity

Laura Di Giacomo[1], Ettore Di Lena[2], Giacomo Patrizi, Livia Pomaranzi, and Federico Sensi[5]

[1] LIX, École Polytechnique, Palaiseau, France digiacomo@lix.polytechnique.fr
[2] Ente Nazionale Energia Elettrica, Rome, Italy ettore.dilena@enel.it
[3] La Sapienza, Università di Roma, Italy g.patrizi@caspur.it
[4] Chefaro Pharma Italia, Rome, Italy livia.pomaranzi@chefaro.it
[5] Enprovia Software Engineering, Italy Federico-Sensi@libero.it

**Summary.** Supply Chain Management (SCM) is an important activity in all producing organizations. To determine efficient policies over a given time interval, it is necessary to consider a simultaneous dynamic estimation and optimization algorithm over the disposable data base. The aim of this paper is to present the Data Driven algorithm, describe its implementation and show through some preliminary applications its potential advantages. To ensure that Certainty Equivalent optimal policies prevail this aspect will be analyzed.

**Key words:** Optimal estimation and control, Dynamic nonlinear, Supply chain management

## 1 Introduction

Positive synergisms can be enhanced by appropriate Supply Chain Management (SCM) policies in an organization. These benefits can be obtained by studying suitable dynamic and nonlinear mathematical models of the activities to be pursued.

These dynamic and nonlinear processes permit a careful monitoring of the changes in the actions in time and the mutual combination of the results of different actions [16]. However, if these decision making mechanisms are not carefully designed, since activities may have greatly differing requirements, instead of achieving greater levels of efficiency, the supply chain policy formulated may incur in severe instabilities, indeterminacy in the process and be technically uncontrollable [8].

Mathematical System Theory [12] permits the formulation of Dynamical systems, under various conditions, as non anticipatory functional forms and can be

L. Bertazzi et al. (eds.), *Innovations in Distribution Logistics*, Lecture Notes in Economics and Mathematical Systems 619, DOI: 10.1007/978-3-540-92944-4,
© 2009 Springer-Verlag Berlin Heidelberg

solved as multi-objective dynamic nonlinear problems, useful in representing generalized SCM structures with many agents. Nonlinear multi functions or point-to-set maps are essential to formulate accurate optimal robust SCM plans [2] and to analyze uncertainty and feedback.

Model relationships and parameters are difficult to determine a priori or by experience, as they evolve in unpredictable ways, so suitable simultaneous estimation and optimization algorithms must be formulated to specify the model. In fact these relationships must be determined as instruments to control accurately the phenomenon, rather than as realist approaches [9]. In such implementations the problem of estimability, stability and certainty equivalents of policies become fundamental [18] [20], but these are often not considered [15]. In large scale models with dynamics, nonlinearities and uncertainties ignoring these aspects may be perilous [10].

Model estimates may be biased, heteroscedastic subject to excessive disturbances, while the residual disturbances tend to be serially correlated [6]. So proper estimation techniques must be effected, to absorb all systematic variation by formulating a simultaneous estimation dynamic nonlinear optimal algorithm, such that disturbances are subject to suitable constraints, which will ensure that these satisfy all the required statistical conditions [8, 17].

The form of the relationships of the model and their coefficients must be determined in the space of the parameters and the values of the control variables must be determined in the space of the variables. Since it can be shown that these two spaces are interrelated, a simultaneous optimization algorithm to solve the combined problem must be defined [8].

The aim of this paper is to present this algorithm, describe its implementation and show through some preliminary applications its advantages. In particular it will be shown how certainty equivalents of optimal policies can be obtained in spite of stochastic disturbances and uncertainties.

The outline of the paper is the following. After the introduction, the formulation of representative dynamic nonlinear stochastic system of supply chains is presented for a multi-level process. In Sect. 3 the proposed dynamic system algorithm and the certainty equivalence of a system are described, while the major statistical properties are indicated, in Sect. 4. Then in Sect. 5 an application will be presented, showing comparative results of various solution strategies. Finally appropriate conclusions will be drawn.


## 2 Analysis of SCM Activity

Optimal production levels in a firm with multiple stages depend on economies of scale and the integrated SCM policy. The determination of an optimal SCM plan is a complex nonlinear dynamic problem with stochastic elements and uncertainties, in which all the various decisions must be evaluated simultaneously.

The aim of this section is to characterize a model of a SCM system integrated with a set of representative stages, such as inventory, production policy and distribution and marketing management stages.

Inventory management policies pervade all the stages of production of a firm and involve management decisions regarding raw-materials, goods in process and manufactured goods, which will be collectively indicated as items. It follows therefore that at each intermediate stage between departments and plants items inventories must be considered [11].

The demand and the cost schedules associated, with inventory management should be generally considered as stochastic processes, changing in time as non-stationary stochastic processes [11]. Without loss of generality, assume that no backlogging occurs. Also a particular lost sales policy is allowed and the limit may be specified.

Multi-level proportional lot sizing and scheduling problems may be proposed with given lead times in production [13] under the general structure of empirical processes.

**Definition 1.** *An Empirical Process is an ordered set of dated Input and Output variables arranged hierarchically in the form of a Gozinto chart or some similar suitable hierarchical scheme, subprocess by subprocess, to formulate a flow model of the potential operations to be carried out on the underlying phenomenon, so as to determine suitable optimal controls, with respect to some merit function [5].*

Limited capacity is also assumed and demand is considered varying, but contrary to many other formulations all the data relevant for the planning process is not assumed deterministic, but must be obtained from the past operations. Machines break down and uncertain events may occur. The operational time for activities is not usually a constant, but depends on a number of environmental aspects.

Dynamic adaptive stochastic diffusion processes are considered for sales models to determine the global SCM plan [3]. The consumption is influenced by a series of events which are random and sequential and depend on environmental factors and Marketing mix plans.

Consumers can be divided into two broad categories:

- Innovators: groups of consumers who are attracted by the novelty of the product and therefore respond well to mass communication through promotion and advertising.
- Imitators: groups of consumers more conservative than the previous group, who react mostly from word of mouth communication.

Without loss of generality assume that the number of individuals who purchase a product at time $t$ is the same as the number of units of the product purchased, since every consumer purchases one unit of product in the period [14] and each group will be characterized by the segment of each group.

Consider the following symbols:

- $f_0^j(q_t^j, t)$ be the non negative setup cost for item j for a quantity $q_t^j$ at period $t$,
- $f^j(I_{t-1}^j, t)$ be the non negative holding cost for holding one item $j$ in stock at time $t$ for one period until time $t + 1$ regarding $I_t^j$ items present at the initial period $t$,

- $h^k(I_t^1, \ldots, I_t^B, t)$   $k = 1, \ldots, K_t$ be the $K_t$ physical space, budget and other capacity constraints which must be satisfied in period t,
- $D_t^j = d^j(n^j(t))$ be the demand for the item $j$ at time $t$,
- $J$ number of items considered in the system,
- $a_{ij}$ is a 'gozinto' factor. Its value is zero if item $j$ is not an immediate successor of item $i$. Otherwise its value is set at $\hat{a}_i^j$, the quantity of item that is directly needed to produce one item $j$,
- $C_t^m$ be the available capacity of machine $m$ in period $t$,
- $I_t^j$ Inventory item $j$ at the end of period $t$,
- $\mathcal{I}_m$ be the set of all items that share the same machine $m$,
- $M$ be the number of machines,
- $m^j$ the machine on which item $j$ is produced,
- $p_m^j$ capacity needed for producing one unit of item $j$ on machine $m$,
- $S_j$ be the set of immediate successors of item $j$,
- $T$ be the number of periods considered,
- $v_j$ Positive and integral lead time in production for item $j$,
- $q_t^j$ acquisition or production quantity of item $j$ at time $t$,
- $x_t^j$ binary variable to indicate whether a setup for item $j$ occurs in period $t$, ($x_t^j = 1$) or not ($x_t^j = 0$),
- $n^j(t)$ number of individuals which will purchase an item $j$ in the interval $(t, t+1)$,
- $Y^j(t)$ marketing and distribution activities in period $t$,
- $a^j\left(t, Y^j(t), Y^j(t-1), \cdots, Y^j(t-k)\right)$ innovation coefficient relative to product $j$ at time $t$, conditioned on the past marketing activities $Y^j(t), Y^j(t-1), \cdots, Y^j(t-k)$, for some $k < t$,
- $b^j(t, Y^j(t, k))$ imitation coefficient relative to product $j$ at time $t$, conditioned on the past marketing activities as above,
- $\tilde{N}^j\left(t, Y^j(t-1), Y^j(t-2)), \cdots, Y^j(t-k)\right)$ total number of potential consumers of item $j$ at time $t$ and subject to marketing and distributional influences $Y^j(t-k)$, for some $k < t$,
- $N^j(t)$ total number of potential consumers that have purchased the item $j$ from the initial period ( 0 ) to the present $t$,
- $r\left(n^j(t)\right)$ be the net revenue function for the sale of $n^j(t)$ units of items $j$ in period $t$ which will include the direct marketing cost $Y^j(t)$ .

The specification of suitable typified representation of the SCM decisions model permits the dynamic system structure to be formulated:

$$Max \quad F = \sum_{j=1}^{J} \left\{ \sum_{t=1}^{T} r\left(n^j(t)\right) - \sum_{t=1}^{T} f_o^j(q_t^j, t) x_t^j + f^j(I_{t-1}^j, t)/2 \right\} \tag{1}$$

$$\text{s.t.} \quad I_t^j = I_{(t-1)}^j + q_t^j - D_t^j - \sum_{i \in S_j} a_{ji} q_t^i \quad \forall j \in J; \quad \forall t = 1, 2, \cdots, T \tag{2}$$

$$I_t^j \geq \sum_{i \in S_j} \sum_{\tau=1}^{min(t+v_j, T)} a_{ji} q_\tau^i \quad \forall j \in J; \quad \forall \tau = 0, 1, \cdots, T-1 \tag{3}$$

$$0 \leq h^k(q_t^1, \ldots, q_t^J, t) \quad \forall k = 1, \ldots, K_t; \quad \forall t = 1, 2, \ldots, T \tag{4}$$

$$p_m^j q_t^j \leq C_t^{m_j}(y_{(t-1)}^j + y_t^j) \quad \forall j \in J; \quad t = 1, \cdots, T \tag{5}$$

$$\sum_{j \in I_m} p_m^j q_t^j \leq C_t^m \quad m = 1, \cdots, M; \quad t = 1, \cdots, T \tag{6}$$

$$\sum_{j \in I_m} y_t^j \leq 1 \quad m = 1, \cdots, M; \quad t = 1, \cdots, T \tag{7}$$

$$(y_t^j - y_{(t-1)}^j) \leq x_t^j \quad \forall j \in J; \quad t = 1, \cdots, T \tag{8}$$

$$n^j(t) = \left[ \hat{a}_j + \hat{b}_j \frac{N^j(t)}{\tilde{N}^j(t, Y^j(t,k))} \right] \left( \tilde{N}^j(t, Y^j(t,k)) - N(t) \right) \forall t = N+1, \cdots, \mathcal{T} \tag{9}$$

$$D_t^j = d^j(n_t^j) \quad \forall j = 1, \cdots, J; \quad \forall t = N+1 \tag{10}$$

$$\hat{a}_j = a_j(t, Y^j(t-1), k), Y^j(t)) \quad \forall j = 1, \cdots, J; \quad \forall t = N+1, \cdots, \mathcal{T} \tag{11}$$

$$\hat{b}_j = b_j(t, Y^j((t-1), k), Y^j(t)) \quad \forall j = 1, \cdots, J; \quad \forall t = N+1, \cdots, \mathcal{T} \tag{12}$$

$$x_t^j \in \{0, 1\} \tag{13}$$

$$y_t^j \in \{0, 1\} \quad \forall j \in J; \quad t = 1, \cdots, T \tag{14}$$

$$0 \leq v_t^j, q_t^j \quad \forall j = 1, 2, \cdots, J; \quad \forall t = 1, 2, \cdots, T \tag{15}$$

$$0 \leq I_t^j, q_t^j, x_t^j \quad \forall j \in J; \quad t = 1, \cdots, T \tag{16}$$

The revenue for the firm is determined from the sales of the items and inventory management costs, including production setup and holding costs in (1) over the time interval. The amounts of items acquired or produced in any period, as relevant, are given by $q_t^j$ and the binary variables $x_t^j$ indicate whether at time $t$ a quantity is ordered or launched in production. The amount of items in stock available at time $t + 1$ for each good $j = 1, 2, ..., J$ are specified by (2) – (3). A set of equations (4) specify the limitations on the inventory, production or physical space limitations.

The set of equations (5) permits the production of item $j$ to be cycled on the machine $m$ at time $t$ to continue the production from the preceding period if $y_{(t-1)} = 1$, without any additional setup cost for the period $t$, as the (8) since if $y_{(t-1)}^j = 1, y_t^j = 1$ then $(y_t^j - y_{(t-1)}^j) = 0$ then the binary variable $x_t^j = 0$ due to the maximization of (1).

The capacity of any machine $m$ must be sufficient to produce the desired items, considering the capacity required for each item, as indicated in (6). In the set of equations (7) the production of an item $j$ is launched only once on a machine, while the (8) manage the eventual continuity of production, as indicated above.

The sales predicted in each period $t$ is indicated by the (9) for each item (which also may be null). In (10) the demand for each item is derived from the sales by decomposing the 'Gozinto' charts, primarily on applying the (2) and (3), but actually the whole system is involved.

The number of items predicted depends on the appropriate consumer segment, depending on complex coefficients. The functional form and the parameters must be determined from the (11) and (12), while the values of the variables of the marketing mix must be determined over the time periods by maximization. The last four sets equations (13)–(16) limit the variation of most of the variables.

Due to the non stationary characteristics of the stochastic process, the cost and demand schedules will depend on the optimal level and other dynamic aspects, as well as the functional form and the parameters which should be determined. The model must be estimated and optimized simultaneously with respect to the historical data time series processes of the firm for the whole Dynamic System and the activities from $t = -T_H, \cdots, 0, 1, \cdots, T$, where the period $(-T_H, \cdots, 0)$ is the known historical period, while the prediction interval is specified as $(1, \cdots, T)$.

## 3 Dynamic System Models

Dynamical Systems are useful to refine concepts and represent applications by appropriate modeling, while whole hierarchies of phenomena may be represented by such systems.

**Definition 2.** *[12]: A Dynamical System is a composite mathematical object defined by the following axioms:*

1. *There is a given time set $T$, a state set $X$, a set of input values $U$, a set of acceptable input functions $\Omega = \omega : \Omega \to U$, a set of output values $Y$ and a set of output functions $\Gamma = \gamma : \Gamma \to Y$.*

2. *(Direction of time). $T$ is an ordered subset of the reals.*

3. *The input space $\Omega$ satisfies the following conditions:*

    *a) (Nontriviality). $\Omega$ is nonempty.*

    *b) (Concatenation of inputs). An input segment $\omega_{(t_1,t_2]}$, $\omega \in \Omega$ restricted to $(t_1, t_2] \cap T$. If $\omega, \omega' \in \Omega$ and $t_1 < t_2 < t_3$ there is an $\omega'' \in \Omega$ such that $\omega''_{(t_1,t_2]} = \omega_{(t_1,t_2]}$ and $\omega''_{(t_2,t_3]} = \omega'_{(t_2,t_3]}$.*

4. *There is a state transition function $\varphi : T \times T \times X \times \Omega \to X$ whose value is the state $x(t) = \varphi(t; \tau, x, \omega) \in X$ resulting at time $t \in T$ from the initial state $x = x(\tau) \in X$ at the initial time $\tau \in T$ under the action of the input $\omega \in \Omega$. $\varphi$ has the following properties:*

a) *(Direction of time). $\varphi$ is defined for all $t \geq \tau$, but not necessarily for all $t < \tau$.*

b) *(Consistency). $\varphi(t; t, x, \omega) = x$ for all $t \in T$, all $x \in X$ and all $\omega \in \Omega$.*

c) *(Composition property). For any $t_1 < t_2 < t_3$ there results:*

$$\varphi(t_3; t_1, x, \omega) = \varphi(t_3; t_2, \varphi(t_2; t_1, x, \omega), \omega)$$

*for all $x \in X$ and all $\omega \in \Omega$.*

d) *(Causality). If $\omega, \omega' \in \Omega$ and $\omega_{(\tau,t]} = \omega'_{(\tau,t]}$ then $\varphi(t; \tau, x, \omega) = \varphi(t; \tau, x, \omega')$.*

5. *There is a given readout map $\eta : T \times X \to Y$ which defines the output $y(t) = \eta(t, x(t))$. The map $(\tau, t] \to Y$ given by $\sigma \mapsto \eta(\sigma, \varphi(\sigma, \tau, x, \omega))$, $\sigma \in (\tau, t]$, is an output segment, that is the restriction $\gamma_{(\tau,t]}$ of some $\gamma \in \Gamma$ to $(\tau, t]$.*

From the accounting data and the 'gozinto' charts, input $U$ and output $Y$ sets of dated quantities are obtained. As the process may be highly nonlinear with marked lags, intermediary vectors, indicated as states $x_t \in X$ are used with an opportune transition function. The effect of activities at an instant $t$ on the state of the system is called an event, specified as $(t, x), t \in T, x \in X$. A trajectory may be understood as the graph of the state as a consequence of the variation in time, given as $(t, \varphi(x_t, u_t))$.

A sufficiently general representation of a dynamic system may be formulated, with a slight abuse of notation, in the following way:

$$x_{t+1} = \varphi(x_t, u_t) \tag{17}$$

$$y_t = \eta(x_t) \tag{18}$$

where:

- $x_t \in X$ is the vector of the state of the dynamical system at period $t$
- $u_t \in U$ is the vector of the control variables of the dynamical system at period $t$
- $y_t \in Y$ is the vector of the output of the dynamical system at period $t$

Dynamical systems are based on intermediary set of states and transition functions, by applying the simultaneous estimation and optimization algorithm to determine the State set $X$ and the transition function [19]. Under appropriate conditions the representation of the system will result unique, as will be indicated below.

**Definition 3.** *Given two states $x_{t_0}$ and $\hat{x}_{t_0}$ belonging to systems $S$ and $\hat{S}$ which may not be identical, but have a common input space $\Omega$ and output space $Y$, the two states are said to be equivalent if and only if for all input segments $\omega_{[t_0,t)} \in \Omega$ the response segment of $S$ starting in state $x_{t_0}$ is identical with the response segment of $\hat{S}$ starting in state $\hat{x}_{t_0}$; that is*

$$x_{t_0} \cong \hat{x}_{t_0} \Leftrightarrow \eta(t, \varphi(x_{t_0}, \omega_{[t_0,t)})) = \hat{\eta}(t, \hat{\varphi}(\hat{x}_{t_0}, \omega_{[t_0,t)})) \quad \forall t \in T, t_0 \leq t, \forall \omega_{[t_0,t)} \in S, \hat{S} \tag{19}$$

The systems $S$ and $\hat{S}$ may be two representations of a SCM system and their policies.

**Definition 4.** *A system is in reduced form if there are no distinct states in its state space which are equivalent to each other.*

Suppose the actual system consists in reduced form, derivable from the firm's data base.

**Definition 5.** *Systems $S$ and $\hat{S}$ are equivalent $S \equiv \hat{S}$ if and only if to every state in the state space of $S$ there corresponds an equivalent state in the state space of $\hat{S}$ and vice versa.*

*Remark 1.* Certain aspects of the SCM representation and the policy that entails should be determined:

1. Can a certain state $s^* \in S$ be reached from the present state, or if the dynamical system attains a given state $x_t$ at time $t = 0$ can it also be made to reach a certain state $x^*$. Evidently it is required to determine the set of states reachable from a specific state $x_t$.
2. Can a dynamical system be driven to a given state by an input $u_t$. Controllability is concerned with the connectedness properties of the system representation.
3. Reachability and controllability lead naturally to the determination of a dynamical system's observability, which provides the conditions to determine the given actual state uniquely.
4. The stability of the system is important since it provides conditions on the way the trajectories will evolve a given perturbation or an admissible control.

Reachability, controllability and stability are seldom formally examined in SCM policy formulation, although at every period, exogenous events can arise to nullify even the best plan formulated, because the controllability of the system was not researched. So these aspects should be studied to ensure the feasibility of the SCM policy.

**Definition 6.** *Simple and multiple experiments involve different sets of input/output pairs:*

- *A simple experiment is an input/output pair $(u_{[t_0,t)}, y_{[t_0,t)})$ that is, given the system in an unknown state an input $u_{[t_0,t)}$ is applied over the interval of time $(t_0, t)$ and the output $y_{[t_0,t)}$ is observed.*
- *A multiple experiment of size M consists of M input/output pairs $(u^i_{[t_0,t)}, y^i_{[t_0,t)}$ $i = 1, 2, ..., M)$ where on applying on the i-th realization of the M systems the input $(u^i_{[t_0,t)})$, the i-th output $y^i_{[t_0,t)}$ is observed.*

While a simple experiment is thought as a stimulus and response experiment, multiple experiments are more complex and define multi-determined reactions, such as synergisms of the system.

**Definition 7.** *A system is simply (multiple) observable at state $x_{t_0}$ if and only if a simple experiment (a multiple experiment) permits the determination of that state uniquely.*

**Definition 8.** *Equivalence of dynamic systems can be distinguished:*

- *Two systems are simply equivalent if it is impossible to distinguish them by any simple experiment.*
- *Two systems are multiply equivalent if it is impossible to distinguish them by any multiple experiment.*

**Theorem 1.** *[12]: If two systems are multiply equivalent then they are equivalent.*

**Definition 9.** *A system is initial-state determinable if the initial state $x_0$ can be determined from an experiment on the system started at $x_0$.*

**Theorem 2.** *[12]: A system is in reduced form if and only if it is initial-state determinable by an infinite multiple experiment.*

The Definitions 6–9 and the results given in Theorems 1–2 formally justify the possibility of defining one or more representations of the dynamical system. The distinction between systems that are simply equivalent and multiply equivalent is crucial, as comparative static or equilibrium models will be simply equivalent, while for the analysis of dynamical systems which are multiple equivalent allow to compare different representations and determine the optimal trajectory for the system.

Thus consider,

**Definition 10.** *An ex ante solution is a solution formed in a given period t based on anticipated outcome of future activities maturing in period t + 1 and consists of the forecast of the optimal values of the control variables.*

**Definition 11.** *An ex post solution is a solution formed in a given period t, regarding outcomes of activities maturing in period t + 1 which are assumed known (with foreknowledge), in period t so as to determine the optimal values of the control variables in period t + 1.*

Let a phenomenon be represented by a multiply equivalent dynamic system in an initial-state determinable and the state in period $t + 1$ be predicted in period $t$. The state at $t + 1$ may be obtained from a representation of another copy of such a system in the period $t + 1$. The state of period $t + 1$ predicted in period $t$ on the basis of the knowledge in the same period, will be equivalent to the state at period $t + 1$ on the basis of the knowledge at period $t + 1$, which is indicated as the ex post state.

**Theorem 3.** *An ex ante multiply equivalent Dynamic system in an initial-state is equivalent to an ex post system, also a multiply equivalent dynamic system in an initial-state.*

**Proof**: By Theorem 1 the ex ante dynamical system is multiply equivalent and its initial state is in reduced form by Theorem 2.

Furthermore by the same reasoning the ex post dynamical system is also a multiply equivalent and its initial state is in reduced form.

By Definitions 4 and 5 they are state equivalent.

**Corollary 1.** *A Certainty Equivalent solution consists of an optimal solution to an optimization problem determined as an ex ante solution in the control variables $u^*$ which are also optimal ex post.*

The state of the systems are equivalent if the state transition functions do not exhibit significant random variation. The eventual random variation will be negligible, but the expected value of the disturbances in both cases will assume null values while the variance of the processes may be positive, but the autocorrelation and cross-correlation must be zero and the processes are stationary.

This formulation is analogous to the properties derived earlier by different methods [18, 20].

No limitations have been enforced for the output equations (18). The random disturbances of these forms may vary. The income resulting, as an output variable may differ, due to disturbances, although optimal solution values of the control variables will be identical in the two processes.

The SCM dynamic estimation problems and optimal policy determination should be formulated by such an approach with a data driven formulation.

# 4 Specification of the Statistical Properties

All the statistical properties of given estimates of the residuals should be fulfilled at every iteration, so suitable constraints must be defined, together with the specification of the model of the phenomenon and the global optimization problem should be solved for all the undetermined variables. Thus a number of independent statistical properties must be satisfied, which may be represented as:

$$\gamma(x_{i+1}, x_t, u_i, y_{t+1}, y_t, w_t, v_t, \theta_1, \theta_2) \geq 0 \quad t = 1, 2, \cdots, T \tag{20}$$

where:

- $w_t, v_t$ are random residual disturbances, to be determined so that a number of conditions specified in the set of equations (20) are satisfied,
- $\theta_1, \theta_2$ are parameters to specify the system, such that the conditions specified by (20) are satisfied.

The set of statistical conditions that the residuals must satisfy to obtain the correct estimates are the following [8]:

1. The parameter estimates are unbiased, this means that:
   - As the size of the data set grows larger, the estimated parameters tend to the asymptotic (true) values.
2. The parameter estimates are consistent, which require the following conditions to be satisfied:
   - The estimated parameters are asymptotically unbiased,
   - The variance of the parameter estimate must tend to zero as the data set tends to infinity.
3. The parameter estimates are asymptotically efficient,
   - The estimated parameters are consistent,
   - The estimated parameters have smaller asymptotic variance as compared to any other consistent estimator.
4. The residuals have the smallest possible variance and to ensure that this is so:
   - The variance of the residuals must be minimum,
   - The residuals must be homoscedastic,
   - The residuals must not be serially correlated.
5. The residuals are unbiased (have zero mean).
6. The residuals have a non informative distribution (usually, the Gaussian distribution). If the distribution of the residuals is informative, the extra information could somehow be exploited by reducing the variance of the residuals, their bias etc. with the result that better estimates are obtained.

Through a correct implementation of statistical estimation techniques, a set of equation (20) can be specified. The parameter estimates are as close as possible to their asymptotic (true) values, all the information that is available is used to determine precise estimates, so the residual uncertainty and the approximation in the data fit is reduced to the maximum extent possible. Thus the estimates of the parameters, which satisfy all these conditions, are the 'best' possible in a 'technical sense' [1].

To the SCM algorithm to determine the optimal policy, indicated by the system (1)–(16) must be added the statistical conditions described above. This can be indicated, for clarity as:

$$Min \quad J = \sum_{t=1}^{T} c(x_t, u_t, y_t) \tag{21}$$

$$x_{t+1} = \varphi(x_t, u_t, y_t, w_t : \theta_1) \qquad t = -T_H, \cdots, 0, 1, 2, \cdots, T - 1 \tag{22}$$

$$y_{t+1} = \eta(x_t, u_t, v_t : \theta_2) \qquad t = -T_H, \cdots, 0, 1, 2, \cdots, T - 1 \tag{23}$$

$$0 \le \gamma(x_{t+1}, x_t, u_t, y_{t+1}, y_t, w_t, v_t, \theta_1, \theta_2), t = -T_H, \cdots, 0, 1, \cdots, T - 1 \tag{24}$$

where the system has been expressed as a coherent dynamical system reformulated from system (1)–(16) and the statistical conditions constrains that have been specified summarily above.

For major clarity, but with no loss of generality, in this representation, opportune terms are added to each expression as residuals $w_t$, $v_t$ and parameters $\theta_1$, $\theta_2$ to be determined to satisfy the conditions.

By recursing on the specifications, i.e. by changing the functional form, and increasing the number of independent variables considered, better and better fits can be obtained, both with regard to the historical data and the predicted optimal control policy, so the optimal control problem (21)–(24) will be solved iteratively.

The mathematical program will be formulated with respect to the residual variables, but it is immediate that for a given functional form, the unknown parameters will be specified and thus the unknowns of the problem will also be defined and available. Thus the mathematical program is fully specified for each functional form to be considered.

In fact the residual terms are given from the (22) and (23) as:

$$w_t = \hat{x}_{t+1} - \varphi_h(\hat{x}_t, \hat{u}_t, \hat{y}_t : \theta_1) \qquad t = -T_H, \cdots, 0, 1, 2, \cdots, T-1, \quad h \in \Omega \quad (25)$$

$$v_t = \hat{y}_{t+1} - \eta_k(x_t, u_t, v_t : \theta_2) \qquad t = -T_H, \cdots, 0, 1, 2, \cdots, T-1, \quad k \in \Gamma \quad (26)$$

where $\hat{\cdot}$, as usual indicates the historical values of a variable and thus suitable values of $\theta_1, \theta_2$ must be determined by the mathematical program, such that all the constraints expressed in terms of $w_i, v_i \; \forall i$ are specified.

The constraints which are added regarding the residuals imply the satisfaction of the moments of the Gaussian distribution function [8].

*Remark 2.* The proposed algorithm dominates the traditional procedures. The algorithm determines always a solution which satisfies all the statistical conditions posed, if the system is feasible, or no solution is determinable.

This solution will ensure that the parameter estimates are unbiased, consistent, asymptotically efficient estimates with minimum variance and the residuals have finite variance and are unbiased. If the estimates determined are sufficiently precise, which can be obtained by appropriate modifications in the model specification, certainty equivalence of the solution will be satisfied by Corollary 1.

## 5 Computational Results

The SCM problem are traditionally formulated as a three stage problem in which the functional form and the parameters estimates are considered known, so that the optimal policy can be determined conditioned on these aspects. Potential uncertainties are not considered ex ante which will render the ex post solution unacceptable, as Theorem 3 is not satisfied.

A Dynamical system formulation of the model, determined by a simultaneous estimation and optimization system as described in Sect. 4, the global objective function will consist of the maximization of the revenue over an interval of time. The state and input output system equations are formulated for each stage consecutively. All the variables must be integrated in a global structure, redefined in a coherent consistent way [8] so that the solution will satisfy all the statistical conditions, as in Remark 2 and the optimal solution will consequently satisfy Theorem 3.

The aim of this section is to show by an application that all the results of this paper are fulfilled. Of course extensive experiments are deemed, but these results are useful to show this objective.

This approach was implemented and solved with the algorithm and for comparison purposes two commercial routines: Lancelot, [7] and Cplex [4] were used with differing estimation procedures.

A simulation of a firm was carried out for illustrative purposes, represented by 48 variables and 18 control variables derived from the system (1)–(16). Also 23 relationships were obtained for the system by considering 2 raw materials, 2 production units and 2 marketing mix variables each consisting of five marketing variables with regard to two manufactured goods.

The 48 variables represent the state variables and the output variables at each period, such as the stock of the two manufactured goods, two raw materials and semi manufactured items, as well as the labor employed on each production unit and the number of individuals who will purchase a manufactured good in that period, the revenue and costs in the period and so on. Instead the 18 control variables for each period include the 10 marketing variables and the price levels chosen and the intermediary six items to carry out the production.

Many non linear and dynamic relationships were introduced in the simulation, as evident in Fig. 1. On the basis the optimal policy for the next six periods the optimal policies were determined and compared.

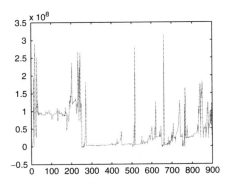

**Fig. 1.** Time series for income in monetary units

The following algorithms were implemented and the results are shown in Tables 1–3 and in Fig. 2:

1. The average values of the coefficients were obtained from the dated values of the perturbed coefficients, used in the simulation, to specify the non linear system to

be solved for the next 6 periods with Lancelot, applying the non linear model. The income obtained was 211,227,602 m.u.(monetary units) over the six periods.

2. The estimation of the parameters of the system was determined by S.P.S.S. based on the simulation data set. The optimal solution over the same is determined by a linear system since it is solved by the Cplex linear program, and an income of 197,167,817 m.u on six periods were predicted.

3. The **C.A.s.S.a.n.D.r.A.** routine was applied on the data set to solve the simultaneous estimation and optimization problem regarding the dynamic relationships for the prediction over the next six periods, with a predicted income of 217,708,547 m.u.

In the first implementation, the model is influenced by random variation and dynamic effects, yet the non linear specification of the system leads to an improvement of 7.13% over the second implementation. The effect of applying dynamic parameters and non linear relationships in **C.A.s.S.a.n.D.r.A.** leads to a further improvement of 4.02% over the first implementation, while the aggregate improvement of this non linear implementation over the linearized three stages implementation was 11.15%.

The Lancelot solution yields a haphazard unstable development path, probably not acceptable to Management, but it reflects the behavior of the income curve in the past. These aspects are consequential to the approach consisting of three stages as indicated in Sect. 1 and Remark (1 sub 4).

The improvement in performance with a simultaneous estimation and optimization approach is relevant, as indicated in Sect. 3, while the consideration of linear modelization leads to relevant inefficiencies.

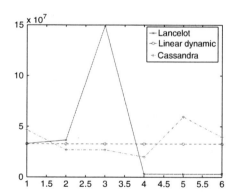

**Fig. 2.** Comparisons of SCM plans for different implementations

The Cplex implementation provides a linear growth of income, due to the use of a linear program as the optimization instrument and an estimation process linear in the parameters, so the ex ante solution may be comforting as an equilibrium growth could be appreciated but this solution is unrealistic and inefficient.

The growth trajectory indicated by the **C.A.s.S.a.n.D.r.A.** routine is stable and provides the best results in apparent adverse market conditions. The ex ante solution

**Table 1.** Prediction of optimal activities for six periods: Lancelot and Cplex algorithms

| Variable | Lancelot | | | | | |
|---|---|---|---|---|---|---|
| | pd.1 | pd.2 | pd.3 | pd.4 | pd.5 | pd. 6 |
| $Y_a^1(t)$ | 185.0 | 191.8 | 192.0 | 192.3 | 192.0 | 192.6 |
| $Y_b^1(t)$ | 199.1 | 199.0 | 199.0 | 199.9 | 199.1 | 199.0 |
| $Y_c^1(t)$ | 199.0 | 199.0 | 199.0 | 199.0 | 199.1 | 199.1 |
| $Y_d^1(t)$ | 199.0 | 199.1 | 199.1 | 199.0 | 199.1 | 199.1 |
| $Y_e^1(t)$ | 177.8 | 181.3 | 198.0 | 199.2 | 199.7 | 199.1 |
| $Y_f^1(t)$ | 14,103.0 | 14,103.0 | 14,103.0 | 14,103.0 | 14,103.0 | 14,103.0 |
| $Y_a^2(t)$ | 186.0 | 194.1 | 200.0 | 204.9 | 204.9 | 214.3 |
| $Y_b^2(t)$ | 186.0 | 196.1 | 202.3 | 206.9 | 206.9 | 214.3 |
| $Y_c^2(t)$ | 199.0 | 199.0 | 199.0 | 199.0 | 199.0 | 199.0 |
| $Y_d^2(t)$ | 161.4 | 162.1 | 180.0 | 212.6 | 212.6 | 308.2 |
| $Y_e^2(t)$ | 1,532.0 | 1,532.0 | 1,532.0 | 1,532.0 | 1,532.0 | 1,532.0 |
| $Y_f^2(t)$ | 15,852.0 | 15,852.0 | 15,852.0 | 15,852.0 | 15,852.0 | 15,852.0 |
| $q_t^1$ | 667.0 | 667.0 | 667.0 | 667.0 | 667.0 | 667.0 |
| $q_t^2$ | 47.2 | 204.0 | 205.0 | 204.0 | 204.9 | 204.9 |
| $q_t^3$ | 47.2 | 504.0 | 205.0 | 204.9 | 204.9 | 204.9 |
| $q_t^4$ | 208.0 | 19.6 | 196.4 | 199.0 | 199.0 | 204.4 |
| $q_t^5$ | 61.7 | 17.0 | 15.2 | 230.7 | 230.7 | 210.3 |
| $q_t^6$ | 0.0 | 14.4 | 10.7 | 0.0 | 0.0 | 75.1 |
| Revenue(000) | 33,135.3 | 36,642.6 | 149,409.0 | 2,982.4 | 3,029.6 | 3,028.7 |

obtained is equivalent to the ex post solution, as Theorem 3 and Corollary 1 hold and is confirmed by numeric values.

Some control variables $Y_f^1(t)$, $Y_f^2(t)$ which are the prices to be levied for the manufactured goods are analogous in the three implementations, while other control variables exhibit pronounced variation between implementations, while for specific control variables variation within an implementation may be large.

For the first two implementations the differences in many of the control variables are large, while incomes are very close, except in one period, which implies that multiple policies can be determined with very different control variables. There also

**Table 2.** Prediction of optimal activities for six periods: Lancelot and Cplex algorithms

| | Cplex | | | | | |
|---|---|---|---|---|---|---|
| Variable | pd.1 | pd.2 | pd.3 | pd.4 | pd.5 | pd. 6 |
| $Y_a^1(t)$ | 7,411.0 | 7,123.0 | 7,491.0 | 7,061.0 | 6,924.0 | 7,023.0 |
| $Y_b^1(t)$ | 0.0 | 0.0 | 0.0 | 0.0 | 0.0 | 0.0 |
| $Y_c^1(t)$ | 0.0 | 0.0 | 0.0 | 0.0 | 0.0 | 0.0 |
| $Y_d^1(t)$ | 9,950.0 | 9,950.0 | 9,950.0 | 9,950.0 | 9,950.0 | 9,950.0 |
| $Y_e^1(t)$ | 4,000.0 | 4,000.0 | 4,000.0 | 4,000.0 | 4,000.0 | 4,000.0 |
| $Y_f^1(t)$ | 14,103.0 | 14,103.0 | 14,103.0 | 14,103.0 | 14,103.0 | 14,103.0 |
| $Y_a^2(t)$ | 60,000.0 | 60,000.0 | 60,000.0 | 60,000.0 | 60,000.0 | 60,000.0 |
| $Y_b^2(t)$ | 1.0 | 1.0 | 1.0 | 1.0 | 1.0 | 1.0 |
| $Y_c^2(t)$ | 2.0 | 2.0 | 2.0 | 2.0 | 2.0 | 2.0 |
| $Y_d^2(t)$ | 4.0 | 4.0 | 4.0 | 4.0 | 4.0 | 4.0 |
| $Y_e^2(t)$ | 22,153.0 | 17,124.0 | 26,210.0 | 14,509.0 | 36,656.0 | 17,565.0 |
| $Y_f^2(t)$ | 15,852.0 | 15,852.0 | 15,852.0 | 15,852.0 | 15,852.0 | 15,852.0 |
| $q_t^1$ | 8,900.0 | 8,900.0 | 8,900.0 | 8,900.0 | 8,900.0 | 8,900.0 |
| $q_t^2$ | 2,220.0 | 2,220.0 | 2,220.0 | 2,220.0 | 2,220.0 | 2,220.0 |
| $q_t^3$ | 8,900.0 | 8,900.0 | 8,900.0 | 8,900.0 | 8,900.0 | 8,900.0 |
| $q_t^4$ | 1,000.0 | 1,000.0 | 1,000.0 | 1,000.0 | 1,000.0 | 1,000.0 |
| $q_t^5$ | 50.0 | 50.0 | 50.0 | 50.0 | 50.0 | 50.0 |
| $q_t^6$ | 5,510.0 | 5,510.0 | 5,510.0 | 5,510.0 | 5,510.0 | 5,510.0 |
| Revenue(000) | 32,862.0 | 32,861.9 | 32,860.8 | 32,861.1 | 32,860.7 | 32,861.1 |

seems to be a multiplicative factor in the values of the control variables for the policy determined for the **C.A.s.S.a.n.D.r.A.** algorithm compared to the Lancelot policy. However, except for one period, which could be considered an outlier, the income of the former policy is an improvement between 20-50%.

The stability of the solution of **C.A.s.S.a.n.D.r.A.** is due to the certainty equivalent requirement, as with differing dynamic lags, the residual variance was always of order $0.8 \times 10^{-9}$ and the estimated model is extremely close to the actual model. This was verified in a number of implementations with different initial points and

**Table 3.** Prediction of optimal activities for six periods: Cassandra algorithm

| | Cassandra | | | | | |
|---|---|---|---|---|---|---|
| Variable | pd.1 | pd.2 | pd.3 | pd.4 | pd.5 | pd. 6 |
| $Y_a^1(t)$ | 8,915.0 | 8,915.0 | 8,915.0 | 0.0 | 8,915.0 | 8,915.0 |
| $Y_b^1(t)$ | 8,915.0 | 8,915.0 | 8,915.0 | 8,195.0 | 8,915.0 | 8,915.0 |
| $Y_c^1(t)$ | 8,985.0 | 8,985.0 | 8,985.0 | 0.0 | 8,985.0 | 8,985.0 |
| $Y_d^1(t)$ | 9,950.0 | 9,950.0 | 9,950.0 | 0.0 | 9,950.0 | 9,950.0 |
| $Y_e^1(t)$ | 4,000.0 | 4,000.0 | 4,000.0 | 1.0 | 4,000.0 | 4,000.0 |
| $Y_f^1(t)$ | 14,103.0 | 14,103.0 | 14,103.0 | 14,103.0 | 14,103.0 | 14,103.0 |
| $Y_a^2(t)$ | 60,000.0 | 60,000.0 | 60,000.0 | 1.0 | 60,000.0 | 60,000.0 |
| $Y_b^2(t)$ | 60,000.0 | 60,000.0 | 60,000.0 | 1.0 | 60,000.0 | 60,000.0 |
| $Y_c^2(t)$ | 58,208.0 | 58,208.0 | 58,208.0 | 2.0 | 58,208.0 | 58,208.0 |
| $Y_d^2(t)$ | 61,628.0 | 61,628.0 | 61,628.0 | 4.0 | 61,628.0 | 61,628.0 |
| $Y_e^2(t)$ | 37,276.0 | 37,276.0 | 37,276.0 | 156.0 | 37,276.0 | 37,276.0 |
| $Y_f^2(t)$ | 15,852.0 | 15,852.0 | 15,852.0 | 15,852.0 | 15,852.0 | 15,852.0 |
| $q_t^1$ | 8,900.0 | 8,900.0 | 8,900.0 | 170.0 | 8,900.0 | 8,900.0 |
| $q_t^2$ | 9,900.0 | 9,900.0 | 9,900.0 | 222.0 | 9,900.0 | 9,900.0 |
| $q_t^3$ | 8,900.0 | 8,900.0 | 8,900.0 | 500.0 | 8,900.0 | 8,900.0 |
| $q_t^4$ | 1,000.0 | 1,000.0 | 1,000.0 | 50.0 | 1,000.0 | 1,000.0 |
| $q_t^5$ | 1,000.0 | 1,000.0 | 1,000.0 | 50.0 | 1,000.0 | 1,000.0 |
| $q_t^6$ | 5,510.0 | 5,510.0 | 5,510.0 | 250.0 | 5,510.0 | 5,510.0 |
| Revenue(000) | 46,591.2 | 26,988.7 | 26,988.7 | 19,904.5 | 59,649.7 | 39,585.7 |

different initial models. In the S.P.S.S. application, the least residual variance was $0.7 \times 10^6$, which is reasonable as the data values vary and are very large.

The result obtained is encouraging and is in line with the methodological results that have been expressed in Corollary 1, so further experimentation seems to be worthwhile [8].

It follows from Theorem 3 that the ex ante solution is optimal compared to the ex post solution, since all the conditions of the theorem and the preceding ones are verified by the **C.A.s.S.a.n.D.r.A.** algorithm.

Experimental comparisons cannot be carried out, since multiple copies of a real firm are never available and balanced experiments performed on random selections of multiple cases cannot be carried out in practice. It would seem that comparisons of different algorithms, as proposed here, could never be carried out by realistic experiments, but the approach formulated in this paper permits the formulation of an instrumentalist approach isomorphically to proposed realistic experiments.

Suppose that an SCM over a number of periods is formulated with **C.A.s.S.a.n.D. r.A.**. The solution is certainty equivalent, as indicated in Corollary 1. Further other systems based on other assumptions may be formulated with similar precision if possible. Depending on their assumed structure simple or multiply equivalent systems will be formulated, as indicated in Definition 5. Various ex ante solutions may be determined which can be compared to the ex ante solution of the **C.A.s.S.a.n.D.r.A.** algorithm, which is a proxy solution to the ex post solution (Theorem 3).

If a single experimental firm is available, the **C.A.s.S.a.n.D.r.A.** ex ante policy can be enacted and this policy can be compared to the expost optimal solution determined at the end of the interval. The latter should be similar to the ex ante solution and the comparative results with other solutions will indicate the efficiency of the different approaches.

Clearly this experiment is isomorphic to a complete multiple experiment carried out on the equivalent of many copies of experimental firms.

# 6 Conclusions

A Supply Chain Management system can be modeled by nonlinear stochastic and dynamical systems, to avoid suboptimization arising from simplifications in the representation. The analysis may be carried out at the highest level of generality with as few a priori assumptions as possible. Data Driven modeling system should be used rather than refer to managerial insights and anecdotal evidence.

Stochastic disturbances and exogenous events may modify the trajectories of development so that the reachability of the goals should be ascertained periodically. Also the controllability of the system must be checked periodically to ensure that the system can still be controlled to achieve the desired aim. As the dynamic SCM system will inevitably be subject to uncertainty, accurate plans may only be defined by considering the properties of the dynamical system in a certainty equivalent context.

Within the Mathematical System framework presented in this paper, all the important aspects described above can be verified and by ensuring, when possible, that multiply equivalent systems are derived, so optimal solutions, both ex ante and ex post are determined, even though the output results may differ somewhat, without influencing the optimal solution.

This methodology allows, multiply equivalent systems to be determined, so that the system can be embedded in an optimal control problem and determine a certainty equivalent optimal solution. Precise alternative experiments can be effected and different exogenous factors may be evaluated precisely.

This approach seems to be useful for the modeling of SCM to determine precise policies.

# References

[1] T. Amemiya. *Advanced Econometrics*. Blackwell, Oxford, 1985.

[2] J.-P. Aubin. *Viability Theory*. Birkhäuser, Boston, 1991.

[3] F. M. Bass, T. V. Krishnan, and D. C. Jain. Why the bass model fits without decision variables. *Marketing Science*, 3:203–223, 1994.

[4] R. E. Bixby. Implementing the simplex method: the initial basis. *Journal of Computing*, 4:267–284, 1992.

[5] E. S. Buffa. *Modern Production Management*. Wiley, New York, 1977.

[6] C. Chatfield. Model uncertainty, data mining and statistical inference. *Journal of the Royal Statistical Soc. Series A*, 158:419 – 466, 1995.

[7] A. R. Conn, N. I. M. Gould, and P. L. Toint. *Trust-Region Methods*. MPS-SIAM, Philadelphia, 2000.

[8] L. Di Giacomo and G. Patrizi. Dynamic nonlinear modelization of operational supply chain systems. *Journal of Global Optimization*, 34:503–534, 2006.

[9] L. Di Giacomo and G. Patrizi. Foundations of supply chain management: towards a normative theory of supply and demand chain management. *submitted for pubblication*, 2007.

[10] J. Gondzio, R. Sarkissian, and J.-Ph. Vial. Parallel implementation of a central decomposition method for solving large scale planning problems. *Computational Optimization and Applications*, 19:5–29, 2001.

[11] G. Hadley and T. M. Whithin. *Analysis of Inventory Systems*. Prentice-Hall, Englewood Cliffs, N. J., 1963.

[12] R. E. Kalman, P. L. Falb, and M. A. Arbib. *Topics in Mathematical System Theory*. McGraw-Hill, New York, 1969.

[13] A. Kimms. *Multi-level Lot Sizing and Scheduling - Methods for Capacitated, Dynamic and Deterministic Models*. Physica Verlag, Heidelberg, 1997.

[14] V. Mahajan, E. Muller, and R. V. Srivastava. Using innovation diffusion models to develop adopter catagories. *Journal of Marketing Research*, 11:37–50, 1990.

[15] A. Nagurney, J. Cruz, and D. Matsypura. Dynamics of global supply chain supernetworks. *Mathematical and Computer Modelling*, 37:963–983, 2003.

[16] A. Nagurney, K. Fe, J. Cruz, K. Hancock, and F. Southworth. Dynamics of supply chains: a multilevel (logistical/informational/financial) network perspective. *Environment and Planning*, B29:795–818, 2000.

[17] G. Patrizi. **S.O.C.R.A.t.E.S.** simultaneous **o**ptimal **c**ontrol by **r**ecursive and **a**daptive **e**stimation system: Problem formulation and computational results. In M. Lassonde, editor, *Optimization and Approximation, Vth International Conference on Approximation and Optimization in the Carribean*. Springer Verlag, Berlin, 2000.

[18] H. A. Simon. Dynamic programming under uncertainty with a quadratic criterion function. *Econometrica*, 24:74–81, 1956.

[19]  T. Söderström and P. Stoica. *System Identification*. Prentice-Hall, Englewood Cliffs, N.J., 1989.

[20]  H. Theil. A note on certainty equivalence in dynamic programming. *Econometrica*, 25:346–349, 1957.

# Ship Routing and Scheduling with Persistence and Distance Objectives

Kjetil Fagerholt[1,2,3], Jarl Eirik Korsvik[3], and Arne Løkketangen[4]

[1] Industrial Economics and Technology Management, Norwegian University of Science and Technology, Norway kjetil.fagerholt@iot.ntnu.no
[2] Norwegian Marine Technology Research Institute (MARINTEK), Norway
[3] Department of Marine Technology, Norwegian University of Science and Technology, Trondheim, Norway korsvik@ntnu.no
[4] Department of Informatics, Molde University College, Norway
Arne.Lokketangen@hiMolde.no

**Summary.** It is well known that decision support systems (DSSs) usually only solve models that simplify and approximate the real problem. The planners might therefore be more interested in a set of diverse high quality solutions to choose from, than in only the optimal (or near-optimal) solution to the model as is usually produced by a DSS. In ship routing and scheduling plans are generated following a rolling horizon principle, where schedules are updated when new relevant information appears. However, the planners have often already made commitments to the customers for the next few voyages, for instance regarding arrival times and which ships that are assigned to service given cargoes. Therefore, the planners are interested in a set of high quality schedules that are close to the current (baseline) schedule in the near future, and diverse from each other in more distant time. We suggest a multi-start heuristic, including a persistence penalty function and distance measures, to produce such schedules. The method has been tested on a set of real-life problems and it provides valuable decision support flexibility for planners in shipping companies.

**Key words:** Ship routing and scheduling, Decision support, Distance measure, Persistence

## 1 Introduction

Ocean shipping is the major transportation mode in international trade and more than 6 billion tons of goods are carried by ships every year [15]. A ship involves a major capital investment, and its daily income and operating costs often amount to several

L. Bertazzi et al. (eds.), *Innovations in Distribution Logistics*, Lecture Notes
in Economics and Mathematical Systems 619, DOI: 10.1007/978-3-540-92944-4,
© 2009 Springer-Verlag Berlin Heidelberg

thousands of dollars. Proper planning of the fleet of vessels is therefore crucial for shipping companies to survive in an increasingly competitive market.

In this paper we focus on decision support for short-term routing and scheduling problems faced by many tramp shipping companies transporting bulk products. A shipping company operating in the tramp market usually has a set of mandatory contract cargoes that they are committed to carry, while trying to increase their profit from optional spot cargoes. The mandatory contract cargoes come from long-term contracts between the shipping company and the cargo owners (shippers). Each cargo (contract and spot) consists of a given quantity to be shipped between a given pair of ports. There are given time windows for loading of the cargoes and sometimes also for unloading. The fleet used for transporting the cargoes is heterogeneous where the ships have different load capacities, speeds, equipment, etc. The ship routing and scheduling problem mainly deals with (1) selecting spot cargoes to service, (2) assigning cargoes to ships in the fleet, and (3) deciding optimal ship routes and schedules. All these tasks must in principle be performed simultaneously. The planners in the shipping companies daily solve this ship routing and scheduling problem, which is basically similar to a multi-vehicle pickup and delivery problem with time windows (m-PDPTW), as described by [8].

TurboRouter [9] is a decision support system (DSS) developed by MARINTEK in Trondheim for solving this type of ship routing and scheduling problems by using local search-based heuristics [2]. However, the model solved by this DSS represents only a simplification of the real-world problem. When implementing and testing the DSS at shipping companies, we have often experienced that there exist constraints that are fuzzy and hard to model. There are sometimes also secondary objectives, which can be unclear, hard to model or to give proper weight to, in addition to the primary one (that usually is measured in monetary terms). The inherent stochastic nature of ocean shipping (for instance sailing times that are influenced by weather conditions) also contributes to make the real-life problem hard to model and to solve. During the work with TurboRouter and from discussing with planners in shipping companies, we realized that it was hardly possible to model all these complicating aspects in a good manner, neither was it desirable. This was something the planners wanted to evaluate themselves, based on their experience. However, a good contribution for the planners would be a DSS that could present a set of diverse high quality solutions to analyze and choose from instead of only one optimal (or near-optimal) solution.

In ship routing and scheduling, planning follows a rolling horizon principle, where plans are updated when new information (new cargoes, ship delays, etc.) appears. Figure 1 illustrates a typical workflow for the planning process.

When the planners perform a complete rescheduling using TurboRouter, the solution after rescheduling is often very different from the solution prior to the changes. However, the planners have often already made commitments to some of the customers, for instance regarding time for start of servicing, and they may have nominated specific ships for loading given cargoes. The planners are therefore interested in new solutions that are in some sense close to the current solution, the *baseline*

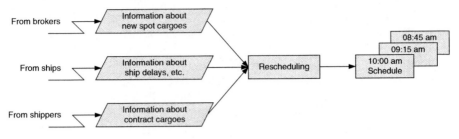

**Fig. 1.** Scheduling workflow at a typical tramp shipping company

*solution,* for cargoes where commitments have been made, usually in the near future. (See the following sections for the explanations of our measures).

The aim of this paper is to develop and present a method that can find and identify such a set of high quality solutions to ship routing and scheduling problems. It is desirable that these solutions are close to the baseline solution in the near future but diverse from each other in the more distant time. In our heuristic search method, we include a persistence penalty function in the objective function for finding solutions close to the baseline solution. For finding a diverse set of high quality solutions we use a structural distance measure.

There exist several references on previous work on similar ship routing and scheduling problems that we study in this paper, see for instance [2, 1] and [11]. A thorough review of ship routing and scheduling problems in general is given by [7]. However, there has been little focus on producing a diverse set of high quality solutions to these problems instead of only the optimal (or near-optimal) solution. Brown et al.[5] consider optimization and persistence in general. They emphasize that new solutions that retain the features of the baseline solution are more acceptable to decision makers than solutions that require more changes. They incorporate this kind of persistence in modelling linear, mixed-integer and integer linear programs. Brown et al. [3] deal with optimizing submarine berthing with a persistence incentive. When the current berthing plan is revised they introduce a penalty in the objective function for moving submarines. In a sample problem the persistence penalty reduced the number of revisions by 75%. Brown et al. [4] describe a problem of scheduling coast guard district cutters. An optimization model is described for solving the problem. In order to avoid major unnecessary changes when revising an accepted schedule they replace the original objective with a surrogate objective that preserves persistence. The cost of the new schedule is constrained to not cost more than the prior schedule.

Within land-based routing, [12] consider various similarity functions for the vehicle routing problem based on structural difference between solutions. The similarity functions or distance measures are used to choose a diverse set of high quality solutions to present to the planner using a decision support system. Their distance measure relies heavily on depot concept, and is thus not directly applicable for ship routing. Sörensen [14] discusses distance measures based on string comparison. The problems are characterized by the fact that the order in which the items appear in the solution is important. The problems discussed include the single-machine and multiple-machine scheduling problems, the traveling salesman problem, vehicle routing problems, and many others. Sörensen [13] considers route stability in vehicle routing decisions. A distance measure is used to create a metaheuristic approach that will find solutions "close" (in the solution space) to a given baseline solution and at the same time have a high quality in the sense that the total distance traveled is small.

The remainder of the paper is organized as follows: Sect. 2 gives a mathematical formulation of the ship routing and scheduling problem. Section 3 deals with the persistence objective to achieve new solutions close to a given baseline solution, while Sect. 4 describes a distance measure used to evaluate the difference between two solutions. A multi-start heuristic is proposed in Sect. 5 to solve the ship routing and scheduling problem with persistence and distance objectives. A computational study is presented in Sect. 6. Finally, Sect. 7 provides a summary and some concluding remarks.

## 2 Mathematical Formulation

This section describes a mathematical formulation of the ship routing and scheduling problem based on a model presented by [6] for a similar problem. Let $N$ be the set of cargoes indexed by $i$. Associated with the loading port of cargo $i$ there is a node $i$, and with the corresponding unloading port a node $N + i$, where $N$ is the number of cargoes (both contract and optional spot cargoes) that may be serviced during the planning horizon. Let $N_P = \{1, ..., N\}$ be the set of loading (or pickup) nodes and $N_D = \{N + 1, ..., 2N\}$ be the set of unloading (or delivery) nodes. The set of loading nodes $N_P$ is divided into two subset, $N_P = N_C \cup N_O$, where $N_C$ is the set of loading nodes associated with the contract cargoes the shipping company is committed to transport, while $N_O$ represents the loading nodes for the optional spot cargoes. Let $V$ be the set of ships in the fleet indexed by $v$. Then, $(N_v, A_v)$ is the network associated with a specific ship $v$. Not all ships can visit all ports and take all cargoes. Denote by $F_v$ the set of feasible nodes for ship $v$. Then, $N_v = F_v \cup \{o(v), d(v)\}$ is the set of nodes that can be visited by ship $v$, where $o(v)$ and $d(v)$ represent the origin and destination node for ship $v$. The origin node $o(v)$ can either be a port or a point at sea, while the destination node $d(v)$ is the last planned unloading node for ship $v$. Here $A_v$ contains the set of all feasible arcs that can be sailed by ship $v$. Further let $N_{Pv} = N_P \cap N_v$ and $N_{Dv} = N_D \cap N_v$ be the sets of loading and unloading node that can be visited by ship $v$, respectively. There is a revenue $R_i$ for transporting cargo $i$. The quantity of cargo $i$ is given by $Q_i$, while the capacity of ship $v$ is given by $V_{CAP_v}$. For each arc, $T_{Sijv}$ represents the calculated sailing time from node $i$ to node $j$ with ship $v$ including service time at node $i$. Let $[T_{MNi}, T_{MXi}]$ denote the time window associated to node $i$. The cost $C_{ijv}$ represents the sailing cost between node $i$ and $j$ for ship $v$ including port cost at node $i$.

We use the following types of variables: the binary flow variable $x_{ijv}, v \in V, (i, j) \in A_v$ equals 1, if ship $v$ sails from node $i$ directly to node $j$, and 0 otherwise. The time variable $t_{iv}, v \in V, i \in N_v$ represents the time at which ship $v$ begins service at node $i$, while variable $l_{iv}, v \in \mathcal{V}, i \in \mathcal{N}_v \backslash d(v)$ gives the total load onboard ship $v$ just after the service is completed at node $i$.

Then, the formulation of the ship routing and scheduling problem can be given as follows:

$$\max \sum_{v \in \mathcal{V}} \sum_{(i,j) \in \mathcal{A}_v | i \in \mathcal{N}_{Pv}} R_i x_{ijv} - \sum_{v \in \mathcal{V}} \sum_{(i,j) \in \mathcal{A}_v} C_{ijv} x_{ijv}, \tag{1}$$

subject to

$$\sum_{v \in \mathcal{V}} \sum_{j \in \mathcal{N}_v} x_{ijv} = 1, \qquad\qquad \forall i \in \mathcal{N}_C, \tag{2}$$

$$\sum_{v \in \mathcal{V}} \sum_{j \in \mathcal{N}_v} x_{ijv} \leq 1, \qquad\qquad \forall i \in \mathcal{N}_O, \tag{3}$$

$$\sum_{j \in \mathcal{N}_{Pv} \cup \{d(v)\}} x_{o(v)jv} = 1, \qquad\qquad \forall v \in \mathcal{V}, \tag{4}$$

$$\sum_{i \in \mathcal{N}_v} x_{ijv} - \sum_{i \in \mathcal{N}_v} x_{jiv} = 0, \qquad\qquad \forall v \in \mathcal{V}, j \in \mathcal{N}_v \backslash \{o(v), d(v)\}, \tag{5}$$

$$\sum_{i \in \mathcal{N}_{Dv} \cup \{o(v)\}} x_{id(v)v} = 1, \qquad\qquad \forall v \in \mathcal{V}, \tag{6}$$

$$x_{ijv}\left(t_{iv} + T_{Sijv} - t_{jv}\right) \leq 1, \qquad\qquad \forall v \in V, (i,j) \in \mathcal{A}_v, \tag{7}$$

$$T_{MNi} \leq t_{iv} \leq T_{MXi}, \qquad\qquad \forall i \in \mathcal{N}_v, \tag{8}$$

$$x_{ijv}\left(l_{iv} + Q_j - l_{jv}\right) = 0, \qquad\qquad \forall v \in \mathcal{V}, (i,j) \in \mathcal{A}_v | j \in \mathcal{N}_{Pv}, \tag{9}$$

$$x_{i,N+j,v}\left(l_{iv} - Q_j - l_{N+j,v}\right) = 0, \qquad\qquad \forall v \in \mathcal{V}, (i, N+j) \in \mathcal{A}_v | j \in \mathcal{N}_{Pv}, \tag{10}$$

$$l_{o(v)} = 0, \qquad\qquad \forall v \in \mathcal{V}, \tag{11}$$

$$\sum_{j \in \mathcal{N}_v} Q_i x_{ijv} \leq l_{iv} \leq \sum_{j \in \mathcal{N}_v} V_{CAPv} x_{ijv}, \qquad\qquad \forall v \in \mathcal{V}, i \in \mathcal{N}_{Pv}, \tag{12}$$

$$0 \leq l_{N+i,v} \leq \sum_{j \in \mathcal{N}_v} (V_{CAPv} - Q_i) x_{N+i,jv}, \qquad\qquad \forall v \in \mathcal{V}, i \in \mathcal{N}_{Pv}, \tag{13}$$

$$\sum_{j \in \mathcal{N}_v} x_{ijv} - \sum_{j \in \mathcal{N}_v} x_{j,N+i,v} = 0, \qquad\qquad \forall v \in \mathcal{V}, i \in \mathcal{N}_{Pv}, \tag{14}$$

$$t_{iv} + T_{Si,N+i,v} - t_{N+i,v} \leq 0, \qquad\qquad \forall v \in \mathcal{V}, i \in \mathcal{N}_{Pv}, \tag{15}$$

$$x_{ijv} \in \{0, 1\}, \qquad\qquad \forall v \in \mathcal{V}, (i,j) \in \mathcal{A}_v. \tag{16}$$

The objective function (1) maximizes the profit of operating the fleet. Constraints (2) ensure that all mandatory contract cargoes are transported exactly once, while (3) guarantee that all optional spot cargoes are transported at most once. Constraints (4)-(6) describe the flow on the sailing route used by ship $v$. Constraints (7) describe the compatibility between routes and schedules. If ship $v$ sails from node $i$ to node $j$ the constraints ensure that the start time of service at node $j$ is greater or equal to the start time of service at node $i$ plus the sailing time from node $i$ to $j$ including service time at node $i$. The time window constraints are given by (8). If ship $v$ is not visiting node $i$, we will get an artificial starting time within the time windows for that $(i, v)$-combination. Constraints (9) and (10) give the relationship between the binary flow variables and the ship load at each loading and unloading port, respectively. The initial load condition for each ship is given by (11), while (12) and (13) represent the ship capacity constraints at loading and unloading nodes, respectively. The coupling constraints (14) and the precedence constraints (15) ensure that both the pickup and delivery node for cargo $i$ belongs to the same route and that the delivery node is visited later than the pickup node. Finally, the formulation includes binary requirements (16) on the flow variables.

## 3 Persistence Penalty Function

As described in Sect. 1, ship routing and scheduling follows a rolling horizon principle, where rescheduling is performed when new information appears. However small modification of the input can often amplify into major changes to the current plan (*baseline solution*). In the baseline solution, especially in the near future, commitments have often been made regarding time for start of servicing and specific ships may have been nominated for transporting given cargoes. Even if it is possible to make changes for cargoes where commitments are made, it is usually not desirable unless there are compelling reasons to do so. For instance, when decisions are made regarding which ship that are planned to service a given cargo, additional equipment or services needed for the port operation, like pilot, tugs and port services may be arranged in advance. The fleet of ships is heterogeneous and if a cargo is rescheduled and assigned to another ship, or is to be serviced at another time, other equipment or services may be required. Then the old arrangements must be cancelled and new ones made. Even if it is possible to make changes it may require a lot of work to do so. When a customer is given a specific time for when he will be served he might be unwilling to accept another time (even if it is within the original time window). The customer might have made his own commitments to other participants in the supply chain, or the customer's storage facilities or production might be dependent on the specified pickup/delivery time.

To achieve solutions that are close to the baseline solution for cargoes where commitments have been made, we introduce a persistence penalty function to penalize solutions deviating from the baseline solution for these cargoes. To describe the persistence penalty function we use the following notation. We denote the baseline solution as $B$. The new solution after rescheduling is denoted by $A$. We let the parameter $U_i^{AB}$ be equal to one if cargo $i$ is served by different ships in solutions $A$ and $B$, and zero otherwise. $T_i^A$ and $T_i^B$ represent the time when service starts at node $i$ in solutions $A$ and $B$, respectively. The *cargo-ship penalty*, $P_{1i}$, is the penalty for servicing cargo $i$ by a different ship in the new solution compared with the baseline solution. Finally the *cargo-time penalty*, $P_{2i}$, is the penalty per time unit difference in service start at node $i$. For cargoes where no commitments have been made, both persistence penalties, $P_{1i}$ and $P_{2i}$, are equal to zero.

The persistence penalty function for solution $A$ can now be given as follows:

$$P(A) = \sum_{i \in N_P} P_{1i} U_i^{AB} + \sum_{i \in N_P \cup N_D} P_{2i} \left| T_i^A - T_i^B \right| \tag{17}$$

This persistence penalty function (17) is included in the objective function when solving the ship routing and scheduling problem.

## 4 Distance Measure

In the ship routing and scheduling problem, the most important difference between two solutions is which ships that service the different cargoes. In our distance measure we therefore calculate the number of cargoes that have changed ship between two solutions. This can be seen as the Hamming distance between the solutions, see [10]. The parameter $U_i^{AB}$ is again equal to one if cargo $i$ is serviced by different ships in solutions $A$ and $B$, and zero otherwise. The normalized Hamming distance (a value between 0 and 1) between solutions $A$ and $B$ is given as follows:

$$D_{AB} = \frac{1}{N} \sum_{i \in N_P} U_i^{AB} \tag{18}$$

The Hamming distance is normalized to make it independent of case. All distance measures can (and should) be normalized to the 0–1 range, see [12]. It can be noted that the penalty function (17) is similar to the normalized Hamming distance (18) if all penalties $P_{1i}$ are equal to $\frac{1}{N}$ and all $P_{2i}$ are zero. However, as mentioned in the previous section, this will not be the case for practical problem solving.

For a general vehicle routing problem (VRP), the Hamming distance is known to be a weak distance measure. In the classical VRP the fleet of vehicles is homogenous and the vehicles start at a given depot. Because of this it makes no difference which vehicle that services a given customer. The focus on distances measure for vehicle routing problems has therefore been to produce routes that are structurally different regarding which customers, and the sequence of the customers, that are serviced together on the same route, see for instance [12, 14]. In ship routing and scheduling problems, we usually have a heterogeneous fleet of ships, and due to the reasons described above the number of different cargo-ship combinations between two solutions is more important for the planner than structurally different routes regarding the sequence of the customers (nodes).

## 5 Solution Method

The solution method is based on the ideas from [2], where a multi-start local search heuristic for the ship routing and scheduling problem is presented. In the multi-start heuristic a number of start solutions are generated by using a constructive heuristic. A part of each solution is randomly generated and the rest is generated by using a deterministic constructive insertion heuristic. The local search operators used are both intra-route and inter-route operators. Intra-route operators try to improve the route of one single ship, while inter-route operators looks for improvement by moving cargoes between ships. The local search heuristic we use consists of three operators:

> *1-resequence*: cargo $i$ (both the loading and unloading node) is removed from the route of ship $v$. The cargo is then reinserted in the best position in the route of the same ship.
>
> *Reassign*: removes a cargo from the route of a ship $v$ and tries to find the best feasible insertion of the cargo into the route of another ship $u$. If there is one or more rejected cargoes in the cargo list, i.e. cargoes for which the heuristic has not found any feasible insertion so far, the heuristic tries to find a place for one of them in the route of ship $v$.
>
> *2-interchange*: two cargoes $i$ and $j$ are removed from the routes of ships $v$ and $u$, respectively, and are then inserted in the best feasible position in the routes of ships $u$ and $v$, respectively.

The persistence penalty function (17) described in Sect. 3 is included in the objective function when evaluating given moves. This will ensure solutions that are close to the baseline solution in near future. The distance measure (18) described in Sect. 4 is included in the multi-start heuristic to find a diverse set of start solutions and in order to choose the final solutions to present to the planner. In the algorithm described below we let $l$ represent the number of start solutions generated, $m$ the number of start solutions to be improved by local-search and $n$ the number of final solutions to present to the planner. We let $T_1$ and $T_2$ represent threshold distance values (between 0 and 1). A pseudo-code of the multi-start heuristic is formulated in Fig. 2. The sub-routine SELECT-SOLUTIONS is called in step 3 and 6 of the MULTI-START algorithm to ensure a diverse set of start solutions and final solutions respectively. In the sub-routine we start by selecting the solution with best objective value in the sorted list of solutions $Y$ to choose from. This solution is included in the list of selected solutions, $X$. We then go through the rest of the solutions in list $Y$. Before including a solution from list $Y$ in list $X$, we calculate the distances to all other solutions selected so far using distance measure (18). If all distances are greater or equal to the threshold distance value, $t$ (equal to $T_1$ and $T_2$, respectively), the solution is selected.

# 6 Computational Study

A computational study has been performed on four test cases based on real data from two different bulk shipping companies operating in the tramp shipping segment. Section 6.1 describes the test cases. In Sect. 6.2 we perform analyses to decide proper values on the penalties in the persistence penalty function (17) described in

MULTI-START $(l, m, n, T_1, T_2)$

$L-$ list of start solutions

$M-$ list of selected start solutions

$N-$ list of final solutions

1  generate $L$ with $l$  start solutions

2  sort $L$ by objective value

3  SELECT-SOLUTIONS $(M, L, m, T_1)$

4  perform local-search on all solutions in $M$ and update objective values

5  sort $M$ by objective value

6  SELECT-SOLUTIONS $(N, M, n, T_2)$

7  present the solutions in list $N$ to the planner (maximum $n$ solutions)

SELECT-SOLUTIONS $(X, Y, no, t)$

$X$.push-back $(Y[1])$

$no\_selected := 1$

**for** $y := 2$ **to** $length[Y]$ **do**

   **if** $no\_selected \geq no$

      **break**

   $selected := true$

   **for** $x := 1$ **to** $length[X]$ **do**

      $d := $ Calculate-Distance $(Y[y], X[x])$

      **if** $d < t$

         $selected := false$

         **break**

**if** $selected$

   $X$.push-back $(Y[y])$

   $no\_selected := no\_selected + 1$

**Fig. 2.** Pseudo-code for the multi-start heuristic

Sect. 3. Section 6.3 presents analyses to decide proper values on the threshold distance values which are included in the multi-start heuristic to get a diverse set of solutions. Section 6.4 presents computational results where different solutions to the test cases are compared regarding solution quality (profit) and distance (closeness) to the baseline solution.

## 6.1  Description of Test Cases

Cases 1 and 2 are collected from a shipping company transporting dry bulk commodities such as rock, iron ore and cement. A chemical commodity shipping company is the basis for cases 3 and 4. The test cases are summarized in Table 1. The number of cargoes includes both the cargoes in the baseline solution and the new optional spot cargoes. The number of spot cargoes is displayed in a separate row.

## 6.2  Setting Persistence Penalties

The cargo-ship penalty $P_{1i}$ and the cargo-time penalties $P_{2i}$ and $P_{2(N+i)}$ in the persistence penalty function (17) given in Sect. 3 have to be decided for each cargo. In Sect. 3 we described that it is desirable when rescheduling that new solutions are close to the baseline solution for some cargoes where certain commitments are made. These cargoes are usually the cargoes that are planned to be serviced in near future of the baseline solution and we will refer to these cargoes as *persistence important*

**Table 1.** Test cases

| Case | 1 | 2 | 3 | 4 |
|---|---|---|---|---|
| Planning horizon [days] | 30 | 30 | 90 | 150 |
| # cargoes | | 24 | 31 | 17 | 40 |
| # optional spot cargoes | 4 | 4 | 4 | 6 |
| # ships | | 7 | 13 | 5 | 6 |

*cargoes.* In the computational study we regard the cargoes that have the ending of the loading time window within the first third of the planning period as persistence important cargoes. This can of course be chosen to another value by the decision maker and the importance of each cargo can also be decided individually. To illustrate that it is often more difficult to make changes early than late in the baseline solution, we use penalties that are linearly decreasing with time, as shown in Fig. 3.

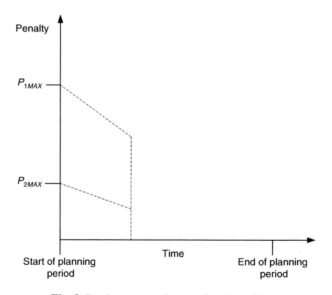

**Fig. 3.** Persistence penalty as a function of time

To decide the proper values of the penalties we use sensitivity analyses. Case 2 is here used as an example illustration. The multi-start heuristic used includes a random element in the generation of start solutions. Therefore, in order to make sensible comparisons we perform ten runs of the algorithm and report average values. Figure 4 shows that the average profit gap from the best solution (regarding profit only) increases with the maximum cargo-ship penalty cost $P_{1MAX}$. The figure also shows the average number of persistence important cargoes that have changed ship after rescheduling for different values on $P_{1MAX}$. From the figure we see that with

$P_{1MAX}$ equal to zero, the average profit gap from best solution is 0% while the average number of cargo-ship changes is 5.1. Increasing the maximum penalty cost will increase the average profit gap while reducing the number of cargo-ship changes. With a maximum penalty cost equal to 0.25% of the baseline objective value the solutions found after rescheduling have on average zero cargo-ship changes and an average profit gap from best solution equal to 0.5%.

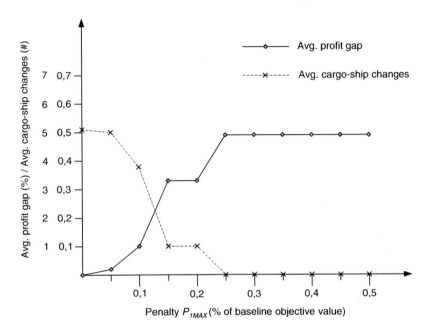

**Fig. 4.** Sensitivity analysis for the cargo-ship penalty for case 2

Similar analysis as for the maximum cargo-ship penalty cost $P_{1MAX}$ is performed for the maximum cargo-time penalty cost and the results are shown in Fig. 5. From the figure we see with $P_{2MAX}$ equal to zero, the average profit gap from best solution is 0% while the total cargo-time difference for the persistence important cargoes is on average 7.9 days between a new solution and the baseline solution. Increasing the maximum cargo-time penalty cost will increase the average profit gap but reduce the total cargo-time difference.

From this we choose the values of the maximum cargo-ship penalty $P_{1MAX}$ and the maximum cargo-time penalty $P_{2MAX}$. For case 2, it seems reasonable to set $P_{1MAX}$ in the range 0.15 – 0.25% of baseline objective value and $P_{2MAX}$ in the range 0.004 – 0.016% of baseline objective value. With these settings, the multi-start heuristic produces high-quality solutions that are close to the baseline solution (for the cargoes in near future). In the rest of the computational study we use for case 2; $P_{1MAX}$ = 0.25% of baseline objective value and $P_{2MAX}$ = 0.016% of baseline objective value.

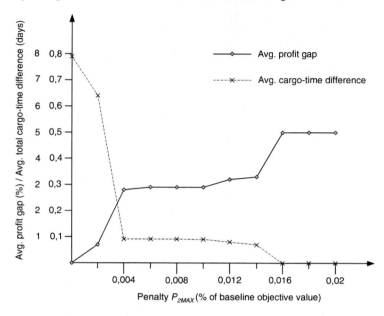

**Fig. 5.** Sensitivity analysis for the cargo-time penalty for case 2

Similar analyses are also performed for the other test cases and corresponding values on the penalties are chosen. The penalties are case specific and must be adapted to the different shipping companies and the problem at hand.

## 6.3 Setting Threshold Distance Values

In the multi-start heuristic described in Sect. 5 we use two acceptation threshold distance values; $T_1$ and $T_2$. Threshold distance value $T_1$ is used to ensure a diverse set of start solutions while threshold distance value $T_2$ is used to ensure a diverse set of final solutions. Figure 6 shows the number of different start solutions for different values on the threshold distance value $T_1$. Case 2 is again used as an example illustration. From the figure we see that if we generate 500 start solutions and use $T_1 = 0.4$ approximately 170 of the start solutions will be considered as sufficiently different to each other.

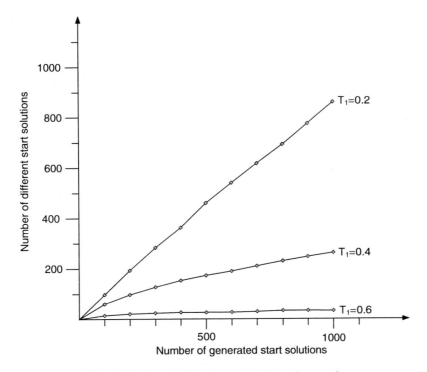

**Fig. 6.** Number of different start solutions for case 2

Figure 6 is produced without persistence penalties in the objective function. Including the persistence penalties presented in Sect. 6.2 in the objective function has only a small impact on the number of sufficiently different start solutions. For example if we generate 500 start solutions and use $T_1 = 0.4$ the number of different start solutions are reduced from 172 to 162 when including persistence penalties.

Figure 7 shows, both with and without persistence penalties in the objective function, the number of final solutions that are considered to be sufficiently different from each other for various values of the threshold distance value $T_2$. Case 2 is again used as an example illustration. We have generated 500 start solutions and use a threshold distance value $T_1 = 0.4$. When the threshold distance value $T_2 = 0$, the number of different final solutions after the multi-start local search is performed is equal to the number of different start solutions (which is 172 (without persistence penalties) and 162 (with persistence penalties)). From figure we see that the number of final solutions considered to be sufficiently different from each other decreases with the threshold distance value. There will also be some fewer different final solutions when including persistence penalties in the objective function. The persistence penalties will direct the local search towards the baseline solution for the persistence important cargoes and more of the final solutions will be similar compared with the solutions we find without persistence penalties in the objective function. Because we seek solutions that are close to the baseline solution for the persistence important cargoes we choose a lower acceptation threshold distance value $T_2$ for solutions that are close to the baseline solution than for solutions that are far from the baseline solution. For case 2, we use acceptation threshold distance value $T_2 = 0.3$ when we generate solutions without persistence penalties in the objective function, while we use $T_2 = 0.2$ when we seek solutions close to the baseline solution. Then, we get 59 and 71 sufficiently different final solutions to choose from, respectively.

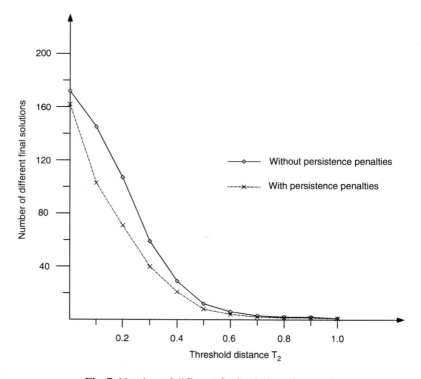

**Fig. 7.** Number of different final solutions for case 2

Similar analyses are done for the other test cases and corresponding values on the acceptation threshold distances are chosen.

## 6.4 Computational Results

We have tested the multi-start heuristic on the test cases described in Sect. 6.1 and the results are summarized in Tables 2 and 3. Table 2 shows the three best solutions to test case 2 when we optimize with respect to profit only, case 2a, and with respect to profit and persistence penalty cost, case 2b. In the table, we report the average number of persistence important cargoes that have changed ship, the average total cargo-time difference for the persistence important cargoes, the average penalty cost and the average profit gap from the best solution (with respect to profit only). The last rows in the table include the distance matrix which reports the average distance between the solutions. From Table 2 we see that including persistence penalties in the objective function has a great effect on finding high quality solutions that are close to the baseline solution. Comparing the best solution for test cases 2a and 2b, we see that the average-ship changes is reduced from 6.2 to 0, while the average total cargo-time difference is reduced from 8.2 to 0 days when including persistence penalties. However, the average profit gap from the best solution is a little bit larger for case 2b. The profit gap for the best solution to test case 2a is 0% from the best solution while for test case 2b the profit gap is 0.5%. The average distances between the solutions are larger for case 2a than for case 2b. The reason for this is that we accept a smaller acceptation threshold distance for solutions that are close to the baseline solution, see Sect. 6.3.

**Table 2.** Computational results for case 2

|  | Case 2a (profit objective) | | | Case 2b (profit incl. persistence penalty) | | |
|---|---|---|---|---|---|---|
| Solution | 1 | 2 | 3 | 1 | 2 | 3 |
| Avg. cargo-ship changes (#) | 6.2 | 5.0 | 6.7 | 0 | 1.0 | 1.1 |
| Avg. total caro-time difference (days) | 8.2 | 7.1 | 9.1 | 0 | 1.7 | 0.7 |
| Avg. penalty cost (% of baseline profit) | 1.7 | 1.4 | 1.8 | 0 | 0.3 | 0.2 |
| Avg. profit gap from best solution (%) | 0 | 0.1 | 0.3 | 0.5 | 0.4 | 0.8 |
| Distance matrix | 1 | 2 | 3 | 1 | 2 | 3 |
| 1 | – | 0.38 | 0.44 | – | 0.26 | 0.23 |
| 2 |  | – | 0.36 |  | – | 0.35 |
| 3 |  |  | – |  |  | – |

Table 3 shows the best solution regarding profit and the best solution regarding profit and penalty cost for each of the four test cases. As in Table 2, we report the average number of persistence important cargoes that have changed ship, the average total cargo-time difference for the persistence important cargoes, the average penalty cost and the average profit gap from the best solution (with respect to profit only). The average value for cases 1-4 a) and for cases 1-4 b) is shown in the two last columns. Remark that for case 4, the average total cargo-time difference is reduced from 86.5 days to 0.3 days when including persistence penalties. The reason for the large decrease is that the time windows are quite wide for this test case, specially the unloading time windows. From the table we see that the profit gap increases, on average from 0.18% to 0.68%, when including the penalty costs in the objective function. In return the persistence penalties direct the multi-start heuristic towards solutions that are very close to the baseline solution with respect to the persistence important cargoes.

**Table 3.** Computational results for cases 1-4

| Case | 1a | 1b | 2a | 2b | 3a | 3b | 4a | 4b | avg. a) | avg. b) |
|---|---|---|---|---|---|---|---|---|---|---|
| Avg. cargo-ship changes (#) | 5.9 | 0.7 | 6.3 | 0 | 3.7 | 1.6 | 12.7 | 1.1 | 7.13 | 0.85 |
| Avg. total cargo-time difference (days) | 12.6 | 2.4 | 8.2 | 0 | 5.4 | 1.4 | 86.5 | 0.3 | 28.18 | 1.03 |
| Avg. penalty cost (% of baseline profit) | 7.2 | 1.1 | 1.7 | 0 | 1.3 | 0.4 | 6.1 | 0.3 | 4.08 | 0.45 |
| Avg. profit gap from best solution (%) | 0.3 | 0.7 | 0 | 0.5 | 0.1 | 0.6 | 0.3 | 0.9 | 0.18 | 0.68 |

Figure 8 shows a new user interface from the DSS TurboRouter [9] for a reduced test case. In the figure, each ship is represented by a colour. The baseline solution is denoted B, while the three new solutions after rescheduling are denoted, S.1, S.2 and S.3. The Gannt diagram shows which ship each cargo is assigned to and the day when service starts (for both loading and unloading), for each of the four solutions. Cargoes 8 and 11 are new spot cargoes and are not serviced in the baseline solution. It is desirable that new solutions are close to the baseline solution for the part to the left of the vertical line in the figure. The new solution, S.2, is the solution closest to the baseline solution. However it should be noted that the solution includes one spot cargo less than solutions S.1 and S.3. The possibility to evaluate a set of solutions, instead of only the optimal (or near-optimal) solution, will in such cases provide valuable decision support flexibility for the planners.

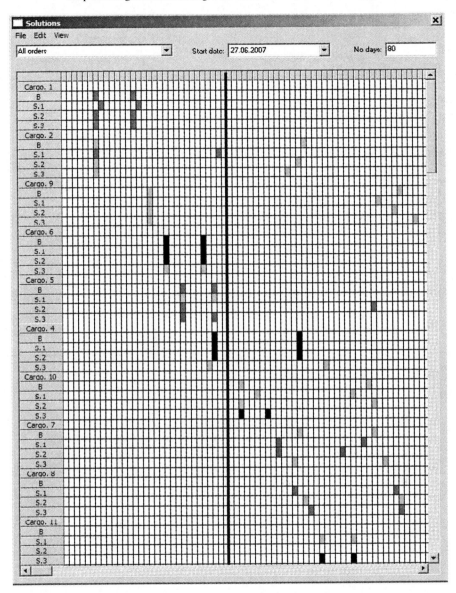

**Fig. 8.** Gannt diagram from TurboRouter showing information about the solutions

# 7 Summary and Concluding Remarks

In this paper we focus on decision support for a short-term routing and scheduling problem faced by many tramp shipping companies. TurboRouter is a decision support system (DSS) developed for solving such problems by using local search-based heuristics, see [2]. However, the model solved by a DSS is only an approximation to the underlying real-life problem. After evaluating the use of TurboRouter in different shipping companies, we realized the need for a DSS that can present a set of good diverse solutions to choose from instead of only one optimal (or near-optimal) solution.

In ship routing and scheduling, planning follows a rolling horizon principle, where plans are updated when new information appears. When planners perform a complete rescheduling using TurboRouter, the solution after rescheduling is often very different from the solution prior to the changes. However, the planners have often already made commitments to some of the customers, for instance regarding time for start of servicing, and they may have nominated specific ships for loading given cargoes. The planners are therefore interested in new solutions close to the previous solution for cargoes where commitments are made, usually in the near future.

This paper presents a method that presents a set of good solutions to ship routing and scheduling problems. It is desirable that these solutions are close to the baseline solution in the near future but diverse from each other in more distant time. The heuristic methods include a persistence penalty function in the objective function, for achieving solutions close to the baseline solution, and a distance measure for finding a diverse set of solutions.

The computational study shows that introducing small persistence penalties in the objective function directs the search towards solutions very close to the baseline solution. The quality (profit) of the solutions is just slightly worse than the solutions found when we optimize with respect to profit only, but the number of cargo-ship changes and the cargo-time difference are considerable reduced. The distance measure is used to find a diverse set of solutions the planner can choose from. A version of the algorithm presented will be implemented in TurboRouter [9]. Introducing persistence and distance objectives when solving ship routing and scheduling problems seem to provide valuable decision support flexibility for planners in shipping companies.

# 8 Acknowledgment

This research was carried out with financial support from the INSUMAR and OPTI-MAR projects, funded by the Research Council of Norway. We are also grateful to the shipping companies that provided real data for the test cases used in the computational study.

# References

[1]  D.O. Bausch, G.G. Brown, and D. Ronen. Scheduling short-term marine transport of bulk products. *Maritime Policy and Management*, 25(4):335–348, 1998.

[2] G. Brønmo, M. Christiansen, K. Fagerholt, and B. Nygreen. A multi-start local search heuristic for ship scheduling - a computational study. *Computers & Operations Research*, 34(3):884–899, 2007.

[3] G.G. Brown, K.J. Cormican, S. Lawphongpanich, and D.B. Widdis. Optimizing submarine berthing with a persistence incentive. *Naval Research Logistics*, 44:301–318, 1997.

[4] G.G. Brown, R.F. Dell, and R.A. Farmer. Scheduling coast guard district cutters. *Interfaces*, 26(2):59–72, 1996.

[5] G.G. Brown, R.F. Dell, and R.K. Wood. Optimization and persistence. *Interfaces*, 27(5):15–37, 1997.

[6] M. Christiansen, K. Fagerholt, B. Nygreen, and D. Ronen. Maritime transportation. In C. Barnhart and G. Laporte, editors, *Transportation*, volume 14 of *Handbooks in Operations Research and Management Science*, pages 189–284. Elsevier Science, 2007.

[7] M. Christiansen, K. Fagerholt, and D. Ronen. Ship routing and scheduling: Status and perspectives. *Transportation Science*, 38(1):1–18, 2004.

[8] J. Desrosiers, Y. Dumas, M.M. Solomon, and F. Soumis. Time constrained routing and scheduling. In M.O. Ball, T.L. Magnanti, C.L. Monma, and G.L. Nemhauser, editors, *Network Routing*, volume 8 of *Handbooks in Operations Research and Management Science*, pages 35–139. Elsevier Science, 1995.

[9] K. Fagerholt. A computer-based decision support system for vessel fleet scheduling - experience and future research. *Decision Support Systems*, 37(1):35–47, 2004.

[10] R.W. Hamming. Error-detecting and error-correcting codes. *Bell System Technical Journal*, 29(2):147–160, 1950.

[11] S.-H. Kim and K.-K. Lee. An optimisation-based decision support system for ship scheduling. *Computers & Industrial Engineering*, 33(3-4):689–692, 1997.

[12] A. Løkketangen and D.L. Woodruff. Similarity functions for VRP decision support. Working paper submitted, 2007.

[13] K. Sörensen. Route stability in vehicle routing decisions: a bi-objective approach using metaheuristics. *Central European Journal of Operations Research*, 14(2):193–207, 2006.

[14] K. Sörensen. Distance measures based on the edit distance for permutation-type representations. *Journal of Heuristics*, 13(1):35–47, 2007.

[15] UNCTAD. Review of maritime transport, 2006. United Nations, New York and Geneva, 2006.

# Logistics in Real-Time: Inventory-Routing Operations Under Stochastic Demand

Ricardo Giesen[1], Hani S. Mahmassani[2], and Patrick Jaillet[3]

[1] Department of Transport Engineering and Logistics, Pontificia Universidad Católica de Chile, Chile Giesen@ing.puc.cl
[2] Transportation Center, Northwestern University, USA Masmah@northwestern.edu
[3] Department of Civil & Environmental Engineering, Massachusetts Institute of Technology, USA Jaillet@MIT.edu

**Summary.** This chapter studies real-time distribution strategies and their associated benefits for a two-level distribution system, from one depot to $N$ retailers, wherein vehicle delivery routes can be updated using real-time information about current inventory levels and vehicle status. Three rolling horizon approaches are proposed in which plans are updated using a mathematical programming formulation. The proposed re-planning strategies are compared against two benchmark policies using a general discrete-event simulation framework. Proposed re-planning strategies are shown to systematically outperform the two benchmark policies.

**Key words:** Logistics, Inventory Routing problem, Vendor managed inventory, Online vehicle routing problem, Dynamic fleet management, Intelligent transportation systems (ITS), Commercial vehicle operations (CVO), Real-time information, On-line problems, Freight transportation

## 1 Introduction

The focus of this Chapter is on formulating Inventory-Routing Problems (IRPs) in a stochastic dynamic environment with real-time information about current inventory levels, as well as delivery vehicle locations and status. The main objectives are to: (a) formulate and analyze the Online Inventory Routing Problem (OIRP), explicitly taking into account real-time information about fleet status and inventory levels at different facilities; (b) develop operational-control strategies to operate a distribution system in which transportation and inventory control are coordinated, tailored to different degrees of real-time information; and (c) evaluate the performance of the

L. Bertazzi et al. (eds.), *Innovations in Distribution Logistics,* Lecture Notes in Economics and Mathematical Systems 619, DOI: 10.1007/978-3-540-92944-4,
© 2009 Springer-Verlag Berlin Heidelberg

proposed real-time strategies and the value of using real-time information in a centrally operated or collaborative distribution system, along with conditions that affect this performance.

This research is motivated by three main considerations: (a) the recognized importance of logistics and distribution systems in the national and local economy (see for example, La Londe, 1994; Ghiani et al., 2004) (b) the current trend to coordinate logistic operations, discussed among others by (Thomas and Griffin, 1996, Campbell et al., 1998, Bertazzi et al., 2007), and (c) the opportunities offered by Information and Communication Technologies (ICT) to operate and control a system in real-time, which allow sharing information between different stages in a supply chain at progressively reduced costs (Rabah and Mahmassani, 2002).

Relevant ICT developments can be divided into three groups: communication and tracking devices that can automate information input to and transmission between computer systems (see, e.g., Blanchard, 2003; Datta, 2003; Masters and La Londe, 1994); Commercial Vehicle Operations (CVO) technologies that allow the control of a fleet of vehicles on a real-time basis (see Regan et al., 1995); and software and Decision-Support Systems (DSS) that provide data processing capabilities at a particular facility (Rutner et al., 2003; Fleischmann and Meyr, 2003; Simchi-Levi et al., 2003). These systems increase the speed and accuracy with which data is entered, gathered, and communicated, and provide real-time visibility into inventory levels throughout the distribution system and better control over a fleet of vehicles.

In summary, with access to real-time information on the current state of the system – i.e. inventory levels at each facility and status of the fleet – managers can react faster to changes in predicted demand patterns or traffic conditions, and make online decisions on a continuing basis to adjust and improve routing plans accordingly. However, the operational decisions are complex, since the underlying problems are combinatorial and unfold in real-time, precluding the evaluation of all possible alternatives by the decision maker. Moreover, the stochastic nature of such systems implies that information about the state of the system is gradually revealed and cannot be accurately predicted in advance. In order to take maximum advantage of real-time information made available by ICTs, supply-chain managers need to use information effectively. That requires the development of models and algorithms that can exploit the full potential of real-time information for distribution-logistic operations.

While previous published contributions on IRPs share some common elements, most of the problems addressed in the literature consider systems that have different characteristics. A detailed review is presented in Baita et al. (1998), Campbell et al. (1998), Campbell and Savelsbergh (2002), Kleywegt et al. (2002), Giesen (2007), and Bertazzi et al. (2007). The main difference between problems found in the literature, with the exception of Giesen et al. (2005), with the problem addressed in the present Chapter, is that vehicle plans are not modified after they leave the depot.

Finally, previous research on real-time fleet management has focused on how to serve load demands for transportation services that are exogenous to the system, in the context of dynamic vehicle-routing problems (Gendreau et al., 1999, Larsen et al., 2002, Larsen et al., 2004, Jehova et al., 2006), pick up and delivery problems with full truck loads (Regan et al., 1995, Regan et al., 1996, Yang et al., 2004, Kim et al., 2002, Kim et al., 2004), and pick up and delivery problems with multiple loads (Mitrovic-Minic et al., 2004, Gendreau et al., 2006).

In this research, routing decisions are coordinated with inventory control. In that fashion, it is expected that monitoring inventory levels would allow improving the forecast and coordination of transportation activities, giving the operator the option to visit a facility earlier than needed to take advantage of transportation savings. That could be particularly useful when demand is highly variable and/or unpredictable, which is normally the case when final consumers are separated by several echelons from the echelon considered, or as a consequence of the phenomenon known as the bullwhip effect (Lee et al., 1997, Fine, 1998, Chen et al., 2000).

The specific distribution system considered is a two-level supply chain, in which a set of geographically dispersed facilities facing stochastic demands have to be repeatedly replenished from a central warehouse (or depot) over a long period of time. The facilities to be replenished could represent final customers, retailers who serve demand from final customers, or distribution centers from which a set of additional facilities are replenished. In this system, products are transported from the depot to the set of retailers by a vehicle with limited capacity, the plans for which can be updated with real-time information about the state of the system, thanks to modern information and communication capabilities. This problem is designated as the Online Inventory Routing Problem (OIRP) under real-time information. The OIRP is formulated and solved considering inventory allocation and transportation decisions together. As such, the OIRP considers the trade-off among transportation, inventory holding, and stock-out costs.

Key features of this OIRP are the presence of uncertainty about future consumption rates at different facilities, and the possibility of updating plans based on accurate real-time information about the complete state of the system. This is in contrast with deterministic environments, in which decisions can be made with perfect hindsight, thus real-time operational capabilities would not modify the nature of the problem. The possibility of updating plans on a quasi-continuous basis, given information on demand realizations, makes possible some additional decisions to update truck-route plans, such as modifying the set and/or the sequence of subsequent customers to be visited; diverting a truck from its current destination to visit a different facility; and adjusting amounts to be delivered to subsequent customers in the route.

Such an operational environment could enable more efficient use of existing resources and increase system reliability. However, the design of effective strategies to operate the system can be extremely difficult. On one hand, the dispatcher faces a fleet-routing and scheduling problem – which is combinatorial – to obtain new operational plans. Since even simplified static and deterministic versions of the inventory-routing problem are computationally hard (Bertazzi et al., 2007), a trade off between quality of solution and speed should be considered in the search for new plans. On the other hand, given that plans can be modified at any time, based on new information, the events and circumstances under which a plan update would be beneficial should be specified.

The rest of the chapter is organized as follows. In the next section, a detailed definition of the problem studied and general approach are presented. Section 3 discusses re-planning strategies for the OIRP under real-time information, presenting a detailed mechanism of how a central decision maker would operate under these policies, and presents a formulation for the routing problem used to update plans corresponding to the problem of interest under real-time information. Section 4 describes the simulation framework developed to evaluate the different strategies proposed, and presents experimental results and analysis from these simulations. Concluding remarks are presented in the final section.

## 2 Problem Definition

### 2.1 Main Assumptions

In this research – unlike the common view in real-time fleet management problems where load demands are exogenous to the system – decisions to replenish inventory, by how much, in what sequence, and by which vehicle, are conducted in an integrated real-time decision framework. In addition a central-planner approach to the problem is assumed. That is, the system is operated and controlled by a central decision maker, with real-time information about the complete state of the system, who seeks to move inventories in the system in such a way as to maximize total expected profit in the long-run for the entire system.

Also, it is assumed that upper hierarchical (strategic and tactical) decisions about the system configuration are given, e.g., the set of facilities to be refilled from a particular distribution center, and the characteristics of the fleet of vehicles assigned to serve those facilities are not directly considered. Moreover, in this initial development, a single-vehicle approach to the problem is assumed.

Demand processes at different facilities are assumed to be the only source of uncertainty; travel times between facilities are assumed to be fixed and known even though, in real-world applications, particularly in urban areas, uncertainties in traffic conditions could lead to significant travel time variation. Moreover, time associated with loading and unloading operations is not considered. In addition, is assumed that demands are known in probability distribution, and that these demand processes at retailers cannot be affected by the central decision maker.

Another important assumption is that daily and weekly cycle operation characteristics are not taken into account; that is, the system is assumed to be operating continuously, without interruption. Moreover, labor-related constraints are not considered. In short, it is assumed that the vehicle and all facilities are always in operation; i.e., deliveries can be scheduled at any time, with neither time windows for particular facilities nor restrictions on the number of hours that a driver can operate a vehicle. Therefore, delivery routes are constrained only by the vehicle's capacity to transport products. Those other types of constraints would cloud the analysis, and interfere with the initial fundamental insights of interest in this investigation; future extensions could readily incorporate such constraints for implementation purposes.

## 2.2 Preliminaries and Problem Parameters

In order to present the OIRP, the following general notation is used. The set of retailers is designated as $\Im$, $\Im = \{1, 2, ..., i, ..., N\}$, and the set of all facilities (depot and retailers) as $\Im_0$, $\Im_0 \equiv \Im \cup \{0\}$. Those $N+1$ facilities are denoted by sub-index $i = 0, 1, 2, ..., N$ (sub-index 0 is for the depot) and are located in a bounded subset in the Euclidean space. The function $d_{ij}$ gives the Euclidean distance between facilities $i$ and $j$, or between a facility $i$ and the vehicle location $j$. Each retailer $i$ has a maximum capacity to store inventory, $\kappa_i$, measured in the units of the single product considered. In addition, the vehicle has limited capacity, $\Upsilon$, measured in the same units, and is assumed to travel at constant speed according to the Euclidean metric. Without loss of generality, the vehicle speed is assumed to be one.

Each retailer $i$ serves an independent demand process. In general, it is assumed that each facility serves a compound Poisson demand process, in which customer arrivals to retailers follow Poisson processes, and customers' demand sizes are independent discrete random variables. Demand processes have associated arrival rates $\lambda_i(t)$ for retailer $i$ at time $t$. In addition, customer demand sizes are assumed to be Poisson distributed with mean $\theta_i(t)$. Thus the expected demand per unit of time at retailer $i$ at time $t$, $\mu_i(t)$, can be calculated as $\mu_i(t) = \lambda_i(t) \cdot \theta_i(t)$.

The state of the system at time $t$, $X(t)$, can be described by the following parameters: (1) inventory levels at time $t$, $\iota(t) = (\iota_1(t), ..., \iota_i(t), ...., \iota_N(t))$, where $\iota_i(t)$ is the inventory level at facility $i$ at time $t$, (2) location of the truck at time $t$, $\ell(t)$, and (3) load remaining in the truck at time $t$, $v(t)$. Hence, the state of the system at time $t$ can be expressed as:

$$X(t) = \left[ \iota(t) \ \ell(t) \ v(t) \right] \tag{1}$$

The decision maker can update plans at any epoch $t$ based on $X(t)$ and past events, but without knowledge of future events. Plan updates are implemented immediately unless the vehicle is moving, in which case a time lag – between the epoch when a decision to update a plan is made and the plan is implemented – is considered. This is modeled using a time projection, which takes into account the time from the moment the decision to update the current plan is made until the new plan begins to be executed. Hence, instead of considering the actual state of the system at time $t$ in the solution procedure, the state of the system is projected to a time $(t + \partial t)$, $\tilde{X}(t + \partial t)$, assuming expected consumption rates and truck current speed and destination, where $\partial t$ is the projection time, which includes any solution procedure used to update plans and the time required for the driver to modify his current destination.

In the OIRP, there are three sets of cost parameters: (1) transportation cost per unit of distance traveled by the vehicle, $TC$, (2) inventory holding costs at each retailer $i$, $h_i$ for retailer $i$, and (3) penalty associated with each unit of demand lost during stock-out, $p_i$ at retailer $i$.

### 2.3 Decision Variables, Main Constraints and Objective Function

In this OIRP system, decisions available to the decision maker are related to truck plans, and consist of when, and how truck plans are updated. A plan or policy $\pi$, can be specified by the sequence of facilities to be visited, amounts to be delivered, and arrival times, to each one of those facilities. In addition, since the state of the system is continuously monitored and plans can be updated at any time, plan update epochs are also decision variables.

The main constraints that must be satisfied in the OIRP are related to the dynamics of the system and could be stated as follows:

(a) Inventory levels at each retailer are always non-negative and less than their capacity, i.e. $0 \leq \iota_i(t) \leq \kappa_i$, for $i \in \mathfrak{I}$, and all $t$.
(b) At consumption epochs, if the demand size is greater than the inventory level at a particular facility, that inventory decrease the demand size, otherwise the difference between the demand size and the inventory remaining is lost demand and the inventory level decreases to zero.
(c) Inventory levels at retailers increase at delivery epochs by an amount equal to the amount delivered.
(d) The load remaining in the truck is always non negative, i.e., $\upsilon(t) \geq 0$, for all $t$.
(e) The amount delivered to a retailer $i$ is not greater than the load remaining in the vehicle at that delivery epoch, and the load remaining in the vehicle after the delivery is decreased by the quantity delivered.
(f) The total amount delivered in a tour does not exceed its capacity.
(g) The location of the truck is modified whenever the truck is not idle, and the truck moves toward the next facility at unit speed.

The objective of the central decision maker is to move the inventories in the system so as to minimize the expected total operating cost, composed of: (1) total transportation costs, (2) total inventory holding costs, and (3) total lost sales penalty costs.

Thus far, the formal definition of the problem studied in this Chapter has been introduced. The next section discusses major sources of complexity, the general approach that is being used to deal with the OIRP, and the proposed re-optimization strategies.

## 3 Proposed Real-Time Strategies for the OIRP

This section presents the general approach to deal with the OIRP, two benchmark policies used to compare the proposed strategies, and the formulation and design of re-optimization based strategies. Under those strategies, the inventory control side of the problem is solved first without considering joint replenishment to different facilities; the resulting fill-up-to levels are then used in a local off-line problem which is then solved in different real-time control strategies, reflecting different degrees of real-time information, to update plans.

## 3.1  General Approach

As previously stated, the OIRP does not seem to be tractable. Among the main difficulties in solving this control problem are:

(a) Simplified static and deterministic versions of the problem are NP-Hard, i.e., given the complete stream of future demands at each retailer, the associated inventory-routing problem, to optimally schedule deliveries to retailers, is very difficult to solve. In addition to the sequencing complexity of the problem, it is difficult to correctly capture the effect of short-term decisions on long-term costs, since deliveries depend on the time and amount reloaded in the previous visit to that facility. An optimal solution to the problem would require a long-term planning horizon; therefore, it is unlikely that the problem could be solved to near optimality in a reasonable time even for small problem-instances. This precludes the use of a complete static and deterministic IRP formulation for re-planning purposes in real-time operations.
(b) In addition to the combinatorial challenge of the static version of the problem, demands are dynamic and stochastic, and decisions can be updated at any time. In fact – in contrast with other real-time fleet operation problems in which requests to the system are clearly decision epochs – in the OIRP, final customer-demand epochs occur so often that it would be infeasible to adjust plans at each one of them. Thus, update epochs are not clearly defined, and obtaining the best update epoch is not trivial.
(c) Because retailer deliveries can be combined on the same route, optimal policies to serve each retailer depend not only upon that retailer's inventory level, but also upon the state of the complete system. In fact, transportation costs to service a particular facility are not fixed, but depend upon the set of facilities served on the same route (Campbell et al., 1998, Bertazzi et al., 2007). Moreover, since a single vehicle serves all retailers, the lead time to replenish a particular retailer might be affected by congestion, in terms of the number of additional deliveries that are scheduled before that visit.
(d) Advances in real-time online combinatorial optimization neither provide tools to solve problems, such as the OIRP, to optimality nor give clear guidance on how to exploit online information in its operation (Grötschel et al., 2001b, Grötschel et al., 2001a).
(e) Finally, as in most real-time combinatorial problems, there is a trade-off between the quality of a new plan and the response time at update epochs.

Those difficulties preclude solving the problem or finding an optimal policy directly from the formulation presented in the previous section. Instead, an approach is proposed wherein the inventory control side of the problem is solved first, taking into account only a simplified version of the routing problem. In this approach, inventory reorder parameters are established for each facility and then used as target levels on a routing problem used to update plans.

For the proposed approach, different operational policies are proposed, tailored for different degrees of sophistication in terms of ICT. Those policies are based on a rolling-horizon framework, wherein new operational plans are repeatedly generated, based on updated information about the state of the system, and are implemented until the next update epoch is reached. In that scheme, operational strategies are defined by when and how plans are updated.

In terms of plan update epochs, three different cases are analyzed, ordered in terms of decreasing ICT requirements: (1) truck routes can be continuously updated, allowing for en-route diversions, (2) truck routes can be updated only at facilities (en-route diversion not allowed), and (3) truck routes cannot be updated after the truck leaves the depot, i.e. truck plans can be updated only upon tour completion. In all cases, full information about the state of the system at plan update epochs is assumed, and a Mixed-Integer Programming (MIP) problem formulation is solved for new plan generation.

In order to evaluate and compare proposed real-time strategies discrete-event simulation experiments are conducted. The proposed policies are compared to each other and against two benchmark policies described next.

### 3.2 Benchmark Policies

There are no accepted benchmarls for evaluating real-time fleet operational strategies. As discussed by Kim (2003), "detailed specifications of the problem have a significant impact on the performance of a policy." Since the OIRP has not been studied before, two benchmark policies are introduced and developed.

The first benchmark policy, BENCH1, emulates what can be achieved operating the system in a decentralized manner with agents following optimal policies. In BENCH1, each retailer manages his own inventory, placing orders to a central supplier who, once a day, schedules deliveries for previous-day orders. In this case, based on the orders received at the end of each day, the supplier creates routes solving a Vehicle-Routing Problem (VRP). Each retailer will follow an optimal continuous-review policy to control his inventory, which in this case corresponds to an $(s, S)$ policy. That is, each retailer will place an order of size $(S-s)$ immediately if his inventory level is below $s$. The optimal parameters for an $(s, S)$ policy can be obtained using Zheng and Federgruen algorithm (1991) or by exhaustive search over the feasible region.

The second benchmark policy, *Most-Urgent-Next* (MUN) is based on a simple greedy decision rule. Under MUN, at each delivery epoch, the vehicle is send next to refill the retailer closest to run out of inventory. To select the next retailer to be refilled, inventory levels are inspected, and the time at which each retailer would run out of inventory, if not visited, is calculated based on average consumption rates. If the vehicle has enough load remaining to refill the selected retailer, it would be sent directly to that location, otherwise it would go first to be refilled at the depot and then to that retailer. In that policy, each retailer $i$ is refilled up to a pre-specified target level, $S_i$, or up to capacity $\kappa_i$. MUN implementation assumes that inventory levels are monitored and that routes are created so that the next delivery is decided upon refilling a retailer. In this strategy, whenever the vehicle is at the depot, a decision on waiting an additional time or departing immediately has to be weighed until vehicle departure. To evaluate the effect of waiting-additional-time decisions, for each facility the expected increment in lost sales costs is compared against transportation costs savings associated with waiting an additional time interval. If the expected increment in lost sale costs are smaller than transportation cost savings an additional time interval is waited, otherwise the vehicle is send to the most urgent facility.

In these benchmark policies, operational control parameters, such as reorder levels, are adjusted based on the specific parameters of each scenario.

## 3.3  Off-Line Optimization Problem Formulation

This subsection presents the local off-line problem used on optimization based control strategies to update plans. First, the general approach to locally update plan is presented. Second, optimization of refilling levels method is presented. Finally, a mathematical formulation of the routing problem used to update delivery plans is presented.

### a) Preliminaries

One of the main difficulties in formulating this problem is to be able to capture the effect of short-term decisions on long-run costs. If the customers were visited in isolation of each other using direct deliveries from the depot, served by independent vehicles, the optimal policy for each customer could be computed. In this case, a well known result on inventory control for single items inventory systems with stochastic consumption rates, constant replenishment lead times, and standard cost assumptions is the optimality of $(s, S)$ policies, see for example Axsäter (2000) and Zipkin (2000). In an $(s, S)$ policy each time the inventory position (inventory on hand plus on order minus backorders) is below $s$ a delivery is scheduled to send a quantity equal to $S$ minus the inventory position, so the inventory position becomes equal to $S$.

However, since only one truck is serving all customers and customer deliveries can be combined on the same route, transportation (delivery) costs are not fixed. Indeed, they would depend on the set of customers that are served together on the same route. Then the optimal policy to serve each customer would depend not only on its inventory level, but also on the complete state of the system.

To deal with this problem, optimal refilling levels for each facility are first specified, assuming that there is no pattern of deliveries. These levels are then kept as targets to refill up to, on each delivery, when plans are generated. However, there are only penalties associated with violating them as they are not included as hard constraints in the off-line routing problem. The off-line routing problem generates a plan that stipulates for each customer the next delivery time and quantity to refill, based on reorder quantities and on the current state of the system, i.e. inventory levels at each facility, and location and load remaining on the truck. This plan is obtained by minimizing the sum of transportation costs, and expected Lost Sales Penalty (*LSP*) costs, subject to visiting all customers once during the planning horizon.

In the next subsection, the method used to compute reorder quantities for each facility is presented, followed by the mathematical formulation of the off-line routing problem.

### b) Optimization of Refilling Levels

To compute the target refilling levels for each facility, the sum of expected costs (per time unit) for all facilities is minimized, assuming there is no pattern of deliveries and the truck visits only one customer per route. That is, the truck goes back to the depot after refilling a customer, and from there would go to the next customer as needed. Since the truck visits only one customer per route, transportation costs associated with serving a particular customer are fixed. However, even though customers are visited in isolation, the possibility of waiting for service due to the (single) truck serving other facilities is incorporated. Additionally, unlike traditional inventory systems where quantities are fixed after orders are placed, quantities can be updated upon arrival at the customer.

Then, using a policy that places orders when the inventory level is $s$ and refills up to level $S$, the expected cost per time unit ($AC$) at steady state at each facility could be calculated using the renewal reward theorem (see for example [35]). An approximation made in the calculation of AC (in (3) below) is to neglect the impact of expected stock-out time during the cycle on the cycle length and holding costs (see Axsäter, 2000 pp. 65), as follows:

$$AC_i = \frac{E[\text{cost per cycle}]}{E[\text{cycle length}]} \tag{2}$$

$$AC_i \approx \frac{FTC_i + h_i\left(\frac{S_i - s_i}{\mu_i} + L_i\right)\left((s_i - L_i\mu_i) + \frac{1}{2}(S_i - s_i + L_i\mu_i)\right) + p_i \cdot \sigma_i \sqrt{L_i} \cdot G\left(\frac{s_i - L_i\mu_i}{\sigma_i \sqrt{L_i}}\right)}{\left(\frac{S_i - s_i}{\mu_i} + L_i\right)} \tag{3}$$

where $FTC_i$ is the fixed transportation cost to serve retailer $i$ (in isolation); $G(x)$ is the loss function that gives the expected number of lost sales at the end of a period with demand distributed $N(0,1)$ given that the initial inventory level is $x$ (see Axsäter, 2000); and $L_i = T + d_{0i} + W_i$, is the sum of the review period, $T$, and the total lead time. The length of the review period, $T$, is the time between plan updates and depends on the policy implemented. The lead time is composed of travel time from the depot to retailer $i$, $d_{0i}$, and the expected waiting time $W_i$ for retailer $i$. This expected waiting time could be expressed, as a function of reorder quantities, using the following recursive expression:

$$W_i = \sum_{j \neq i} \left\{ \underbrace{\text{Pr\{ret. } j \text{ is in service\}}}_{\beta_j} \cdot \left(d_{0j}\right) + \underbrace{\text{Pr\{ret. } j \text{ is waiting for service\}}}_{\gamma_j} \cdot \left(2d_{0j}\right) \right\} \tag{4}$$

$$\beta_j = \frac{2d_{0j}}{\left(\frac{S_j - s_j}{\mu_j}\right) + L_j}, \tag{5}$$

and

$$\gamma_j = \frac{W_j}{\left(\frac{S_j - s_j}{\mu_j}\right) + L_j} \tag{6}$$

where $\beta$ and $\gamma$ are the proportion of the cycle in which a retailer is being served and waits for service, respectively. To evaluate the waiting times for all facilities, given a vector of reorder quantities, a bisection procedure is used iteratively until the waiting times for all facilities are consistent.

Finally, to obtain the optimal reorder quantities, the sum of average cost for all facilities is minimized, subject to $L_i = T + d(0, i) + W_i$; equations (4), (5), and (6); and $\sum_i 2d_{0i} \cdot (\mu_i/(S_i - s_i)) < 1$. The first four are definitional constraints, and the fourth implies that truck utilization rate should be less than 100%. This problem is solved using a steepest decent numerical procedure, in which at each step the gradient is evaluated numerically. Then, the solutions found in this step are used as input parameters every time the off-line routing problem presented in the next subsection is called.

### c) Mathematical Formulation of the Problem

In this off-line routing problem, the current inventory levels at all facilities are considered as given, as are the load remaining and the distance to all facilities for the truck. When the truck is at the depot the load remaining is equal to the truck capacity. Additional input parameters are the transportation cost $TC$ [\$/hr]; inventory holding cost $h_i$ [\$/unit-day]; lost sales cost $p_i$ [\$/unit]; and order up to level $S_i$[units].

It is assumed that the central decision maker would try to follow the optimal reorder up to $S$ policy for each customer. However, since patterns of deliveries are not considered, he/she would deviate from that policy to take advantage of transportation savings. In order to measure the impact on transportation and inventory cost of deviating from the reorder up to $S$ policy, Incremental Inventory Costs ($IIC$) for each facility are computed. These $IIC$ are calculated as a one time deviation from the reorder up to $S$ policy, assuming that after this deviation the optimal policy is resumed. These $IIC$ can be expressed as the sum of expected Incremental Transportation Costs ($ITC$), and expected Lost Sales Penalty costs ($LSP$). Notice that the impacts on holding costs are only considered through the specification of reorder up to levels. To compute $ITC$, first notice that if each retailer is considered in isolation, for a given consumption rate $\mu$, the minimum transportation costs are achieved when deliveries arrive when the inventory level is zero and the quantity delivered is $S$. In this case transportation costs per unit of time are $FTC \cdot (\mu/S)$. Then $ITC$, associated with scheduling a delivery of size $q$ units at time $t$ after the current time, given that the current inventory level is $\iota$, could be expressed as

$$ITC(q,t/\iota) = \left( \frac{FTC}{\left(\frac{q}{\mu}\right)} - \frac{FTC}{\left(\frac{S}{\mu}\right)} \right)\left(\frac{q}{\mu}\right) = FTC \cdot \left( \frac{\mu}{q} - \frac{\mu}{S} \right)\left(\frac{q}{\mu}\right) = FTC \cdot \left( 1 - \frac{q}{S} \right) \quad (7)$$

where $0 \le q \le (S - \iota + \mu \cdot t)$, and $S$ is the optimal reorder up to level. On the other hand, expected Lost Sale Penalty (LSP) costs, associated with scheduling a delivery at time $t$ after the current time, given that the current inventory level is $\iota$, could be computed, approximating the distribution of total demand during $t$ as $N(t\mu, t\sigma^2)$, as:

$$LSP(t / \iota) = p \cdot \int_{\iota}^{\infty} (u - \iota)f_{D(t)}(u)du \quad (8)$$

$$LSP(t/\iota) = p \cdot \sigma \sqrt{t} \int_{\frac{\iota-t\mu}{\sigma\sqrt{t}}}^{\infty} \left( v - \left( \frac{\iota - t\mu}{\sigma\sqrt{t}} \right) \right) \phi(v)dv = p \cdot \sigma \sqrt{t} \cdot G\left( \frac{\iota - t\mu}{\sigma\sqrt{t}} \right) \quad (9)$$

Based on these IIC, an off-line problem could be formulated similarly to a vehicle routing problem (VRP), where the next visit to each customer is scheduled based on its current inventory level, but adding in the objective function the IIC. This static off-line problem is formulated as minimizing the sum of IIC for all retailers and total transportation costs for the next delivery, subject to visiting all customers once during the planning period (next week) and inventory levels not exceeding order – up-to levels, $S$, for each retailer.

Thus, the variables of this problem are:

$q_i^r$ : Quantity to be delivered to retailer $i$ by the truck in its $r^{th}$ tour, where tours are numbered from 0 (0 is the current tour).

$$x_{ij}^r = \begin{cases} 1 \text{ If facility } j \text{ is visited immediately after facility } i \text{ by the truck} \\ \quad \text{in its } r^{th} \text{ tour} \\ 0 \text{ Otherwise} \end{cases}$$

$$y_i^r = \begin{cases} 1 \text{ If facility } i \text{ is served by the truck in its } r^{th} \text{ tour} \\ 0 \text{ Otherwise} \end{cases}$$

$t_i$ : Arrival time to retailer $i$. ($i \in \mathfrak{I}$).

$t_0^r$: Arrival time to the depot by the truck in its $r^{th}$ tour. $t_0^0$ is the truck arrival time to the depot in its current tour.

In addition, the parameters of this model are:

$\iota_i$ : Retailer $i$ current inventory level.

$\kappa_i$ : Retailer $i$ capacity to store inventory.

$\Upsilon$ : Truck capacity.

$v$ : Load remaining in the truck, which is equal to $\Upsilon$ when the truck is at the depot.

$TC$ : Transportation cost per unit of distance traveled by the truck. This is measured in [\$/hr], since the truck moves at constant speed.

$h_i$ : Retailer $i$ inventory holding cost [\$/unit-day].

$p_i$ : Retailer $i$ lost sales cost per unit of demand not satisfied [\$/unit].

$S_i$ : Retailer $i$ order up to level [units].

$s_i$ : Retailer $i$ reorder level [units].

$\ell$: Facility where the truck is currently located. $\ell \in \{0, 1, 2, ..., N, N+1\}$, it is equal to $N+1$ when truck is en-route. In this case, a dummy node, $N+1$, is created at the projected position.

$d_{ij}$: Distance from facility $i$ to facility $j$. Notice that when the truck is en-route distance from the dummy node $N+1$ to all facilities should be included.

$\mathfrak{R} \equiv \{0, 1, 2, ..., R\}$ : Set of tours (routes) for the truck in the planning horizon, where $R$ is the maximum number of tours not considering the current tour ($r = 0$). Thus, $r \in \mathfrak{R}$.

H : Length of planning horizon (maximum number of hours of operation).

The Mixed Integer Programming (MIP) formulation is presented below:

$$Min. \sum_{i \in \mathfrak{I}} LS P_i(t_i/\iota_i) + \sum_{i \in \mathfrak{I}} \sum_{r \in \mathfrak{R}} ITC_i(q_i^r, t_i/\iota_i) \cdot y_i^r + TC \cdot \sum_{i \in \mathfrak{I}_0} \sum_{j \in \mathfrak{I}_0 : j \neq i} d_{ij} \cdot \left( \sum_{r \in \mathfrak{R}} x_{ij}^r \right) \quad (10)$$

$$= \sum_{i \in \mathfrak{I}} \left\{ p_i \cdot \sigma_i \sqrt{t_i} \cdot G \left( \frac{\iota_i - t_i \mu_i}{\sigma_i \sqrt{t_i}} \right) \right\} + \sum_{i \in \mathfrak{I}} \sum_{r \in \mathfrak{R}} FTC_i \cdot y_i^r - \sum_{i \in \mathfrak{I}} \sum_{r \in \mathfrak{R}} FTC_i \cdot \left( \frac{q_i^r}{S_i} \right) +$$
$$TC \cdot \sum_{i \in \mathfrak{I}_0} \sum_{j \in \mathfrak{I}_0 : j \neq i} d_{ij} \cdot \left( \sum_{r \in \mathfrak{R}} x_{ij}^r \right)$$
$$(11)$$

Subject to:

$$\sum_{j \in \mathfrak{I}_0 : j \neq i} \sum_{r \in \mathfrak{R}} x_{ij}^r = 1 \quad , \text{for } i \in \mathfrak{I} \quad (12)$$

$$\sum_{i \in \mathfrak{I}} x_{i0}^r = 1 \quad , \text{for } r = 1, 2, \ldots, R \quad (13)$$

$$\sum_{i \in \mathfrak{I}} x_{i0}^0 + x_{N+1,0}^0 = 1 \quad (14)$$

$$\sum_{j \in \mathfrak{I}_0 : j \neq \ell} x_{\ell j}^0 = 1 \quad (15)$$

$$\sum_{j \in \mathfrak{I}_0 : j \neq i} x_{ij}^0 = \sum_{j \in \mathfrak{I}_0 : j \neq i} x_{ji}^0 + \mathbf{1} \{i = \ell\} \quad , \text{if } \ell \neq N+1; \text{ for } i \in \mathfrak{I} \quad (16)$$

$$\sum_{j \in \mathfrak{I}_0 : j \neq i} x_{ij}^0 = \sum_{j \in \mathfrak{I}_0 : j \neq i} x_{ji}^0 + x_{N+1,i}^0 \quad , \text{if } \ell = N+1; \text{for } i \in \mathfrak{I} \quad (17)$$

$$\sum_{j \in \mathfrak{I}_0 : j \neq i} x_{ij}^r = \sum_{j \in \mathfrak{I}_0 : j \neq i} x_{ji}^r \quad , \text{for } i \in \mathfrak{I}_0; r = 1, 2, \ldots, R \quad (18)$$

$$\sum_{j\in\mathfrak{I}} x_{0j}^1 \le \sum_{i=1}^{N+1} x_{i0}^0 \tag{19}$$

$$\sum_{j\in\mathfrak{I}} x_{0j}^r \le \sum_{i\in\mathfrak{I}} x_{i0}^{(r-1)} \quad , \text{for } r = 2, 3, \ldots, R \tag{20}$$

$$\sum_{i\in\mathfrak{I}} x_{\ell i}^0 \ge \sum_{i\in\mathfrak{I}:i\ne j} x_{ij}^0 \quad , \text{for } j \in \mathfrak{I} \tag{21}$$

$$\sum_{i\in\mathfrak{I}} x_{0i}^r \ge \sum_{i\in\mathfrak{I}:i\ne j} x_{ij}^r \quad , \text{for } j \in \mathfrak{I}; r = 1, 2, \ldots, R \tag{22}$$

$$t_j \ge t_i + d_{ij} - M\left(1 - x_{ij}^r\right) \quad , \text{for } i \in \mathfrak{I}; \ j \in \mathfrak{I}; r \in \mathfrak{R} \tag{23}$$

$$t_j \le t_i + d_{ij} + M\left(1 - x_{ij}^r\right) \quad , \text{for } i \in \mathfrak{I}; \ j \in \mathfrak{I}; r \in \mathfrak{R} \tag{24}$$

$$t_j \ge d_{\ell j} - M\left(1 - x_{\ell j}^0\right) \quad , \text{for } j \in \mathfrak{I} \tag{25}$$

$$t_j \le d_{\ell j} + M\left(1 - x_{\ell j}^0\right) \quad , \text{if } \ell \ne 0, \text{ for } j \in \mathfrak{I} \tag{26}$$

$$t_j \ge t_0^r + d_{0j} - M\left(1 - x_{0j}^{(r+1)}\right) \quad , \text{for } j \in \mathfrak{I}, r \in \mathfrak{R} \tag{27}$$

$$t_\ell = 0 \quad , \text{if } \ell \ne N + 1 \tag{28}$$

$$t_0^0 \ge d_{N+1,0} - M\left(1 - x_{N+1,0}^0\right) \quad , \text{if } \ell = N + 1 \tag{29}$$

$$t_0^0 \le d_{N+1,0} + M\left(1 - x_{N+1,0}^0\right) \quad , \text{if } \ell = N + 1 \tag{30}$$

$$t_0^r \ge t_i + d_{i0} - M\left(1 - x_{i0}^r\right) \quad , \text{for } i \in \mathfrak{I}; r \in \mathfrak{R} \tag{31}$$

$$t_0^r \le t_i + d_{i0} + M\left(1 - x_{i0}^r\right) \quad , \text{for } i \in \mathfrak{I}; r \in \mathfrak{R} \tag{32}$$

$$t_0^r \ge t_0^{(r-1)} \quad , \text{for } r = 1, 2, \ldots, R \tag{33}$$

$$t_0^R \le \mathrm{H} \tag{34}$$

$$\sum_{j\in\mathfrak{I}_0:j\ne i} x_{ij}^r = y_i^r \quad , \text{for } i \in \mathfrak{I}; r \in \mathfrak{R} \tag{35}$$

$$0 \le q_i^r \le \Upsilon \cdot y_i^r \quad , \text{for } i \in \mathfrak{I}; r \in \mathfrak{R} \tag{36}$$

$$\sum_{i \in \mathfrak{I}} q_i^r \leq \varUpsilon \quad , \text{for } r = 1, 2, \ldots, R \tag{37}$$

$$\sum_{i \in \mathfrak{I}} q_i^0 \leq v \tag{38}$$

$$s_i - \iota_i + \mu_i t_i \leq \sum_{r \in \mathfrak{R}} q_i^r \leq S_i - \iota_i + \mu_i t_i \quad , \text{for } i \in \mathfrak{I} : i \neq \ell \tag{39}$$

$$\min\{(s_\ell - \iota_\ell), v\} \leq q_\ell^0 \leq S_\ell - \iota_\ell \quad , \text{if } \ell \neq N + 1 \tag{40}$$

$$x_{ij}^r \in \{0, 1\} \text{ and } y_i^r \in \{0, 1\} \quad , \text{for } i \in \mathfrak{I}_0; \ j \in \mathfrak{I}_0; r \in \mathfrak{R} \tag{41}$$

$$x_{N+1,j}^0 \in \{0, 1\} \quad , \text{if } \ell = N+1, \text{for } j \in \mathfrak{I}_0 \tag{42}$$

The objective function (10)–(11) minimizes the sum of the total expected lost sale penalties at each facility for the next scheduled visit, and the total transportation costs.

All tours start at the depot with a full truck, with the exception of the current tour, $r = 0$, in which the truck starts at any facility or en-route (at node $N+1$), and the truck might not be full (its current load is $v$).

Constraint (12) ensures that the next visit for each retailer is programmed. Equations (13) and (14) ensure that the truck returns to the depot in all its tours. Constraint (15) dictates that the truck should leave from its current location. Constraints (16)–(17)–(18) give continuity of flow ensuring that the number of arrivals equals the number of departures at each node. Constraints (19)–(20) ensure that subsequent routes could be traveled only if the previous route is completed. Constraints (21)–(22) ensure that current route leave the initial node and subsequent routes leave the depot to visit retailers. Constraints (23) through (33) ensure that the arrival times at each facility are consistent with travel times between them and the initial conditions, where $M$ is a big number. Constraint (34) dictates that the last route should be completed before the end of the planning horizon H. Constraint (35) relates facilities served on each route with its links. Constraints (36) guarantee that only customers visited from a particular route could receive deliveries from it. Constraints (37)–(38) ensure that the truck capacities are not exceeded and that the quantity delivered cannot exceed the load remaining in the vehicle. Constraints (39)–(40) guarantee that inventory levels should not exceed the order up to level, $S_i$ and inventory level should be greater than $s_i$ after refilling; however an exception is allowed at the current facility if the load remaining in the truck is insufficient (40).

As mentioned, the purpose of this formulation is to update truck plans making use of updated information about the state of the system. The next section describes three strategies that solve this formulation in a rolling horizon framework.

## 3.4 Re-optimization Based Real-Time Strategies

The off-line problem described in the previous section will be used to determine how to update truck routes and inventory allocations in a rolling horizon framework. Different policies could be devised based on how often the off-line problem is solved and/or how many steps of the current solution are implemented before solving a new instance with updated system state information. Three policies, ordered in terms of increasing ICT requirement, are presented and tested. These are Replan at Tour Completion (RTC), Replan at Delivery Epochs (RDE), and Replan at Delivery Epochs with possible en-route diversions (RDE+div).

*a) Replan at Tour Completion (RTC) Strategy*

In Replan at Tour Completion (RTC), the off-line IRP is solved each time the truck returns to the depot, i.e. completes a tour, and only the first route of the current solution is implemented. In this policy, the review period, $T$, used to compute the optimal refill levels, is obtained as the expected distance on a tour over the set of retailers.

*b) Replan at Delivery Epochs (RDE) Strategy*

In Replan at Delivery Epochs (RDE), the off-line problem is called at delivery epochs. Each time a truck arrives to a facility, either a retailer or the depot (delivery epoch), an off-line IRP is solved and the solution implemented until the next delivery epoch. That is the amount specified by the solution is delivered at the current facility, and the truck is sent to the next facility specified by the solution. In this policy, the review period, $T$, used to compute the optimal refilling levels, is obtained as the expected distance between two retailers.

*c) Replan at Delivery Epochs with Possible En-Route Diversion (RDE+div) Strategy*

In Replan at Delivery Epochs with possible en-route diversion (RDE+div), plans are updated at delivery epochs, as in RDE, as well as when demand disruptions occur. In this case, inventory levels are continuously monitored while the vehicle is traveling; whenever a facility's consumption since the last plan update falls below or above 3 standard deviations from its expected demand, the current plan is updated. To update the plan, the state of the system (i.e. the location of the truck and inventory levels assuming expected consumption rates) is first projected. Then, based on the projected state of the system, an off-line routing problem is solved and the next step implemented. In this strategy, the truck could be diverted if in the new plan the next facility to be visited differs from the current destination.

In order to solve the off-line IRP formulation used in these strategies, the first term in equation is piecewise linearized, so that small instances can be efficiently solved using CPLEX 10.0 with default settings. This problem can be solved in a few seconds for most instances with less than six facilities and a few minutes for instance with less than nine facilities. The design of heuristics to solve larger size instances is beyond the scope of this Chapter, and is left as a future extension.

# 4 Simulation Experiments and Results

This section describes the experiments designed to evaluate and compare proposed real-time policies, and discusses the results. Simulation runs were performed with all three re-optimization strategies, RTC, RDE and RDE+div, as well as the two benchmark policies, BENCH1 and MUN presented in the previous section.

## 4.1 Simulation Scenarios

This section presents the main elements and defining parameters of the simulated scenarios. First, the set of fixed parameters used in all simulations is introduced. Second, parameters in scenarios with steady-state demand processes are defined. Third, inventory reorder levels are obtained for each combination of strategy and scenario simulated. Finally, experiments with demand disruption at one facility are presented.

Because of limited computational resources (each simulation run takes hours of computer time – even days for some re-optimization strategies –), a full factorial design was not practical, so not all parameter combinations were applied. For each strategy, simulations were performed for four cases of facility layouts, and 12 parameter sets, representing typical cost settings, probabilistic scenarios, and constraints.

Distances between facilities are Euclidean and are measured in units of time [hours], because it is assumed, without loss of generality, that the vehicle moves at unit speed. All facilities are located in a square region, with side length of 4 h, with the depot in the center of the square region, i.e. the depot is located at (2, 2). In all cases simulated, seven retailers and one depot are considered. In case 0, retailers are symmetrically distributed around the depot at 1.2 h apart, and in cases 1 to 3 they are randomly distributed in the region. Figures 1 through 4 show the locations of facilities for each case. Facilities were renumbered to coincide with their position in the TSP tour.

In addition to the location of facilities for each case studied, the following set of parameters is considered as fixed: the vehicle capacity, $\Upsilon = 400$ [units]; the length of the planning horizon used on the off-line problem for re-optimization strategies, $H = 100$ [hrs] which is also assumed to be the amount of working hours per week; lost sales penalty costs, $p_i = 100$ [\$/unit] for all retailers; and the fixed transportation costs used to obtain refill levels for re-optimization strategies, $FTC_i = 2 \cdot d_{0i} \cdot TC$, which is computed as twice the cost of a tour from the depot only to that retailer. In addition, the time projection used in the case of diverting the vehicle is 6 min, i.e. $\partial t = 6$ min.

Two sets of scenarios were studied: (1) products with high inventory-holding costs and no capacity constraints at retailers' sites, and (2) products with low inventory-holding costs and capacity constraints at retailers' sites. Scenarios with low inventory holding costs and no capacity constraints at retailers were not considered, since for those scenarios the best policy would be full-truckload deliveries.

For each set of scenarios, a base case was considered. Parameter set 1 is the base case for high inventory-holding cost scenarios, in which $TC = 100$ [\$/hr], $h_i = 10$ [\$/unit-day] for all retailers, and demand parameters are the same for all retailers, and

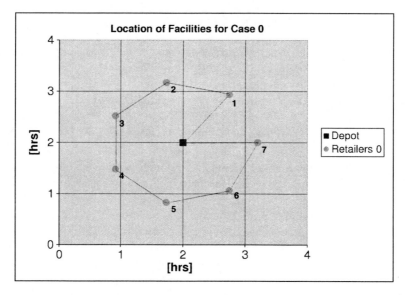

**Fig. 1.** Location of facilities for Case 0 (Symmetric case)

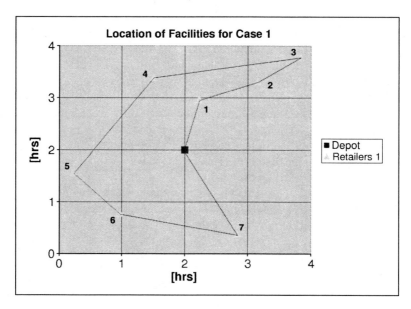

**Fig. 2.** Location of facilities for Case 1

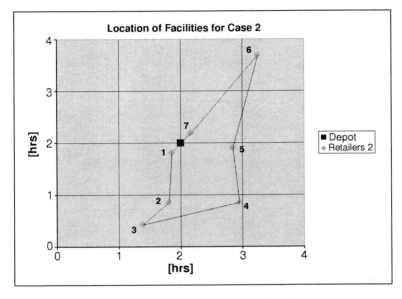

**Fig. 3.** Location of facilities for Case 2

equal to $\lambda_i = 50$ [arrivals/Day] and $\theta_i = 1$ [units], for all $i$. This demand process can be approximated as $N(50, 10^2)$ for daily periods. For low inventory-holding cost scenarios, Parameter set 7 is the base case, in which $h_i = 1$ [\$/unit-day], $\kappa_i = 100$ [units] for all retailers, and the remaining parameters are the same as in Parameter set 1.

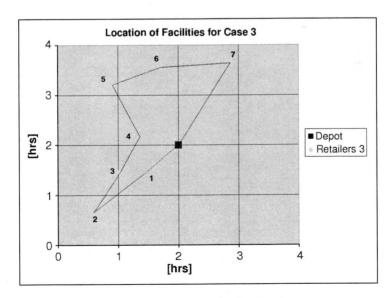

**Fig. 4.** Location of facilities for Case 3

In order to analyze the impact of transportation costs and demand variability on the proposed policies, ten additional scenarios were designed, as shown in Table 1. In scenarios with high (low) inventory holding costs and without (with) retailer capacity constraints, parameter sets 2 and 3 (8 and 9) capture variation in transportation costs, whereas parameter sets 4 to 6 (10 to 12) reflect differing levels of demand variability.

In the experimental design, the magnitude of each cost parameter is less relevant than the relationship among them. For that reason, the scenarios studied vary the ratio between transportation and inventory costs.

The reorder parameters of the $(s, S)$ policy used in BENCH1 were obtained using Zheng and Federgruen algorithm (1991). RTC and RDE reorder parameters were obtained using the procedure described in Sect. 3.3.2. The only difference between RTC and RDE is the review period considered. In RTC the expected TSP length was used, and in RDE the expected distance between two facilities was used. For the MUN strategy, RDE refilling up to levels, $S$, were used.

**Table 1.** Simulation scenarios

| Parameter Set | TC [\$/hr] | h [\$/day] | $\kappa$ [units] | $\lambda$ [arrivals/day] | $\theta$ [units] | Approx. $N(\mu,\sigma^2)$ |
|---|---|---|---|---|---|---|
| 1 | 100 | 10 | $\infty$ | 50 | 1 | $N(50,10^2)$ |
| 2 | 33 | 10 | $\infty$ | 50 | 1 | $N(50,10^2)$ |
| 3 | 300 | 10 | $\infty$ | 50 | 1 | $N(50,10^2)$ |
| 4 | 100 | 10 | $\infty$ | 10.5 | 4.8 | $N(50,17^2)$ |
| 5 | 100 | 10 | $\infty$ | 4.35 | 11.5 | $N(50,25^2)$ |
| 6 | 100 | 10 | $\infty$ | 2.4 | 20.8 | $N(50,33^2)$ |
| 7 | 100 | 1 | 100 | 50 | 1 | $N(50,10^2)$ |
| 8 | 33 | 1 | 100 | 50 | 1 | $N(50,10^2)$ |
| 9 | 300 | 1 | 100 | 50 | 1 | $N(50,10^2)$ |
| 10 | 100 | 1 | 100 | 10.5 | 4.8 | $N(50,17^2)$ |
| 11 | 100 | 1 | 100 | 4.35 | 11.5 | $N(50,25^2)$ |
| 12 | 100 | 1 | 100 | 2.4 | 20.8 | $N(50,33^2)$ |

## 4.2 Simulation Results

For every combination of strategy and set of parameters studied, simulations were carried for 30 replication runs of 100 h (1 week) each, and all four facility layout cases. For each parameter set, the different strategies were simulated with common random numbers (for demand generation), and the same initial conditions. The initial conditions for the first replication were the same for all strategies in the same scenario, starting with the vehicle at the depot, and the same initial inventory levels. The effect of initial conditions is only relevant up to the first visit to each facility, which is small compared to the length of each run to have significant effects. Moreover, those initial conditions were only used in the first replication, and results suggest that transient-state effects are negligible.

For each simulation run, the following measures of performance were examined: average transportation costs; average inventory holding costs; and average lost-sale penalty costs. For each of those measures, interval estimates were obtained. These results are presented in Appendix.

## 4.3 Analysis of Results

Simulations were carried only under typical system parameters, and run under idealized probabilistic distributions. Thus, the results presented are valid only for the range of values studied, which are nonetheless intended to be representative of real-world applications, and hence adequate to provide general insight for them.

For all parameter sets considered, the three proposed online strategies systematically outperformed benchmark strategies. The best proposed strategies achieved reductions in average total costs of approximately 30 and 15% compared against benchmark policies BENCH1 and MUN, respectively. The average cost improvements were computed as the average of:

$$\frac{(\text{Avg. Total Cost of Strategy} - \text{Avg. Total Cost of BENCH})}{\text{Avg. Total Cost of BENCH}} \cdot 100\% \quad (43)$$

For all sets of parameters and cases considered. Moreover, the optimal decentralized benchmark policy, BENCH1, was systematically outperformed by centralized strategies. This can be explained in part by the fact that BENCH1 tends to carry more inventory to protect against longer lead times. In addition, all proposed strategies achieved less variability in average costs than the BENCH1 strategy.

Among re-optimization strategies, those that update plans at delivery epochs, RDE and RDE+div, were the best strategies for the set of parameters considered. The possibility of diversion – either en-route or when the vehicle is idle at the depot – improves system performance in scenarios with low inventory-holding costs and high demand variability. However, further research is needed to better delineate scenarios in which en-route diversion could be beneficial, since in RDE vehicle idle time at the depot is set upon arrival and not updated, even when that might be profitable. The benefits of re-planning at delivery epochs tend to be higher in cases where there are clusters of facilities close to each other and/or near to the depot, such as in case 2.

As noted, experiments were carried out for two sets of scenarios: (1) products with high inventory-holding costs, and (2) products with low inventory-holding costs and retailer capacity constraints. As shown in Fig. 5, a comparison of the two sets of scenarios illustrates that re-planning strategies tended to increase their benefits relative to the benchmark policies when applied to scenarios with low inventory-holding costs.

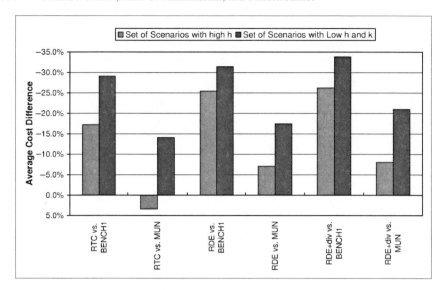

**Fig. 5.** Impact of inventory holding cost

In Figs. 6 and 7, reductions in average total costs for proposed re-planning strategies vs. the two benchmark policies are illustrated. In scenarios with high inventory-holding costs (Fig. 6), the benefits of the proposed strategies tend to decrease (increase) as transportation costs increase, when compared with BENCH1 (MUN). That again can be explained by the longer lead times in BENCH1, requiring facilities to carry more inventory. Thus, the higher the inventory-holding costs, the worse the performance of BENCH1.

In scenarios with low inventory-holding costs and retailer capacities (Fig. 7), the benefit of the proposed strategies tends to increase, as transportation costs increase when compared to any of the benchmark policies. Thus, with the exception of BENCH1, for scenarios with high inventory-holding costs, the benefits of the proposed re-planning strategies tend to increase as a function of the proportion of transportation costs in the total cost function.

Among re-planning strategies when transportation costs are more significant in the total costs, re-planning at delivery epochs is less beneficial, compared with re-planning only at tour completions. A comparison between RDE and RTC shows that RDE reduced by approximately 11, 8, and 4% the average total costs in scenarios with high inventory-holding costs, i.e. under parameter sets 2 ($TC = 33$), 1 ($TC = 100$), and 3 ($TC = 300$), respectively. In scenarios with low inventory-holding costs, the differences are less dramatic and remain relatively constant (around 2.5%) with respect to changes in transportation costs. Those differences could be explained mainly by the higher inventory levels maintained under RTC, which amplify the differences between RDE and RTC when inventory-holding costs are predominant in the total cost function.

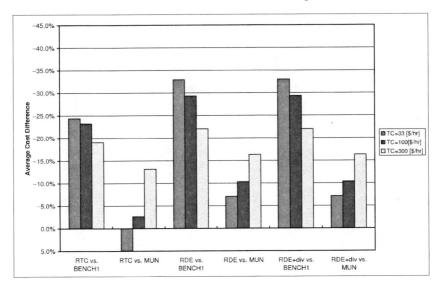

**Fig. 6.** Impact of increments in transportation costs in scenarios with high inventory holding cost, parameter sets 1, 2 and 3

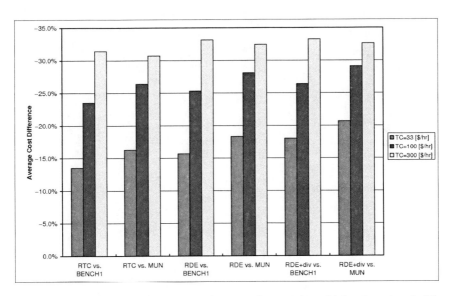

**Fig. 7.** Impact of increments in transportation costs in scenarios with low inventory holding cost, parameter sets 7, 8 and 9

As shown in Fig. 8, with high inventory-holding costs, as demand variability increases, the benefits of the proposed strategies tend to decrease, compared to benchmark strategies. Conversely, in scenarios with low inventory-holding costs and capacity constraints (Fig. 9), the benefits of the proposed strategies tend to increase vs. BENCH1 and decrease vs. MUN, as demand variability increases. As expected, MUN becomes more competitive in scenarios with very high demand variability. In scenarios with increased demand variability, i.e. those with parameter sets 5, 6, 11, and 12, the advantage of re-planning at delivery epochs vs. only at tour completion tends to be slightly higher than in scenarios with less demand variability.

The possibility of diversion – either en-route or when the vehicle is idle at the depot – improves system performance in scenarios with low inventory-holding costs and high demand variability. However, further research is needed to identify scenarios in which en-route diversion would be beneficial.

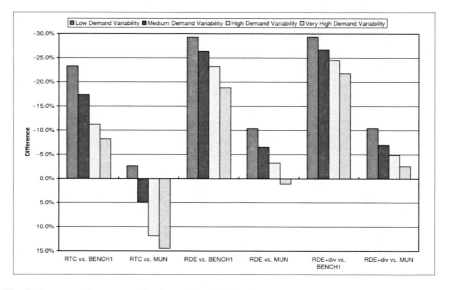

**Fig. 8.** Impact of increments in demand variability in scenarios with high inventory-holding costs, parameter sets 1, 4, 5 and 6

## 5 Conclusion

In this Chapter the OIRP was introduced and three rolling horizon strategies were proposed, corresponding to different degrees of sophistication in terms of ICT. The central idea of the proposed approach (which is then applied online with varying triggers depending on the operational control strategy followed), is to establish inventory reorder parameters for each facility, and use those as target levels in the routing

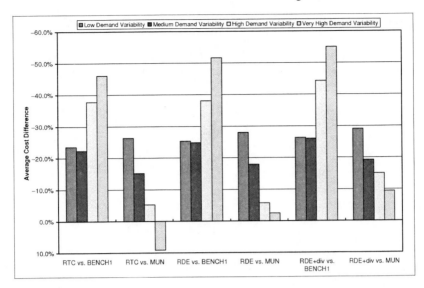

**Fig. 9.** Impact of increments in demand variability in scenarios with low inventory-holding costs, parameter sets 7, 10, 11 and 12

problem solved to update the plans. These target levels are obtained by solving the inventory control side of the problem first, assuming known transport costs corresponding to a simplified version of the routing problem. The routing problem used to update plans includes both transportation and inventory-related costs in its objective function, and is formulated as a MIP, solved for small instances in the experiments presented in the paper using a commercial solver.

The proposed strategies were compared against two benchmark policies using a general discrete-event simulation framework. The proposed re-planning strategies are shown to systematically outperform the two benchmark policies. Among proposed strategies, those that update plans at delivery epochs exhibited superior performance for the set of parameters considered in the experiments.

The work presented here represents a first step in addressing the complex problem of optimally operating logistics processes in real-time. It has been previously established that combining inventory planning with fleet operation could result in potentially considerable savings relative to operating each process separately. Operating the two processes jointly in real-time can help realize those advantages, and amplify them substantially by responding more effectively to demand fluctuations. The specific problem addressed in this Chapter can be extended along several dimensions, including multiple-vehicle fleets, additional echelons, multiple product types, among many others. For systems with non-affiliated retailers, the problem and approach presented here provide an example of collaborative logistics, which give rise to interesting problem classes and open significant opportunities for novel approaches and mechanisms for achieving mutual gains through online collaboration.

## Acknowledgments

The first author would like to thank the Chilean National Fund for the Development of Scientific and Technological Research (FONDECYT Project 1107218) for partially financing this research.

## References

[1] Axaäter, S. (2000) *Inventory control*, Boston, Mass.; London, Kluwer Academic.
[2] Baita, F., Ukovich, W., Pesenti, R. & Favaretto, D. (1998) Dynamic routing-and-inventory problems: A review. *Transportation Research Part a-Policy and Practice*, 32, 585-598.
[3] Bertazzi, L., Savelsbergh, M. & Speranza, M. G. (2007) Inventory Routing.
[4] Blanchard, D. (2003) Fears of "Big Brother" sidetrack Benetton's smart tag initiative. *Transportation and Distribution*, 44, 20.
[5] Campbell, A. M., Clarke, L. W. & Savelsbergh, M. (1998) The inventory routing problem. In, Crainic T. G. & Laporte, G. (Eds.) *Fleet Management and Logistics*. Dordrecht, The Netherlands, Kluwer Academic Publishers.
[6] Campbell, A. M. & Savelsbergh, M. (2002) Inventory Routing in Practice. In Toth, P. & Vigo, D. (Eds.) *The Vehicle Routing Problem*. SIAM Monographs on Discrete Mathematics and Applications.
[7] Chen, F., Drezner, Z., Ryan, J. K. & Simchi-Levi, D. (2000) Quantifying the bullwhip effect in a simple supply chain: The impact of forecasting, lead times, and information. *Management Science*, 46, 436-443.
[8] Datta, S. (2003) Radio frequency identification (RFID) made easy. MIT forum for supply chain innovation.
[9] Fine, C. H. (1998) *Clockspeed: winning industry control in the age of temporary advantage*, Reading, Mass., Perseus Books.
[10] Fleischmann, B. & Meyr, H. (2003) Planning Hierarchy, Modeling and Advance Planning Systems. IN Kok, A. G. D. & Graves, S. C. (Eds.) *Supply chain management: design, coordination and operation*. Elsevier.
[11] Gendreau, M., Guertin, F., Potvin, J. Y. & Seguin, R. (2006) Neighborhood search heuristics for a dynamic vehicle dispatching problem with pick-ups and deliveries. *Transportation Research Part C-Emerging Technologies*, 14, 157-174.
[12] Gendreau, M., Guertin, F., Potvin, J. Y. & Taillard, E. (1999) Parallel tabu search for real-time vehicle routing and dispatching. *Transportation Science*, 33, 381-390.
[13] Ghiani, G., Musmanno, R. & Laporte, G. (2004) *Introduction to logistics systems planning and control*, Chichester, West Sussex; Hoboken, NJ, USA, J. Wiley.

[14] Giesen, R. (2007) Online Inventory Replenishment and Fleet Routing Decisions Under Real-Time Information. *Ph.D. Dissertation, University of Maryland, College Park.*

[15] Giesen, R., Mahmassani, H. S. & Jaillet, P. (2005) Strategies for online inventory routing problem under real-time information. *Transportation Research Record 1923, Network Modeling 2005,* 164-179.

[16] Grötschel, M., Krumke, S. O. & Rambau, J. (2001a) Online Optimization of Complex Transportation Systemes. In Grötschel, M., Krumke, S. O. & Rambau, J. (Eds.) *Online optimization of large scale systems.* Berlin; New York, Springer.

[17] Grötshel, M., Krumke, S. O., Rambau, J., Winter, T. & Zimmermann, U. W. (2001b) Combinatorial Online Optimization in Real-Time. In Grötschel, M., Krumke, S. O. & Rambau, J. (Eds.) *Online Optimization of Large Scale Systems.* Berlin; New York, Springer.

[18] Ichoua, S., Gendreau, M. & Potvin, J. Y. (2006) Exploiting knowledge about future demands for real-time vehicle dispatching. *Transportation Science,* 40, 211-225.

[19] Kim, Y. (2003) Hybrid Approaches To Solve Dynamic Fleet Management Problems. *Ph D dissertation, The University of Texas at Austin.*

[20] Kim, Y., Mahmassani, H. S. & Jaillet, P. (2002) Dynamic truckload truck routing and scheduling in oversaturated demand situations. *Transportation Network Modeling 2002,* 66-71.

[21] Kim, Y. J., Mahmassani, H. S. & Jaillet, P. (2004) Dynamic truckload routing, scheduling, and load acceptance for large fleet operation with priority demands. *Transportation Network Modeling 2004,* 120-128.

[22] Kleywegt, A. J., Nori, V. S. & Savelsbergh, M. W. P. (2002) The Stochastic inventory routing problem with direct deliveries. *Transportation Science,* 36, 94-118.

[23] La Londe, B. J. (1994) Evolution of the integrated logistics concept. In Robeson, J. F., Copacino, W. C. & Howe, R. E. (Eds.) *The Logistics Handbook.* New York, The Free Press.

[24] Larsen, A., Madsen, O. & Solomon, M. (2002) Partially dynamic vehicle routing - models and algorithms. *Journal of the Operational Research Society,* 53, 637-646.

[25] Larsen, A., Madsen, O. B. G. & Solomon, M. M. (2004) The A priori dynamic traveling salesman problem with time windows.*Transportation Science,* 38, 459-472.

[26] Lee, H. L., Padmanbhan, V. & Whang, S. (1997) The bullwhip effect in supply chains. *Sloan Management Review,* 38, 93-102.

[27] Masters, J. M. & La Londe, B. J. (1994) The role of new information technology in the practice of traffic management. In Robeson, J. F., Copacino, W. C. & Howe, R. E. (Eds.) *The Logistics Handbook.* New York, The Free Press.

[28] Mitrovic-Minic, S., Krishnamurti, R. & Laporte, G. (2004) Double-horizon based heuristics for the dynamic pickup and delivery problem with time windows. *Transportation Research Part B-Methodological,* 38, 669-685.

[29] Rabah, M. & Mahmassani, H. S. (2002) Impact of information and communication technologies on logistics and freight transportation - Example of vendor-managed inventories. *Freight Transportation 2002*, 10-19.

[30] Regan, A. C., Mahmassani, H. S. & Jaillet, P. (1995) Improving the efficiency of commercial vehicle operations using real-time information: potential uses and assignment strategies. *Transportation Research Record*, 188-198.

[31] Regan, A. C., Mahmassani, H. S. & Jaillet, P. (1996) Dynamic decision making for commercial fleet operations using real-time information. *Transportation Research Record*, 91-97.

[32] Rutner, S. M., Gibson, B. J. & Williams, S. R. (2003) The impacts of the integrated logistics systems on electronic commerce and enterprise resource planning systems. *Transportation Research Part E-Logistics and Transportation Review*, 39, 83-93.

[33] Simchi-Levi, D., Kaminsky, P. & Simchi-Levi, E. (2003) *Designing and managing the supply chain: concepts, strategies, and case studies*, Boston, McGraw-Hill/Irwin.

[34] Thomas, D. J. & Griffin, P. M. (1996) Coordinated supply chain management. *European Journal of Operational Research*, 94, 1-15.

[35] Wolff, R. W. (1989) *Stochastic modeling and the theory of queues*, Englewood Cliffs, N.J., Prentice Hall.

[36] Yang, J., Jaillet, P. & Mahmassani, H. (2004) Real-time multivehicle truckload pickup and delivery problems. *Transportation Science*, 38, 135-148.

[37] Zheng, Y. S. & Federgruen, A. (1991) Finding Optimal (S, S) Policies Is About as Simple as Evaluating a Single Policy. *Operations Research*, 39, 654-665.

[38] Zipkin, P. H. (2000) *Foundations of inventory management*, Boston, McGraw-Hill.

# Appendix

**Fig. 10.** Simulation results: parameter set 1 TC = 100 [\$/hr], $h_i$ = 50[\$/week], $\lambda_i$ = 50[arrivals/day], $\theta_i$ = 1 [units] for all i

| Strategy | | Inv. Holding Cost | Lost Sales Cost | Total Inv. Cost | Transp. Cost | Total Cost |
|---|---|---|---|---|---|---|
| | | **Case 0 (Symmetric case)** | | | | |
| BENCH1 | Mean | 22,569 | 2,041 | 24,610 | 3,428 | 28,038 |
| | St Dev | 862 | 1,474 | 1,244 | 164 | 1,240 |
| | CV | 0.04 | 0.72 | 0.05 | 0.05 | 0.04 |
| MUN | Mean | 12,843 | 1,901 | 14,744 | 6,415 | 21,159 |
| | St Dev | 539 | 1,798 | 1,572 | 494 | 1,633 |
| | CV | 0.04 | 0.95 | 0.11 | 0.08 | 0.08 |
| RTC | Mean | 16,221 | 167 | 16,388 | 4,621 | 21,008 |
| | St Dev | 646 | 278 | 591 | 241 | 561 |
| | CV | 0.04 | 1.67 | 0.04 | 0.05 | 0.03 |
| RDE | Mean | 13,997 | 312 | 14,308 | 4,943 | 19,252 |
| | St Dev | 227 | 387 | 486 | 258 | 543 |
| | CV | 0.02 | 1.24 | 0.03 | 0.05 | 0.03 |
| RDE+div | Mean | 14,158 | 228 | 14,386 | 4,980 | 19,366 |
| | St Dev | 234 | 304 | 307 | 248 | 397 |
| | CV | 0.02 | 1.33 | 0.02 | 0.05 | 0.02 |
| | | **Case 1** | | | | |
| BENCH1 | Mean | 22,812 | 2,928 | 25,739 | 4,363 | 30,103 |
| | St Dev | 841 | 2,302 | 1,852 | 432 | 1,963 |
| | CV | 0.04 | 0.79 | 0.07 | 0.10 | 0.07 |
| MUN | Mean | 16,361 | 1,763 | 18,124 | 7,437 | 25,561 |
| | St Dev | 725 | 1,672 | 1,393 | 545 | 1,541 |
| | CV | 0.04 | 0.95 | 0.08 | 0.07 | 0.06 |
| RTC | Mean | 18,895 | 321 | 19,215 | 5,308 | 24,523 |
| | St Dev | 703 | 498 | 886 | 330 | 911 |
| | CV | 0.04 | 1.55 | 0.05 | 0.06 | 0.04 |
| RDE | Mean | 17,287 | 307 | 17,594 | 5,536 | 23,130 |
| | St Dev | 316 | 410 | 498 | 317 | 602 |
| | CV | 0.02 | 1.34 | 0.03 | 0.06 | 0.03 |
| RDE+div | Mean | 17,238 | 239 | 17,477 | 5,606 | 23,083 |
| | St Dev | 327 | 271 | 417 | 354 | 592 |
| | CV | 0.02 | 1.13 | 0.02 | 0.06 | 0.03 |
| | | **Case 2** | | | | |
| BENCH1 | Mean | 22,721 | 1,432 | 24,153 | 3,041 | 27,194 |
| | St Dev | 683 | 1,691 | 1,576 | 216 | 1,665 |
| | CV | 0.03 | 1.18 | 0.07 | 0.07 | 0.06 |
| MUN | Mean | 12,725 | 612 | 13,337 | 7,366 | 20,703 |
| | St Dev | 381 | 1,202 | 1,104 | 525 | 1,372 |
| | CV | 0.03 | 1.97 | 0.08 | 0.07 | 0.07 |
| RTC | Mean | 14,903 | 660 | 15,563 | 4,367 | 19,930 |
| | St Dev | 362 | 755 | 693 | 264 | 701 |
| | CV | 0.02 | 1.14 | 0.04 | 0.06 | 0.04 |
| RDE | Mean | 12,647 | 680 | 13,327 | 4,922 | 18,249 |
| | St Dev | 250 | 574 | 636 | 230 | 640 |
| | CV | 0.02 | 0.84 | 0.05 | 0.05 | 0.04 |
| RDE+div | Mean | 12,658 | 443 | 13,101 | 4,919 | 18,019 |
| | St Dev | 252 | 592 | 667 | 235 | 739 |
| | CV | 0.02 | 1.34 | 0.05 | 0.05 | 0.04 |
| | | **Case 3** | | | | |
| BENCH1 | Mean | 23,221 | 1,821 | 25,041 | 3,355 | 28,396 |
| | St Dev | 901 | 1,620 | 1,410 | 279 | 1,488 |
| | CV | 0.04 | 0.89 | 0.06 | 0.08 | 0.05 |
| MUN | Mean | 14,696 | 1,046 | 15,742 | 6,650 | 22,392 |
| | St Dev | 616 | 1,234 | 981 | 555 | 1,244 |
| | CV | 0.04 | 1.18 | 0.06 | 0.08 | 0.06 |
| RTC | Mean | 16,938 | 534 | 17,473 | 4,466 | 21,939 |
| | St Dev | 528 | 667 | 855 | 282 | 851 |
| | CV | 0.03 | 1.25 | 0.05 | 0.06 | 0.04 |
| RDE | Mean | 14,960 | 257 | 15,217 | 4,727 | 19,943 |
| | St Dev | 337 | 576 | 535 | 222 | 627 |
| | CV | 0.02 | 2.24 | 0.04 | 0.05 | 0.03 |
| RDE+div | Mean | 15,020 | 212 | 15,232 | 4,805 | 20,037 |
| | St Dev | 308 | 326 | 410 | 198 | 478 |
| | CV | 0.02 | 1.54 | 0.03 | 0.04 | 0.02 |

30 replication with common random numbers
Results in [\$/week]

**Fig. 11.** Simulation results: parameter set 2 TC = 33[$/hr], $h_i = 50$ [$/week], $\lambda_i = 50$ [arrivals/day], $\theta_i = 1$ [units] for all i

| Strategy | | Case 0 (Symmetric case) | | | | |
|---|---|---|---|---|---|---|
| | | Inv. Holding Cost | Lost Sales Cost | Total Inv. Cost | Transp. Cost | Total Cost |
| BENCH1 | Mean | 21,344 | 181 | 21,525 | 1,408 | 22,932 |
| | St Dev | 461 | 502 | 544 | 38 | 536 |
| | CV | 0.02 | 2.78 | 0.03 | 0.03 | 0.02 |
| MUN | Mean | 11,904 | 1,748 | 13,651 | 2,476 | 16,128 |
| | St Dev | 489 | 1,579 | 1,290 | 186 | 1,278 |
| | CV | 0.04 | 0.90 | 0.09 | 0.08 | 0.08 |
| RTC | Mean | 14,906 | 315 | 15,220 | 1,689 | 16,910 |
| | St Dev | 477 | 670 | 775 | 72 | 770 |
| | CV | 0.03 | 2.13 | 0.05 | 0.04 | 0.05 |
| RDE | Mean | 12,649 | 187 | 12,837 | 1,890 | 14,727 |
| | St Dev | 231 | 282 | 281 | 72 | 301 |
| | CV | 0.02 | 1.50 | 0.02 | 0.04 | 0.02 |
| RDE+div | Mean | 12,635 | 236 | 12,871 | 1,901 | 14,772 |
| | St Dev | 248 | 313 | 312 | 84 | 293 |
| | CV | 0.02 | 1.33 | 0.02 | 0.04 | 0.02 |

| Strategy | | Case 1 | | | | |
|---|---|---|---|---|---|---|
| | | Inv. Holding Cost | Lost Sales Cost | Total Inv. Cost | Transp. Cost | Total Cost |
| BENCH1 | Mean | 20,641 | 1,886 | 22,527 | 1,920 | 24,447 |
| | St Dev | 699 | 1,950 | 1,512 | 72 | 1,535 |
| | CV | 0.03 | 1.03 | 0.07 | 0.04 | 0.06 |
| MUN | Mean | 15,338 | 2,274 | 17,612 | 2,632 | 20,244 |
| | St Dev | 802 | 2,434 | 2,076 | 189 | 2,081 |
| | CV | 0.05 | 1.07 | 0.12 | 0.07 | 0.10 |
| RTC | Mean | 17,971 | 432 | 18,402 | 1,873 | 20,275 |
| | St Dev | 703 | 801 | 869 | 119 | 891 |
| | CV | 0.04 | 1.86 | 0.05 | 0.06 | 0.04 |
| RDE | Mean | 16,246 | 282 | 16,527 | 1,980 | 18,507 |
| | St Dev | 313 | 330 | 412 | 124 | 399 |
| | CV | 0.02 | 1.17 | 0.02 | 0.06 | 0.02 |
| RDE+div | Mean | 16,212 | 243 | 16,455 | 1,987 | 18,442 |
| | St Dev | 297 | 367 | 415 | 91 | 403 |
| | CV | 0.02 | 1.51 | 0.03 | 0.05 | 0.02 |

| Strategy | | Case 2 | | | | |
|---|---|---|---|---|---|---|
| | | Inv. Holding Cost | Lost Sales Cost | Total Inv. Cost | Transp. Cost | Total Cost |
| BENCH1 | Mean | 20,559 | 787 | 21,345 | 1,300 | 22,646 |
| | St Dev | 418 | 859 | 835 | 55 | 848 |
| | CV | 0.02 | 1.09 | 0.04 | 0.04 | 0.04 |
| MUN | Mean | 11,087 | 985 | 12,072 | 2,576 | 14,648 |
| | St Dev | 405 | 1,006 | 873 | 177 | 864 |
| | CV | 0.04 | 1.02 | 0.07 | 0.07 | 0.06 |
| RTC | Mean | 13,720 | 490 | 14,210 | 1,635 | 15,845 |
| | St Dev | 348 | 632 | 671 | 81 | 707 |
| | CV | 0.03 | 1.29 | 0.05 | 0.05 | 0.04 |
| RDE | Mean | 11,243 | 518 | 11,761 | 1,903 | 13,664 |
| | St Dev | 203 | 611 | 619 | 81 | 639 |
| | CV | 0.02 | 1.18 | 0.05 | 0.04 | 0.05 |
| RDE+div | Mean | 11,275 | 386 | 11,661 | 1,952 | 13,613 |
| | St Dev | 193 | 411 | 439 | 109 | 452 |
| | CV | 0.02 | 1.07 | 0.04 | 0.06 | 0.03 |

| Strategy | | Case 3 | | | | |
|---|---|---|---|---|---|---|
| | | Inv. Holding Cost | Lost Sales Cost | Total Inv. Cost | Transp. Cost | Total Cost |
| BENCH1 | Mean | 21,317 | 493 | 21,811 | 1,389 | 23,199 |
| | St Dev | 482 | 745 | 843 | 43 | 842 |
| | CV | 0.02 | 1.51 | 0.04 | 0.03 | 0.04 |
| MUN | Mean | 13,287 | 909 | 14,195 | 2,316 | 16,511 |
| | St Dev | 504 | 1,452 | 1,146 | 181 | 1,151 |
| | CV | 0.04 | 1.60 | 0.08 | 0.08 | 0.07 |
| RTC | Mean | 15,735 | 247 | 15,981 | 1,604 | 17,585 |
| | St Dev | 464 | 480 | 666 | 72 | 697 |
| | CV | 0.03 | 1.95 | 0.04 | 0.04 | 0.04 |
| RDE | Mean | 13,682 | 363 | 14,045 | 1,751 | 15,795 |
| | St Dev | 209 | 501 | 516 | 75 | 512 |
| | CV | 0.02 | 1.38 | 0.04 | 0.04 | 0.03 |
| RDE+div | Mean | 13,766 | 238 | 14,004 | 1,790 | 15,794 |
| | St Dev | 205 | 374 | 410 | 90 | 426 |
| | CV | 0.01 | 1.57 | 0.03 | 0.05 | 0.03 |

30 replication with common random numbers
Results in [$/week]

**Fig. 12.** Simulation results: parameter Set 3 TC = 300 [$/hr], $h_i = 50$[$/week], $\lambda_i = 50$ [arrivals/day], $\theta_i = 1$ [units] for all i

| Strategy | | Case 0 (Symmetric case) | | | | |
|---|---|---|---|---|---|---|
| | | Inv. Holding Cost | Lost Sales Cost | Total Inv. Cost | Transp. Cost | Total Cost |
| BENCH1 | Mean | 29,895 | 488 | 30,383 | 8,441 | 38,825 |
| | St Dev | 642 | 731 | 761 | 688 | 1,156 |
| | CV | 0.02 | 1.50 | 0.03 | 0.08 | 0.03 |
| MUN | Mean | 19,248 | 829 | 20,077 | 15,671 | 35,748 |
| | St Dev | 707 | 1,386 | 1,114 | 1,459 | 2,103 |
| | CV | 0.04 | 1.67 | 0.06 | 0.09 | 0.06 |
| RTC | Mean | 20,780 | 368 | 21,147 | 9,666 | 30,814 |
| | St Dev | 628 | 577 | 743 | 749 | 1,050 |
| | CV | 0.03 | 1.57 | 0.04 | 0.08 | 0.03 |
| RDE | Mean | 19,086 | 448 | 19,534 | 10,027 | 29,562 |
| | St Dev | 470 | 547 | 517 | 682 | 931 |
| | CV | 0.02 | 1.22 | 0.03 | 0.07 | 0.03 |
| RDE+div | Mean | 19,167 | 452 | 19,619 | 10,063 | 29,682 |
| | St Dev | 437 | 483 | 538 | 709 | 914 |
| | CV | 0.02 | 1.07 | 0.03 | 0.07 | 0.03 |
| Strategy | | Case 1 | | | | |
| | | Inv. Holding Cost | Lost Sales Cost | Total Inv. Cost | Transp. Cost | Total Cost |
| BENCH1 | Mean | 30,680 | 1,703 | 32,383 | 10,581 | 42,965 |
| | St Dev | 980 | 1,742 | 1,813 | 845 | 2,226 |
| | CV | 0.03 | 1.02 | 0.06 | 0.08 | 0.05 |
| MUN | Mean | 21,104 | 1,529 | 22,633 | 16,244 | 38,877 |
| | St Dev | 1,063 | 1,710 | 1,426 | 1,885 | 2,254 |
| | CV | 0.05 | 1.12 | 0.06 | 0.12 | 0.06 |
| RTC | Mean | 24,095 | 609 | 24,705 | 11,183 | 35,888 |
| | St Dev | 874 | 934 | 1,132 | 897 | 1,555 |
| | CV | 0.04 | 1.53 | 0.05 | 0.08 | 0.04 |
| RDE | Mean | 22,833 | 529 | 23,362 | 11,523 | 34,886 |
| | St Dev | 536 | 624 | 800 | 1,062 | 1,364 |
| | CV | 0.02 | 1.18 | 0.03 | 0.09 | 0.04 |
| RDE+div | Mean | 22,800 | 505 | 23,305 | 11,578 | 34,883 |
| | St Dev | 738 | 682 | 966 | 1,012 | 1,448 |
| | CV | 0.03 | 1.35 | 0.04 | 0.09 | 0.04 |
| Strategy | | Case 2 | | | | |
| | | Inv. Holding Cost | Lost Sales Cost | Total Inv. Cost | Transp. Cost | Total Cost |
| BENCH1 | Mean | 28,522 | 727 | 29,250 | 7,470 | 36,719 |
| | St Dev | 661 | 1,167 | 1,231 | 716 | 1,562 |
| | CV | 0.02 | 1.60 | 0.04 | 0.10 | 0.04 |
| MUN | Mean | 16,594 | 4,798 | 21,393 | 15,627 | 37,020 |
| | St Dev | 781 | 3,346 | 2,759 | 2,391 | 2,722 |
| | CV | 0.05 | 0.70 | 0.13 | 0.15 | 0.07 |
| RTC | Mean | 19,246 | 771 | 20,017 | 8,978 | 28,995 |
| | St Dev | 655 | 1,172 | 1,023 | 796 | 1,247 |
| | CV | 0.03 | 1.52 | 0.05 | 0.09 | 0.04 |
| RDE | Mean | 17,627 | 553 | 18,180 | 9,578 | 27,758 |
| | St Dev | 428 | 765 | 888 | 819 | 1,218 |
| | CV | 0.02 | 1.38 | 0.05 | 0.09 | 0.04 |
| RDE+div | Mean | 17,561 | 469 | 18,031 | 9,566 | 27,597 |
| | St Dev | 460 | 417 | 537 | 663 | 869 |
| | CV | 0.03 | 0.89 | 0.03 | 0.07 | 0.03 |
| Strategy | | Case 3 | | | | |
| | | Inv. Holding Cost | Lost Sales Cost | Total Inv. Cost | Transp. Cost | Total Cost |
| BENCH1 | Mean | 29,530 | 1,009 | 30,539 | 8,368 | 38,907 |
| | St Dev | 905 | 1,053 | 1,118 | 745 | 1,399 |
| | CV | 0.03 | 1.04 | 0.04 | 0.09 | 0.04 |
| MUN | Mean | 20,035 | 449 | 20,484 | 14,654 | 35,138 |
| | St Dev | 818 | 555 | 949 | 1,645 | 2,134 |
| | CV | 0.04 | 1.24 | 0.05 | 0.11 | 0.06 |
| RTC | Mean | 22,267 | 372 | 22,639 | 9,184 | 31,823 |
| | St Dev | 636 | 444 | 689 | 821 | 1,207 |
| | CV | 0.03 | 1.19 | 0.03 | 0.09 | 0.04 |
| RDE | Mean | 20,555 | 561 | 21,116 | 9,563 | 30,678 |
| | St Dev | 416 | 806 | 822 | 832 | 1,286 |
| | CV | 0.02 | 1.44 | 0.04 | 0.09 | 0.04 |
| RDE+div | Mean | 20,594 | 573 | 21,167 | 9,585 | 30,751 |
| | St Dev | 533 | 592 | 673 | 803 | 1,089 |
| | CV | 0.03 | 1.03 | 0.03 | 0.08 | 0.04 |

30 replication with common random numbers
Results in [$/week]

**Fig. 13.** Simulation results: parameter set 4 TC = 100 [$/hr], $h_i = 50$[$/week], $\lambda_i = 10.5$ [arrivals/day], $\theta_i = 4.8$ [units] for all i

| Strategy | | Case 0 (Symmetric case) | | | | |
|---|---|---|---|---|---|---|
| | | Inv. Holding Cost | Lost Sales Cost | Total Inv. Cost | Transp. Cost | Total Cost |
| BENCH1 | Mean | 26,668 | 1,924 | 28,592 | 3,293 | 31,885 |
| | St Dev | 833 | 1,779 | 1,870 | 238 | 1,922 |
| | CV | 0.03 | 0.92 | 0.07 | 0.07 | 0.06 |
| MUN | Mean | 16,535 | 923 | 17,458 | 6,903 | 24,360 |
| | St Dev | 531 | 996 | 1,118 | 514 | 1,339 |
| | CV | 0.03 | 1.08 | 0.06 | 0.07 | 0.05 |
| RTC | Mean | 20,424 | 863 | 21,287 | 4,757 | 26,044 |
| | St Dev | 636 | 989 | 1,109 | 331 | 1,169 |
| | CV | 0.03 | 1.15 | 0.05 | 0.07 | 0.04 |
| RDE | Mean | 17,050 | 1,018 | 18,068 | 5,197 | 23,265 |
| | St Dev | 355 | 855 | 798 | 303 | 820 |
| | CV | 0.02 | 0.84 | 0.04 | 0.06 | 0.04 |
| RDE+div | Mean | 16,934 | 727 | 17,660 | 5,326 | 22,987 |
| | St Dev | 436 | 804 | 769 | 333 | 831 |
| | CV | 0.03 | 1.11 | 0.04 | 0.06 | 0.04 |

| Strategy | | Case 1 | | | | |
|---|---|---|---|---|---|---|
| | | Inv. Holding Cost | Lost Sales Cost | Total Inv. Cost | Transp. Cost | Total Cost |
| BENCH1 | Mean | 26,615 | 2,797 | 29,412 | 4,326 | 33,738 |
| | St Dev | 1,055 | 2,524 | 2,131 | 401 | 2,200 |
| | CV | 0.04 | 0.90 | 0.07 | 0.09 | 0.07 |
| MUN | Mean | 19,530 | 1,282 | 20,812 | 7,222 | 28,034 |
| | St Dev | 957 | 1,953 | 1,806 | 543 | 1,972 |
| | CV | 0.05 | 1.52 | 0.09 | 0.08 | 0.07 |
| RTC | Mean | 23,530 | 524 | 24,054 | 5,388 | 29,442 |
| | St Dev | 908 | 839 | 1,185 | 317 | 1,225 |
| | CV | 0.04 | 1.60 | 0.05 | 0.06 | 0.04 |
| RDE | Mean | 20,716 | 649 | 21,365 | 5,867 | 27,231 |
| | St Dev | 430 | 690 | 701 | 480 | 882 |
| | CV | 0.02 | 1.06 | 0.03 | 0.08 | 0.03 |
| RDE+div | Mean | 20,810 | 307 | 21,118 | 5,872 | 26,990 |
| | St Dev | 428 | 555 | 611 | 431 | 686 |
| | CV | 0.02 | 1.81 | 0.03 | 0.07 | 0.03 |

| Strategy | | Case 2 | | | | |
|---|---|---|---|---|---|---|
| | | Inv. Holding Cost | Lost Sales Cost | Total Inv. Cost | Transp. Cost | Total Cost |
| BENCH1 | Mean | 26,427 | 1,817 | 28,244 | 2,953 | 31,196 |
| | St Dev | 1,117 | 1,751 | 1,602 | 254 | 1,662 |
| | CV | 0.04 | 0.96 | 0.06 | 0.09 | 0.05 |
| MUN | Mean | 15,341 | 1,651 | 16,992 | 7,352 | 24,343 |
| | St Dev | 540 | 1,621 | 1,570 | 563 | 1,590 |
| | CV | 0.04 | 0.98 | 0.09 | 0.08 | 0.07 |
| RTC | Mean | 19,364 | 545 | 19,909 | 4,340 | 24,249 |
| | St Dev | 751 | 596 | 982 | 287 | 989 |
| | CV | 0.04 | 1.09 | 0.05 | 0.07 | 0.04 |
| RDE | Mean | 15,385 | 699 | 16,084 | 4,976 | 21,060 |
| | St Dev | 273 | 641 | 617 | 311 | 673 |
| | CV | 0.02 | 0.92 | 0.04 | 0.06 | 0.03 |
| RDE+div | Mean | 15,458 | 812 | 16,270 | 5,125 | 21,395 |
| | St Dev | 269 | 784 | 806 | 310 | 817 |
| | CV | 0.02 | 0.96 | 0.05 | 0.06 | 0.04 |

| Strategy | | Case 3 | | | | |
|---|---|---|---|---|---|---|
| | | Inv. Holding Cost | Lost Sales Cost | Total Inv. Cost | Transp. Cost | Total Cost |
| BENCH1 | Mean | 27,080 | 2,313 | 29,393 | 3,318 | 32,711 |
| | St Dev | 849 | 2,482 | 2,267 | 255 | 2,309 |
| | CV | 0.03 | 1.07 | 0.08 | 0.08 | 0.07 |
| MUN | Mean | 17,761 | 860 | 18,621 | 6,730 | 25,351 |
| | St Dev | 682 | 965 | 1,089 | 550 | 1,238 |
| | CV | 0.04 | 1.12 | 0.06 | 0.08 | 0.05 |
| RTC | Mean | 21,753 | 1,165 | 22,918 | 4,513 | 27,431 |
| | St Dev | 943 | 1,582 | 1,557 | 294 | 1,565 |
| | CV | 0.04 | 1.36 | 0.07 | 0.07 | 0.06 |
| RDE | Mean | 18,207 | 911 | 19,118 | 4,893 | 24,011 |
| | St Dev | 439 | 1,205 | 1,020 | 305 | 1,031 |
| | CV | 0.02 | 1.32 | 0.05 | 0.06 | 0.04 |
| RDE+div | Mean | 18,213 | 578 | 18,791 | 4,949 | 23,740 |
| | St Dev | 510 | 672 | 627 | 322 | 764 |
| | CV | 0.03 | 1.16 | 0.03 | 0.07 | 0.03 |

30 replication with common random numbers
Results in [$/week]

**Fig. 14.** Simulation results: parameter set 5 TC = 100 [\$/hr], $h_i = 50$[\$/week], $\lambda_i = 4.35$ [arrivals/day], $\theta_i = 11.5$ [units] for all i

| Strategy | | Case 0 (Symmetric case) | | | | |
| --- | --- | --- | --- | --- | --- | --- |
| | | Inv. Holding Cost | Lost Sales Cost | Total Inv. Cost | Transp. Cost | Total Cost |
| BENCH1 | Mean | 30,862 | 2,273 | 33,135 | 3,105 | 36,241 |
| | St Dev | 1,292 | 2,761 | 2,590 | 243 | 2,615 |
| | CV | 0.04 | 1.21 | 0.08 | 0.08 | 0.07 |
| MUN | Mean | 19,999 | 1,509 | 21,508 | 7,251 | 28,759 |
| | St Dev | 1,043 | 1,693 | 1,803 | 598 | 2,105 |
| | CV | 0.05 | 1.12 | 0.08 | 0.08 | 0.07 |
| RTC | Mean | 25,127 | 2,488 | 27,615 | 4,981 | 32,596 |
| | St Dev | 989 | 2,867 | 2,857 | 437 | 2,860 |
| | CV | 0.04 | 1.15 | 0.10 | 0.09 | 0.09 |
| RDE | Mean | 19,988 | 1,895 | 21,883 | 5,358 | 27,241 |
| | St Dev | 419 | 1,540 | 1,473 | 436 | 1,615 |
| | CV | 0.02 | 0.81 | 0.07 | 0.08 | 0.06 |
| RDE+div | Mean | 20,272 | 1,383 | 21,654 | 5,558 | 27,212 |
| | St Dev | 485 | 1,386 | 1,370 | 381 | 1,499 |
| | CV | 0.02 | 1.00 | 0.06 | 0.07 | 0.06 |

| Strategy | | Case 1 | | | | |
| --- | --- | --- | --- | --- | --- | --- |
| | | Inv. Holding Cost | Lost Sales Cost | Total Inv. Cost | Transp. Cost | Total Cost |
| BENCH1 | Mean | 30,842 | 4,103 | 34,945 | 4,083 | 39,028 |
| | St Dev | 1,354 | 4,971 | 4,560 | 375 | 4,586 |
| | CV | 0.04 | 1.21 | 0.13 | 0.09 | 0.12 |
| MUN | Mean | 23,346 | 1,259 | 24,606 | 7,219 | 31,825 |
| | St Dev | 1,117 | 1,514 | 1,478 | 853 | 2,139 |
| | CV | 0.05 | 1.20 | 0.06 | 0.12 | 0.07 |
| RTC | Mean | 28,961 | 1,647 | 30,607 | 5,616 | 36,223 |
| | St Dev | 1,605 | 1,640 | 2,242 | 472 | 2,197 |
| | CV | 0.06 | 1.00 | 0.07 | 0.08 | 0.06 |
| RDE | Mean | 24,254 | 1,248 | 25,502 | 6,386 | 31,888 |
| | St Dev | 637 | 1,519 | 1,367 | 571 | 1,487 |
| | CV | 0.03 | 1.22 | 0.05 | 0.09 | 0.05 |
| RDE+div | Mean | 24,592 | 959 | 25,551 | 6,327 | 31,878 |
| | St Dev | 709 | 1,350 | 1,375 | 584 | 1,455 |
| | CV | 0.03 | 1.41 | 0.05 | 0.09 | 0.05 |

| Strategy | | Case 2 | | | | |
| --- | --- | --- | --- | --- | --- | --- |
| | | Inv. Holding Cost | Lost Sales Cost | Total Inv. Cost | Transp. Cost | Total Cost |
| BENCH1 | Mean | 30,787 | 2,473 | 33,260 | 2,919 | 36,179 |
| | St Dev | 1,061 | 2,254 | 2,035 | 274 | 2,165 |
| | CV | 0.03 | 0.91 | 0.06 | 0.09 | 0.06 |
| MUN | Mean | 18,397 | 2,570 | 20,967 | 7,416 | 28,384 |
| | St Dev | 599 | 2,245 | 2,082 | 770 | 2,188 |
| | CV | 0.03 | 0.87 | 0.10 | 0.10 | 0.08 |
| RTC | Mean | 24,345 | 1,885 | 26,229 | 4,363 | 30,593 |
| | St Dev | 963 | 2,152 | 2,193 | 482 | 2,229 |
| | CV | 0.04 | 1.14 | 0.08 | 0.11 | 0.07 |
| RDE | Mean | 18,311 | 2,872 | 21,182 | 5,361 | 26,543 |
| | St Dev | 418 | 2,646 | 2,545 | 501 | 2,645 |
| | CV | 0.02 | 0.92 | 0.12 | 0.09 | 0.10 |
| RDE+div | Mean | 18,589 | 1,596 | 20,184 | 5,312 | 25,496 |
| | St Dev | 344 | 1,633 | 1,741 | 501 | 1,848 |
| | CV | 0.02 | 1.02 | 0.09 | 0.09 | 0.07 |

| Strategy | | Case 3 | | | | |
| --- | --- | --- | --- | --- | --- | --- |
| | | Inv. Holding Cost | Lost Sales Cost | Total Inv. Cost | Transp. Cost | Total Cost |
| BENCH1 | Mean | 31,057 | 3,478 | 34,536 | 3,098 | 37,634 |
| | St Dev | 1,391 | 3,501 | 3,377 | 252 | 3,371 |
| | CV | 0.04 | 1.01 | 0.10 | 0.08 | 0.09 |
| MUN | Mean | 21,483 | 999 | 22,482 | 6,897 | 29,378 |
| | St Dev | 1,187 | 1,158 | 1,457 | 848 | 2,098 |
| | CV | 0.06 | 1.16 | 0.06 | 0.12 | 0.07 |
| RTC | Mean | 26,585 | 1,774 | 28,359 | 4,703 | 33,062 |
| | St Dev | 1,204 | 2,158 | 2,348 | 303 | 2,343 |
| | CV | 0.05 | 1.22 | 0.08 | 0.06 | 0.07 |
| RDE | Mean | 21,582 | 1,767 | 23,348 | 5,591 | 28,940 |
| | St Dev | 494 | 1,567 | 1,615 | 551 | 1,651 |
| | CV | 0.02 | 0.89 | 0.07 | 0.10 | 0.06 |
| RDE+div | Mean | 21,587 | 905 | 22,492 | 5,639 | 28,131 |
| | St Dev | 525 | 1,315 | 1,209 | 544 | 1,384 |
| | CV | 0.02 | 1.45 | 0.05 | 0.10 | 0.05 |

30 replication with common random numbers
Results in [\$/week]

**Fig. 15.** Simulation results: parameter set 6 TC = 100 [$/hr], $h_i = 50$[$/week], $\lambda_i = 2.4$ [arrivals/day], $\theta_i = 20.8$ [units] for all i

| Strategy | | Case 0 (Symmetric case) | | | | |
|---|---|---|---|---|---|---|
| | | Inv. Holding Cost | Lost Sales Cost | Total Inv. Cost | Transp. Cost | Total Cost |
| BENCH1 | Mean | 34,425 | 4,670 | 39,096 | 3,044 | 42,140 |
| | St Dev | 1,691 | 4,308 | 3,834 | 325 | 3,758 |
| | CV | 0.05 | 0.92 | 0.10 | 0.11 | 0.09 |
| MUN | Mean | 23,539 | 1,954 | 25,493 | 7,302 | 32,795 |
| | St Dev | 1,197 | 1,728 | 1,783 | 881 | 1,941 |
| | CV | 0.05 | 0.88 | 0.07 | 0.12 | 0.06 |
| RTC | Mean | 29,653 | 4,261 | 33,914 | 5,281 | 39,195 |
| | St Dev | 1,614 | 3,799 | 3,574 | 523 | 3,630 |
| | CV | 0.05 | 0.89 | 0.11 | 0.10 | 0.09 |
| RDE | Mean | 23,151 | 4,040 | 27,190 | 5,717 | 32,907 |
| | St Dev | 633 | 2,860 | 2,841 | 479 | 2,775 |
| | CV | 0.03 | 0.71 | 0.10 | 0.08 | 0.08 |
| RDE+div | Mean | 23,482 | 2,630 | 26,112 | 5,793 | 31,906 |
| | St Dev | 524 | 2,704 | 2,699 | 417 | 2,749 |
| | CV | 0.02 | 1.03 | 0.10 | 0.07 | 0.09 |
| Strategy | | Case 1 | | | | |
| | | Inv. Holding Cost | Lost Sales Cost | Total Inv. Cost | Transp. Cost | Total Cost |
| BENCH1 | Mean | 34,452 | 5,503 | 39,955 | 4,050 | 44,005 |
| | St Dev | 1,673 | 4,789 | 4,108 | 414 | 4,211 |
| | CV | 0.05 | 0.87 | 0.10 | 0.10 | 0.10 |
| MUN | Mean | 26,848 | 2,233 | 29,081 | 7,225 | 36,306 |
| | St Dev | 1,119 | 2,763 | 2,382 | 776 | 2,527 |
| | CV | 0.04 | 1.24 | 0.08 | 0.11 | 0.07 |
| RTC | Mean | 33,758 | 2,504 | 36,262 | 5,824 | 42,086 |
| | St Dev | 1,463 | 3,145 | 3,157 | 479 | 3,215 |
| | CV | 0.04 | 1.26 | 0.09 | 0.08 | 0.08 |
| RDE | Mean | 27,965 | 3,227 | 31,192 | 7,002 | 38,194 |
| | St Dev | 841 | 2,572 | 2,350 | 560 | 2,517 |
| | CV | 0.03 | 0.80 | 0.08 | 0.08 | 0.07 |
| RDE+div | Mean | 28,194 | 1,324 | 29,518 | 6,785 | 36,303 |
| | St Dev | 833 | 1,997 | 2,050 | 720 | 2,134 |
| | CV | 0.03 | 1.51 | 0.07 | 0.11 | 0.06 |
| Strategy | | Case 2 | | | | |
| | | Inv. Holding Cost | Lost Sales Cost | Total Inv. Cost | Transp. Cost | Total Cost |
| BENCH1 | Mean | 34,610 | 3,952 | 38,563 | 2,734 | 41,297 |
| | St Dev | 1,464 | 2,876 | 2,592 | 331 | 2,660 |
| | CV | 0.04 | 0.73 | 0.07 | 0.12 | 0.06 |
| MUN | Mean | 21,356 | 3,562 | 24,918 | 7,523 | 32,441 |
| | St Dev | 841 | 2,225 | 1,930 | 857 | 1,944 |
| | CV | 0.04 | 0.62 | 0.08 | 0.11 | 0.06 |
| RTC | Mean | 28,890 | 1,965 | 30,855 | 4,419 | 35,274 |
| | St Dev | 1,271 | 2,334 | 2,813 | 498 | 2,829 |
| | CV | 0.04 | 1.19 | 0.09 | 0.11 | 0.08 |
| RDE | Mean | 21,304 | 4,612 | 25,916 | 5,247 | 31,163 |
| | St Dev | 587 | 3,785 | 3,640 | 548 | 3,844 |
| | CV | 0.03 | 0.82 | 0.14 | 0.10 | 0.12 |
| RDE+div | Mean | 21,667 | 2,631 | 24,298 | 5,800 | 30,098 |
| | St Dev | 453 | 2,452 | 2,386 | 646 | 2,430 |
| | CV | 0.02 | 0.93 | 0.10 | 0.11 | 0.08 |
| Strategy | | Case 3 | | | | |
| | | Inv. Holding Cost | Lost Sales Cost | Total Inv. Cost | Transp. Cost | Total Cost |
| BENCH1 | Mean | 35,055 | 3,169 | 38,224 | 3,119 | 41,343 |
| | St Dev | 1,409 | 2,945 | 3,135 | 290 | 3,112 |
| | CV | 0.04 | 0.93 | 0.08 | 0.09 | 0.08 |
| MUN | Mean | 25,234 | 1,656 | 26,891 | 7,029 | 33,920 |
| | St Dev | 1,309 | 1,979 | 1,978 | 789 | 2,347 |
| | CV | 0.05 | 1.19 | 0.07 | 0.11 | 0.07 |
| RTC | Mean | 31,198 | 2,420 | 33,618 | 4,926 | 38,544 |
| | St Dev | 1,359 | 2,973 | 2,462 | 423 | 2,527 |
| | CV | 0.04 | 1.23 | 0.07 | 0.09 | 0.07 |
| RDE | Mean | 24,729 | 4,071 | 28,800 | 6,065 | 34,865 |
| | St Dev | 762 | 3,254 | 2,833 | 634 | 2,964 |
| | CV | 0.03 | 0.80 | 0.10 | 0.10 | 0.09 |
| RDE+div | Mean | 25,215 | 2,420 | 27,635 | 6,213 | 33,848 |
| | St Dev | 609 | 2,029 | 1,893 | 642 | 1,937 |
| | CV | 0.02 | 0.84 | 0.07 | 0.10 | 0.06 |

30 replication with common random numbers
Results in [$/week]

**Fig. 16.** Simulation results: parameter set 7 TC = 100 [$/hr], $h_i$ = 5[$/week], $\lambda_i$ =50[arrivals/day], $\theta_i$ = 1 [units] for all i

| Strategy | | Case 0 (Symmetric case) | | | | |
|---|---|---|---|---|---|---|
| | | Inv. Holding Cost | Lost Sales Cost | Total Inv. Cost | Transp. Cost | Total Cost |
| BENCH1 | Mean | 2,178 | 440 | 2,618 | 4,158 | 6,776 |
| | St Dev | 41 | 605 | 594 | 173 | 588 |
| | CV | 0.02 | 1.38 | 0.23 | 0.04 | 0.09 |
| MUN | Mean | 2,066 | 762 | 2,828 | 4,648 | 7,476 |
| | St Dev | 75 | 1,357 | 1,325 | 484 | 1,355 |
| | CV | 0.04 | 1.78 | 0.47 | 0.10 | 0.18 |
| RTC | Mean | 2,153 | 183 | 2,336 | 3,215 | 5,551 |
| | St Dev | 58 | 350 | 351 | 172 | 414 |
| | CV | 0.03 | 1.92 | 0.15 | 0.05 | 0.07 |
| RDE | Mean | 2,098 | 220 | 2,318 | 3,134 | 5,452 |
| | St Dev | 46 | 246 | 232 | 211 | 328 |
| | CV | 0.02 | 1.12 | 0.10 | 0.07 | 0.06 |
| RDE+div | Mean | 2,104 | 93 | 2,197 | 3,183 | 5,380 |
| | St Dev | 42 | 227 | 233 | 177 | 309 |
| | CV | 0.02 | 2.44 | 0.11 | 0.06 | 0.06 |

| Strategy | | Case 1 | | | | |
|---|---|---|---|---|---|---|
| | | Inv. Holding Cost | Lost Sales Cost | Total Inv. Cost | Transp. Cost | Total Cost |
| BENCH1 | Mean | 2,004 | 2,128 | 4,132 | 5,872 | 10,004 |
| | St Dev | 67 | 1,633 | 1,590 | 196 | 1,643 |
| | CV | 0.03 | 0.77 | 0.38 | 0.03 | 0.16 |
| MUN | Mean | 2,024 | 1,299 | 3,323 | 6,397 | 9,720 |
| | St Dev | 87 | 1,892 | 1,845 | 680 | 1,800 |
| | CV | 0.04 | 1.46 | 0.56 | 0.11 | 0.19 |
| RTC | Mean | 2,135 | 177 | 2,311 | 4,360 | 6,671 |
| | St Dev | 74 | 341 | 341 | 306 | 548 |
| | CV | 0.03 | 1.93 | 0.15 | 0.07 | 0.08 |
| RDE | Mean | 2,098 | 307 | 2,405 | 4,208 | 6,613 |
| | St Dev | 44 | 479 | 467 | 334 | 606 |
| | CV | 0.02 | 1.56 | 0.19 | 0.08 | 0.09 |
| RDE+div | Mean | 2,087 | 184 | 2,271 | 4,286 | 6,557 |
| | St Dev | 45 | 292 | 286 | 339 | 508 |
| | CV | 0.02 | 1.59 | 0.13 | 0.08 | 0.08 |

| Strategy | | Case 2 | | | | |
|---|---|---|---|---|---|---|
| | | Inv. Holding Cost | Lost Sales Cost | Total Inv. Cost | Transp. Cost | Total Cost |
| BENCH1 | Mean | 2,196 | 603 | 2,799 | 3,931 | 6,731 |
| | St Dev | 56 | 781 | 768 | 200 | 753 |
| | CV | 0.03 | 1.30 | 0.27 | 0.05 | 0.11 |
| MUN | Mean | 1,991 | 429 | 2,420 | 4,729 | 7,149 |
| | St Dev | 63 | 648 | 644 | 490 | 813 |
| | CV | 0.03 | 1.51 | 0.27 | 0.10 | 0.11 |
| RTC | Mean | 2,056 | 444 | 2,500 | 3,059 | 5,559 |
| | St Dev | 61 | 476 | 442 | 249 | 472 |
| | CV | 0.03 | 1.07 | 0.18 | 0.08 | 0.08 |
| RDE | Mean | 1,955 | 323 | 2,278 | 3,062 | 5,339 |
| | St Dev | 41 | 650 | 647 | 196 | 718 |
| | CV | 0.02 | 2.02 | 0.28 | 0.06 | 0.13 |
| RDE+div | Mean | 1,945 | 175 | 2,119 | 3,071 | 5,191 |
| | St Dev | 42 | 266 | 273 | 236 | 394 |
| | CV | 0.02 | 1.53 | 0.13 | 0.08 | 0.08 |

| Strategy | | Case 3 | | | | |
|---|---|---|---|---|---|---|
| | | Inv. Holding Cost | Lost Sales Cost | Total Inv. Cost | Transp. Cost | Total Cost |
| BENCH1 | Mean | 2,165 | 1,108 | 3,273 | 4,215 | 7,489 |
| | St Dev | 60 | 1,241 | 1,214 | 205 | 1,271 |
| | CV | 0.03 | 1.12 | 0.37 | 0.05 | 0.17 |
| MUN | Mean | 2,083 | 424 | 2,507 | 5,073 | 7,579 |
| | St Dev | 77 | 647 | 616 | 526 | 757 |
| | CV | 0.04 | 1.53 | 0.25 | 0.10 | 0.10 |
| RTC | Mean | 2,151 | 264 | 2,416 | 3,185 | 5,600 |
| | St Dev | 61 | 488 | 494 | 252 | 637 |
| | CV | 0.03 | 1.85 | 0.20 | 0.08 | 0.11 |
| RDE | Mean | 2,115 | 188 | 2,303 | 3,156 | 5,458 |
| | St Dev | 38 | 311 | 305 | 225 | 407 |
| | CV | 0.02 | 1.66 | 0.13 | 0.07 | 0.07 |
| RDE+div | Mean | 2,114 | 105 | 2,218 | 3,205 | 5,424 |
| | St Dev | 48 | 227 | 232 | 249 | 314 |
| | CV | 0.02 | 2.17 | 0.10 | 0.08 | 0.06 |

30 replication with common random numbers
Results in [$/week]

**Fig. 17.** Simulation results: parameter set 8  TC = 33  [$/hr],  $h_i = 5$[$/week],  $\lambda_i = 50$ [arrivals/day], $\theta_i = 1$ [units] for all i

| Strategy | | Case 0 (Symmetric case) | | | | |
|---|---|---|---|---|---|---|
| | | Inv. Holding Cost | Lost Sales Cost | Total Inv. Cost | Transp. Cost | Total Cost |
| BENCH1 | Mean | 2,228 | 107 | 2,335 | 1,411 | 3,746 |
| | St Dev | 46 | 333 | 324 | 34 | 330 |
| | CV | 0.02 | 3.11 | 0.14 | 0.02 | 0.09 |
| MUN | Mean | 2,086 | 364 | 2,450 | 1,575 | 4,025 |
| | St Dev | 97 | 619 | 576 | 158 | 612 |
| | CV | 0.05 | 1.70 | 0.24 | 0.10 | 0.15 |
| RTC | Mean | 2,153 | 126 | 2,279 | 1,062 | 3,341 |
| | St Dev | 53 | 315 | 292 | 89 | 286 |
| | CV | 0.02 | 2.50 | 0.13 | 0.08 | 0.09 |
| RDE | Mean | 2,106 | 133 | 2,240 | 1,049 | 3,288 |
| | St Dev | 37 | 324 | 317 | 68 | 322 |
| | CV | 0.02 | 2.43 | 0.14 | 0.06 | 0.10 |
| RDE+div | Mean | 2,109 | 93 | 2,202 | 1,053 | 3,255 |
| | St Dev | 36 | 168 | 172 | 77 | 202 |
| | CV | 0.02 | 1.80 | 0.08 | 0.07 | 0.06 |

| Strategy | | Case 1 | | | | |
|---|---|---|---|---|---|---|
| | | Inv. Holding Cost | Lost Sales Cost | Total Inv. Cost | Transp. Cost | Total Cost |
| BENCH1 | Mean | 2,082 | 727 | 2,809 | 1,989 | 4,799 |
| | St Dev | 59 | 859 | 839 | 48 | 842 |
| | CV | 0.03 | 1.18 | 0.30 | 0.02 | 0.18 |
| MUN | Mean | 2,028 | 860 | 2,888 | 2,142 | 5,030 |
| | St Dev | 80 | 1,184 | 1,142 | 160 | 1,184 |
| | CV | 0.04 | 1.38 | 0.40 | 0.07 | 0.24 |
| RTC | Mean | 2,148 | 322 | 2,470 | 1,440 | 3,910 |
| | St Dev | 58 | 439 | 458 | 109 | 461 |
| | CV | 0.03 | 1.36 | 0.19 | 0.08 | 0.12 |
| RDE | Mean | 2,064 | 358 | 2,423 | 1,427 | 3,850 |
| | St Dev | 51 | 549 | 541 | 112 | 577 |
| | CV | 0.02 | 1.53 | 0.22 | 0.08 | 0.15 |
| RDE+div | Mean | 2,068 | 213 | 2,281 | 1,448 | 3,729 |
| | St Dev | 50 | 338 | 340 | 96 | 373 |
| | CV | 0.02 | 1.59 | 0.15 | 0.07 | 0.10 |

| Strategy | | Case 2 | | | | |
|---|---|---|---|---|---|---|
| | | Inv. Holding Cost | Lost Sales Cost | Total Inv. Cost | Transp. Cost | Total Cost |
| BENCH1 | Mean | 2,231 | 308 | 2,539 | 1,359 | 3,897 |
| | St Dev | 40 | 593 | 591 | 42 | 588 |
| | CV | 0.02 | 1.92 | 0.23 | 0.03 | 0.15 |
| MUN | Mean | 1,845 | 90 | 1,934 | 1,797 | 3,731 |
| | St Dev | 49 | 184 | 173 | 174 | 259 |
| | CV | 0.03 | 2.05 | 0.09 | 0.10 | 0.07 |
| RTC | Mean | 1,874 | 377 | 2,251 | 1,053 | 3,304 |
| | St Dev | 38 | 635 | 615 | 71 | 621 |
| | CV | 0.02 | 1.68 | 0.27 | 0.07 | 0.19 |
| RDE | Mean | 1,745 | 421 | 2,166 | 1,092 | 3,258 |
| | St Dev | 37 | 447 | 440 | 70 | 435 |
| | CV | 0.02 | 1.06 | 0.20 | 0.06 | 0.13 |
| RDE+div | Mean | 1,750 | 236 | 1,986 | 1,089 | 3,075 |
| | St Dev | 48 | 363 | 347 | 88 | 360 |
| | CV | 0.03 | 1.54 | 0.17 | 0.08 | 0.12 |

| Strategy | | Case 3 | | | | |
|---|---|---|---|---|---|---|
| | | Inv. Holding Cost | Lost Sales Cost | Total Inv. Cost | Transp. Cost | Total Cost |
| BENCH1 | Mean | 2,215 | 194 | 2,409 | 1,417 | 3,826 |
| | St Dev | 51 | 534 | 542 | 35 | 532 |
| | CV | 0.02 | 2.76 | 0.22 | 0.02 | 0.14 |
| MUN | Mean | 2,027 | 259 | 2,286 | 1,753 | 4,039 |
| | St Dev | 83 | 458 | 419 | 158 | 429 |
| | CV | 0.04 | 1.77 | 0.18 | 0.09 | 0.11 |
| RTC | Mean | 2,101 | 280 | 2,381 | 1,079 | 3,460 |
| | St Dev | 57 | 501 | 496 | 74 | 526 |
| | CV | 0.03 | 1.79 | 0.21 | 0.07 | 0.15 |
| RDE | Mean | 2,007 | 170 | 2,177 | 1,099 | 3,276 |
| | St Dev | 43 | 273 | 277 | 70 | 299 |
| | CV | 0.02 | 1.61 | 0.13 | 0.06 | 0.09 |
| RDE+div | Mean | 2,009 | 98 | 2,107 | 1,121 | 3,228 |
| | St Dev | 40 | 228 | 230 | 71 | 247 |
| | CV | 0.02 | 2.33 | 0.11 | 0.06 | 0.08 |

30 replication with common random numbers
Results in [$/week]

**Fig. 18.** Simulation results: parameter set 9 TC = 300 [\$/hr], $h_i = 5$[\$/week], $\lambda_i = 50$[arrivals/day], $\theta_i = 1$ [units] for all i

| Strategy | | Case 0 (Symmetric case) | | | | |
|---|---|---|---|---|---|---|
| | | Inv. Holding Cost | Lost Sales Cost | Total Inv. Cost | Transp. Cost | Total Cost |
| BENCH1 | Mean | 2,007 | 3,317 | 5,324 | 11,593 | 16,917 |
| | St Dev | 63 | 2,482 | 2,445 | 550 | 2,587 |
| | CV | 0.03 | 0.75 | 0.46 | 0.05 | 0.15 |
| MUN | Mean | 2,077 | 426 | 2,503 | 14,269 | 16,772 |
| | St Dev | 95 | 442 | 427 | 1,532 | 1,648 |
| | CV | 0.05 | 1.04 | 0.17 | 0.11 | 0.10 |
| RTC | Mean | 2,132 | 172 | 2,304 | 9,507 | 11,811 |
| | St Dev | 63 | 279 | 283 | 776 | 826 |
| | CV | 0.03 | 1.62 | 0.12 | 0.08 | 0.07 |
| RDE | Mean | 2,096 | 192 | 2,288 | 9,358 | 11,646 |
| | St Dev | 36 | 301 | 290 | 491 | 642 |
| | CV | 0.02 | 1.57 | 0.13 | 0.05 | 0.06 |
| RDE+div | Mean | 2,109 | 140 | 2,249 | 9,466 | 11,715 |
| | St Dev | 44 | 251 | 251 | 597 | 621 |
| | CV | 0.02 | 1.80 | 0.11 | 0.06 | 0.05 |

| Strategy | | Case 1 | | | | |
|---|---|---|---|---|---|---|
| | | Inv. Holding Cost | Lost Sales Cost | Total Inv. Cost | Transp. Cost | Total Cost |
| BENCH1 | Mean | 1,799 | 7,725 | 9,524 | 15,720 | 25,244 |
| | St Dev | 62 | 2,998 | 2,977 | 1,000 | 3,075 |
| | CV | 0.03 | 0.39 | 0.31 | 0.06 | 0.12 |
| MUN | Mean | 2,009 | 929 | 2,937 | 18,889 | 21,826 |
| | St Dev | 67 | 1,205 | 1,168 | 1,384 | 1,469 |
| | CV | 0.03 | 1.30 | 0.40 | 0.07 | 0.07 |
| RTC | Mean | 2,060 | 1,001 | 3,060 | 12,338 | 15,399 |
| | St Dev | 62 | 837 | 852 | 853 | 1,310 |
| | CV | 0.03 | 0.84 | 0.28 | 0.07 | 0.09 |
| RDE | Mean | 2,029 | 873 | 2,902 | 12,072 | 14,974 |
| | St Dev | 39 | 745 | 741 | 945 | 1,097 |
| | CV | 0.02 | 0.85 | 0.26 | 0.08 | 0.07 |
| RDE+div | Mean | 2,031 | 493 | 2,524 | 12,172 | 14,696 |
| | St Dev | 47 | 635 | 638 | 783 | 1,046 |
| | CV | 0.02 | 1.29 | 0.25 | 0.06 | 0.07 |

| Strategy | | Case 2 | | | | |
|---|---|---|---|---|---|---|
| | | Inv. Holding Cost | Lost Sales Cost | Total Inv. Cost | Transp. Cost | Total Cost |
| BENCH1 | Mean | 2,056 | 3,174 | 5,229 | 10,616 | 15,845 |
| | St Dev | 69 | 2,274 | 2,233 | 748 | 2,359 |
| | CV | 0.03 | 0.72 | 0.43 | 0.07 | 0.15 |
| MUN | Mean | 2,022 | 2,323 | 4,345 | 12,921 | 17,266 |
| | St Dev | 101 | 2,007 | 1,966 | 1,470 | 2,024 |
| | CV | 0.05 | 0.86 | 0.45 | 0.11 | 0.12 |
| RTC | Mean | 2,089 | 685 | 2,774 | 8,874 | 11,647 |
| | St Dev | 61 | 703 | 678 | 753 | 1,110 |
| | CV | 0.03 | 1.03 | 0.24 | 0.08 | 0.10 |
| RDE | Mean | 2,058 | 430 | 2,488 | 8,747 | 11,234 |
| | St Dev | 46 | 563 | 549 | 619 | 871 |
| | CV | 0.02 | 1.31 | 0.22 | 0.07 | 0.08 |
| RDE+div | Mean | 2,065 | 380 | 2,445 | 8,881 | 11,326 |
| | St Dev | 55 | 1,100 | 1,085 | 662 | 1,236 |
| | CV | 0.03 | 2.89 | 0.44 | 0.07 | 0.11 |

| Strategy | | Case 3 | | | | |
|---|---|---|---|---|---|---|
| | | Inv. Holding Cost | Lost Sales Cost | Total Inv. Cost | Transp. Cost | Total Cost |
| BENCH1 | Mean | 1,991 | 3,514 | 5,505 | 11,643 | 17,148 |
| | St Dev | 66 | 2,453 | 2,412 | 667 | 2,464 |
| | CV | 0.03 | 0.70 | 0.44 | 0.06 | 0.14 |
| MUN | Mean | 2,050 | 880 | 2,930 | 14,527 | 17,457 |
| | St Dev | 89 | 1,169 | 1,116 | 1,470 | 2,068 |
| | CV | 0.04 | 1.33 | 0.38 | 0.10 | 0.12 |
| RTC | Mean | 2,096 | 721 | 2,816 | 9,178 | 11,995 |
| | St Dev | 62 | 807 | 831 | 689 | 1,194 |
| | CV | 0.03 | 1.12 | 0.30 | 0.08 | 0.10 |
| RDE | Mean | 2,072 | 617 | 2,688 | 9,036 | 11,724 |
| | St Dev | 48 | 637 | 632 | 733 | 1,041 |
| | CV | 0.02 | 1.03 | 0.23 | 0.08 | 0.09 |
| RDE+div | Mean | 2,056 | 493 | 2,550 | 9,121 | 11,670 |
| | St Dev | 46 | 569 | 570 | 570 | 821 |
| | CV | 0.02 | 1.15 | 0.22 | 0.06 | 0.07 |

30 replication with common random numbers
Results in [\$/week]

**Fig. 19.** Simulation results: parameter set 10 TC = 100 [$/hr], $h_i = 5$[$/week], $\lambda_i = 10.5$ [arrivals/day], $\theta_i = 4.8$ [units] for all i

| Strategy | | Case 0 (Symmetric case) | | | | |
|---|---|---|---|---|---|---|
| | | Inv. Holding Cost | Lost Sales Cost | Total Inv. Cost | Transp. Cost | Total Cost |
| BENCH1 | Mean | 2,173 | 2,082 | 4,255 | 4,119 | 8,374 |
| | St Dev | 58 | 2,259 | 2,229 | 200 | 2,281 |
| | CV | 0.03 | 1.09 | 0.52 | 0.05 | 0.27 |
| MUN | Mean | 2,227 | 808 | 3,035 | 5,007 | 8,042 |
| | St Dev | 107 | 914 | 915 | 600 | 1,130 |
| | CV | 0.05 | 1.13 | 0.30 | 0.12 | 0.14 |
| RTC | Mean | 2,395 | 658 | 3,054 | 3,782 | 6,836 |
| | St Dev | 94 | 786 | 768 | 237 | 801 |
| | CV | 0.04 | 1.19 | 0.25 | 0.06 | 0.12 |
| RDE | Mean | 2,273 | 652 | 2,925 | 3,603 | 6,528 |
| | St Dev | 57 | 1,264 | 1,250 | 297 | 1,307 |
| | CV | 0.02 | 1.94 | 0.43 | 0.08 | 0.20 |
| RDE+div | Mean | 2,281 | 602 | 2,883 | 3,620 | 6,503 |
| | St Dev | 52 | 877 | 859 | 321 | 913 |
| | CV | 0.02 | 1.46 | 0.30 | 0.09 | 0.14 |

| Strategy | | Case 1 | | | | |
|---|---|---|---|---|---|---|
| | | Inv. Holding Cost | Lost Sales Cost | Total Inv. Cost | Transp. Cost | Total Cost |
| BENCH1 | Mean | 2,000 | 3,985 | 5,986 | 5,721 | 11,707 |
| | St Dev | 68 | 2,999 | 2,958 | 257 | 2,971 |
| | CV | 0.03 | 0.75 | 0.49 | 0.04 | 0.25 |
| MUN | Mean | 2,183 | 1,023 | 3,206 | 6,661 | 9,867 |
| | St Dev | 106 | 1,241 | 1,199 | 571 | 1,162 |
| | CV | 0.05 | 1.21 | 0.37 | 0.09 | 0.12 |
| RTC | Mean | 2,419 | 726 | 3,145 | 5,181 | 8,326 |
| | St Dev | 114 | 901 | 917 | 269 | 953 |
| | CV | 0.05 | 1.24 | 0.29 | 0.05 | 0.11 |
| RDE | Mean | 2,273 | 634 | 2,907 | 5,212 | 8,119 |
| | St Dev | 52 | 906 | 900 | 553 | 1,074 |
| | CV | 0.02 | 1.43 | 0.31 | 0.11 | 0.13 |
| RDE+div | Mean | 2,292 | 637 | 2,929 | 5,080 | 8,009 |
| | St Dev | 49 | 932 | 927 | 357 | 993 |
| | CV | 0.02 | 1.46 | 0.32 | 0.07 | 0.12 |

| Strategy | | Case 2 | | | | |
|---|---|---|---|---|---|---|
| | | Inv. Holding Cost | Lost Sales Cost | Total Inv. Cost | Transp. Cost | Total Cost |
| BENCH1 | Mean | 2,199 | 2,081 | 4,280 | 3,842 | 8,122 |
| | St Dev | 75 | 1,643 | 1,630 | 196 | 1,613 |
| | CV | 0.03 | 0.79 | 0.38 | 0.05 | 0.20 |
| MUN | Mean | 2,161 | 541 | 2,702 | 4,971 | 7,674 |
| | St Dev | 110 | 707 | 663 | 487 | 926 |
| | CV | 0.05 | 1.31 | 0.25 | 0.10 | 0.12 |
| RTC | Mean | 2,408 | 466 | 2,874 | 3,459 | 6,334 |
| | St Dev | 110 | 578 | 591 | 271 | 645 |
| | CV | 0.05 | 1.24 | 0.21 | 0.08 | 0.10 |
| RDE | Mean | 2,190 | 690 | 2,881 | 3,586 | 6,466 |
| | St Dev | 43 | 946 | 949 | 484 | 1,120 |
| | CV | 0.02 | 1.37 | 0.33 | 0.13 | 0.17 |
| RDE+div | Mean | 2,195 | 364 | 2,559 | 3,431 | 5,990 |
| | St Dev | 54 | 520 | 504 | 284 | 596 |
| | CV | 0.02 | 1.43 | 0.20 | 0.08 | 0.10 |

| Strategy | | Case 3 | | | | |
|---|---|---|---|---|---|---|
| | | Inv. Holding Cost | Lost Sales Cost | Total Inv. Cost | Transp. Cost | Total Cost |
| BENCH1 | Mean | 2,159 | 2,412 | 4,571 | 4,126 | 8,697 |
| | St Dev | 72 | 2,049 | 2,017 | 225 | 2,051 |
| | CV | 0.03 | 0.85 | 0.44 | 0.05 | 0.24 |
| MUN | Mean | 2,233 | 570 | 2,804 | 5,155 | 7,959 |
| | St Dev | 101 | 737 | 729 | 533 | 836 |
| | CV | 0.05 | 1.29 | 0.26 | 0.10 | 0.11 |
| RTC | Mean | 2,442 | 710 | 3,152 | 3,805 | 6,957 |
| | St Dev | 108 | 939 | 919 | 288 | 910 |
| | CV | 0.04 | 1.32 | 0.29 | 0.08 | 0.13 |
| RDE | Mean | 2,297 | 468 | 2,765 | 3,653 | 6,418 |
| | St Dev | 50 | 758 | 746 | 264 | 790 |
| | CV | 0.02 | 1.62 | 0.27 | 0.07 | 0.12 |
| RDE+div | Mean | 2,296 | 558 | 2,854 | 3,744 | 6,598 |
| | St Dev | 45 | 864 | 868 | 298 | 899 |
| | CV | 0.02 | 1.55 | 0.30 | 0.08 | 0.14 |

30 replication with common random numbers
Results in [$/week]

**Fig. 20.** Simulation results: parameter set 11 TC $= 100$ [$/hr], $h_i = 5$[$/week], $\lambda_i = 4.35$ [arrivals/day], $\theta_i = 11.5$ [units] for all i

| Strategy | | Case 0 (Symmetric case) | | | | |
|---|---|---|---|---|---|---|
| | | Inv. Holding Cost | Lost Sales Cost | Total Inv. Cost | Transp. Cost | Total Cost |
| BENCH1 | Mean | 2,155 | 7,635 | 9,790 | 4,017 | 13,807 |
| | St Dev | 112 | 5,935 | 5,883 | 297 | 5,990 |
| | CV | 0.05 | 0.78 | 0.60 | 0.07 | 0.43 |
| MUN | Mean | 2,432 | 985 | 3,417 | 5,626 | 9,042 |
| | St Dev | 123 | 1,166 | 1,159 | 560 | 1,302 |
| | CV | 0.05 | 1.18 | 0.34 | 0.10 | 0.14 |
| RTC | Mean | 2,704 | 1,200 | 3,903 | 4,612 | 8,515 |
| | St Dev | 139 | 1,811 | 1,743 | 360 | 1,969 |
| | CV | 0.05 | 1.51 | 0.45 | 0.08 | 0.23 |
| RDE | Mean | 2,438 | 2,003 | 4,441 | 4,102 | 8,543 |
| | St Dev | 59 | 1,963 | 1,958 | 359 | 2,036 |
| | CV | 0.02 | 0.98 | 0.44 | 0.09 | 0.24 |
| RDE+div | Mean | 2,467 | 1,476 | 3,943 | 4,199 | 8,142 |
| | St Dev | 57 | 1,778 | 1,758 | 355 | 1,965 |
| | CV | 0.02 | 1.20 | 0.45 | 0.08 | 0.24 |

| Strategy | | Case 1 | | | | |
|---|---|---|---|---|---|---|
| | | Inv. Holding Cost | Lost Sales Cost | Total Inv. Cost | Transp. Cost | Total Cost |
| BENCH1 | Mean | 1,997 | 11,111 | 13,108 | 5,463 | 18,571 |
| | St Dev | 112 | 5,915 | 5,856 | 334 | 5,932 |
| | CV | 0.06 | 0.53 | 0.45 | 0.06 | 0.32 |
| MUN | Mean | 2,365 | 1,828 | 4,193 | 7,101 | 11,294 |
| | St Dev | 115 | 2,188 | 2,141 | 679 | 2,086 |
| | CV | 0.05 | 1.20 | 0.51 | 0.10 | 0.18 |
| RTC | Mean | 2,703 | 1,454 | 4,157 | 6,321 | 10,479 |
| | St Dev | 134 | 1,850 | 1,793 | 417 | 1,933 |
| | CV | 0.05 | 1.27 | 0.43 | 0.07 | 0.18 |
| RDE | Mean | 2,462 | 1,752 | 4,214 | 6,097 | 10,311 |
| | St Dev | 66 | 2,302 | 2,289 | 596 | 2,526 |
| | CV | 0.03 | 1.31 | 0.54 | 0.10 | 0.24 |
| RDE+div | Mean | 2,488 | 921 | 3,409 | 6,006 | 9,414 |
| | St Dev | 64 | 1,253 | 1,225 | 521 | 1,543 |
| | CV | 0.03 | 1.36 | 0.36 | 0.09 | 0.16 |

| Strategy | | Case 2 | | | | |
|---|---|---|---|---|---|---|
| | | Inv. Holding Cost | Lost Sales Cost | Total Inv. Cost | Transp. Cost | Total Cost |
| BENCH1 | Mean | 2,199 | 6,968 | 9,167 | 3,689 | 12,857 |
| | St Dev | 84 | 4,807 | 4,759 | 282 | 4,894 |
| | CV | 0.04 | 0.69 | 0.52 | 0.08 | 0.38 |
| MUN | Mean | 2,406 | 1,310 | 3,716 | 5,258 | 8,975 |
| | St Dev | 133 | 2,227 | 2,155 | 756 | 2,089 |
| | CV | 0.06 | 1.70 | 0.58 | 0.14 | 0.23 |
| RTC | Mean | 2,728 | 1,732 | 4,460 | 4,306 | 8,766 |
| | St Dev | 131 | 1,873 | 1,841 | 374 | 1,947 |
| | CV | 0.05 | 1.08 | 0.41 | 0.09 | 0.22 |
| RDE | Mean | 2,398 | 1,974 | 4,372 | 4,129 | 8,501 |
| | St Dev | 58 | 1,550 | 1,561 | 620 | 1,679 |
| | CV | 0.02 | 0.79 | 0.36 | 0.15 | 0.20 |
| RDE+div | Mean | 2,436 | 653 | 3,088 | 4,073 | 7,161 |
| | St Dev | 61 | 940 | 944 | 577 | 1,060 |
| | CV | 0.03 | 1.44 | 0.31 | 0.14 | 0.15 |

| Strategy | | Case 3 | | | | |
|---|---|---|---|---|---|---|
| | | Inv. Holding Cost | Lost Sales Cost | Total Inv. Cost | Transp. Cost | Total Cost |
| BENCH1 | Mean | 2,144 | 7,516 | 9,659 | 3,960 | 13,620 |
| | St Dev | 88 | 4,136 | 4,083 | 197 | 4,123 |
| | CV | 0.04 | 0.55 | 0.42 | 0.05 | 0.30 |
| MUN | Mean | 2,415 | 946 | 3,361 | 5,691 | 9,052 |
| | St Dev | 85 | 1,372 | 1,337 | 654 | 1,289 |
| | CV | 0.04 | 1.45 | 0.40 | 0.11 | 0.14 |
| RTC | Mean | 2,715 | 1,144 | 3,859 | 4,678 | 8,537 |
| | St Dev | 140 | 1,398 | 1,353 | 398 | 1,444 |
| | CV | 0.05 | 1.22 | 0.35 | 0.09 | 0.17 |
| RDE | Mean | 2,459 | 1,823 | 4,282 | 4,458 | 8,740 |
| | St Dev | 71 | 2,178 | 2,142 | 530 | 2,402 |
| | CV | 0.03 | 1.19 | 0.50 | 0.12 | 0.27 |
| RDE+div | Mean | 2,502 | 784 | 3,287 | 4,562 | 7,849 |
| | St Dev | 74 | 1,138 | 1,130 | 376 | 1,162 |
| | CV | 0.03 | 1.45 | 0.34 | 0.08 | 0.15 |

30 replication with common random numbers
Results in [$/week]

**Fig. 21.** Simulation results: parameter set 12 (TC = 100 [$/hr], $h_i = 5$[$/week], $\lambda_i = 2.4$ [arrivals/day], $\theta_i = 20.8$ [units] for all i)

| Strategy | | Case 0 (Symmetric case) | | | | |
|---|---|---|---|---|---|---|
| | | Inv. Holding Cost | Lost Sales Cost | Total Inv. Cost | Transp. Cost | Total Cost |
| BENCH1 | Mean | 2,137 | 18,784 | 20,921 | 3,758 | 24,679 |
| | St Dev | 145 | 8,613 | 8,512 | 363 | 8,639 |
| | CV | 0.07 | 0.46 | 0.41 | 0.10 | 0.35 |
| MUN | Mean | 2,673 | 1,174 | 3,847 | 6,441 | 10,288 |
| | St Dev | 118 | 1,487 | 1,460 | 671 | 1,474 |
| | CV | 0.04 | 1.27 | 0.38 | 0.10 | 0.14 |
| RTC | Mean | 2,906 | 4,250 | 7,156 | 5,436 | 12,592 |
| | St Dev | 149 | 4,277 | 4,239 | 471 | 4,401 |
| | CV | 0.05 | 1.01 | 0.59 | 0.09 | 0.35 |
| RDE | Mean | 2,626 | 3,741 | 6,367 | 4,641 | 11,008 |
| | St Dev | 82 | 3,575 | 3,563 | 385 | 3,475 |
| | CV | 0.03 | 0.96 | 0.56 | 0.08 | 0.32 |
| RDE+div | Mean | 2,660 | 2,214 | 4,874 | 4,779 | 9,653 |
| | St Dev | 66 | 2,021 | 2,034 | 517 | 2,128 |
| | CV | 0.02 | 0.91 | 0.42 | 0.11 | 0.22 |
| Strategy | | Case 1 | | | | |
| | | Inv. Holding Cost | Lost Sales Cost | Total Inv. Cost | Transp. Cost | Total Cost |
| BENCH1 | Mean | 2,017 | 17,912 | 19,928 | 5,131 | 25,059 |
| | St Dev | 127 | 7,455 | 7,373 | 349 | 7,482 |
| | CV | 0.06 | 0.42 | 0.37 | 0.07 | 0.30 |
| MUN | Mean | 2,571 | 2,220 | 4,791 | 7,824 | 12,615 |
| | St Dev | 127 | 2,385 | 2,345 | 733 | 2,368 |
| | CV | 0.05 | 1.07 | 0.49 | 0.09 | 0.19 |
| RTC | Mean | 2,866 | 3,121 | 5,986 | 7,362 | 13,349 |
| | St Dev | 129 | 3,558 | 3,512 | 560 | 3,604 |
| | CV | 0.05 | 1.14 | 0.59 | 0.08 | 0.27 |
| RDE | Mean | 2,652 | 1,920 | 4,572 | 7,253 | 11,826 |
| | St Dev | 68 | 1,667 | 1,654 | 605 | 1,739 |
| | CV | 0.03 | 0.87 | 0.36 | 0.08 | 0.15 |
| RDE+div | Mean | 2,660 | 1,946 | 4,606 | 7,451 | 12,057 |
| | St Dev | 71 | 1,832 | 1,814 | 543 | 1,889 |
| | CV | 0.03 | 0.94 | 0.39 | 0.07 | 0.16 |
| Strategy | | Case 2 | | | | |
| | | Inv. Holding Cost | Lost Sales Cost | Total Inv. Cost | Transp. Cost | Total Cost |
| BENCH1 | Mean | 2,232 | 13,941 | 16,173 | 3,398 | 19,571 |
| | St Dev | 122 | 8,261 | 8,195 | 303 | 8,219 |
| | CV | 0.05 | 0.59 | 0.51 | 0.09 | 0.42 |
| MUN | Mean | 2,617 | 2,381 | 4,997 | 5,639 | 10,636 |
| | St Dev | 132 | 2,314 | 2,324 | 711 | 2,297 |
| | CV | 0.05 | 0.97 | 0.47 | 0.13 | 0.22 |
| RTC | Mean | 2,926 | 3,228 | 6,154 | 4,816 | 10,970 |
| | St Dev | 130 | 3,391 | 3,354 | 486 | 3,479 |
| | CV | 0.04 | 1.05 | 0.54 | 0.10 | 0.32 |
| RDE | Mean | 2,616 | 2,712 | 5,328 | 4,683 | 10,011 |
| | St Dev | 74 | 2,434 | 2,426 | 759 | 2,389 |
| | CV | 0.03 | 0.90 | 0.46 | 0.16 | 0.24 |
| RDE+div | Mean | 2,669 | 1,805 | 4,474 | 4,667 | 9,142 |
| | St Dev | 78 | 1,884 | 1,861 | 604 | 2,051 |
| | CV | 0.03 | 1.04 | 0.42 | 0.13 | 0.22 |
| Strategy | | Case 3 | | | | |
| | | Inv. Holding Cost | Lost Sales Cost | Total Inv. Cost | Transp. Cost | Total Cost |
| BENCH1 | Mean | 2,168 | 14,369 | 16,537 | 3,766 | 20,303 |
| | St Dev | 97 | 7,092 | 7,049 | 289 | 7,109 |
| | CV | 0.04 | 0.49 | 0.43 | 0.08 | 0.35 |
| MUN | Mean | 2,711 | 1,360 | 4,071 | 6,624 | 10,695 |
| | St Dev | 136 | 1,965 | 1,967 | 792 | 2,176 |
| | CV | 0.05 | 1.44 | 0.48 | 0.12 | 0.20 |
| RTC | Mean | 2,917 | 2,907 | 5,824 | 5,410 | 11,235 |
| | St Dev | 121 | 4,036 | 4,015 | 386 | 4,119 |
| | CV | 0.04 | 1.39 | 0.69 | 0.07 | 0.37 |
| RDE | Mean | 2,668 | 2,066 | 4,734 | 5,480 | 10,214 |
| | St Dev | 78 | 2,432 | 2,423 | 650 | 2,565 |
| | CV | 0.03 | 1.18 | 0.51 | 0.12 | 0.25 |
| RDE+div | Mean | 2,710 | 1,175 | 3,886 | 5,405 | 9,291 |
| | St Dev | 69 | 1,753 | 1,732 | 744 | 2,099 |
| | CV | 0.03 | 1.49 | 0.45 | 0.14 | 0.23 |

30 replication with common random numbers
Results in [$/week]

# On Optimizing Production Nodes in Supply Chain Systems

Davide Giglio[1], Riccardo Minciardi[2], Simona Sacone[3], and Silvia Siri[4]

[1] Department of Communications, Computers and Systems Science, University of Genova,
Italy davide.giglio@unige.it
[2] Department of Communications, Computers and Systems Science, University of Genova,
Italy riccardo.minciardi@unige.it
[3] Department of Communications, Computers and Systems Science, University of Genova,
Italy simona.sacone@unige.it
[4] Department of Communications, Computers and Systems Science, University of Genova,
Italy silvia.siri@unige.it

**Summary.** In this chapter a supply chain system is considered, focusing in particular on a single production node. For the single node we propose a hybrid model that combines a continuous-time dynamics, associated with the production process, with a discrete-event dynamics, related to arrivals of raw materials and departures of finite products. The system state variables are given by the inventory levels of raw materials and finite products, while the decisions to be taken concern the arrival process (in terms of number of orders, quantities and ordering time instants), the production process (definition of the production function) and the departure process (determination of the delivery time instants). The resulting optimization problem is nonlinear and functional and it is decomposed into two subproblems that are studied separately (the former is related to the production and the delivery process, the latter concerns the replenishment policy). Some properties of the optimal solution of these subproblems are shown and, in some cases, the optimal control strategies are found. In the final part of the chapter, the proposed optimization approach is extended to consider the case of multi-site schemes, by introducing two simple cases of cooperative and competitive structures.

**Key words:** Supply chain systems, Hybrid model, Optimization, Replenishment problem

L. Bertazzi et al. (eds.), *Innovations in Distribution Logistics*, Lecture Notes
in Economics and Mathematical Systems 619, DOI: 10.1007/978-3-540-92944-4,
© 2009 Springer-Verlag Berlin Heidelberg

# 1 Introduction

Analysis, planning and control of distributive logistic systems represent nowadays a major research field, which has aroused great interest of companies as well. Actually, because of the ever increasing competition, companies must always be efficient and productive, with ambitious objectives such as maximizing service levels, complying with required due-dates, minimizing inventory levels at each stage of the network, minimizing transportation and infrastructure costs. For this reason, the integration of production, distribution and inventory management results to be a crucial aspect to be considered. In this context, different groups of researchers have devoted their attention to the design, analysis, optimization and management of distributed production systems, in order to define optimal decisions or coordination schemes for the different decision makers acting in such networks. All these aspects are summarized in the common expression *supply chain management*, which emphasizes the view of the company as a part of a chain composed of different stages, such as suppliers, manufacturers, assemblers, warehouses, customers [1, 2, 3].

From a modelling point of view, a supply chain can be represented in different ways, corresponding to centralized or decentralized structures, to analytical or simulation models, and so on. For the single node of a distributed production system, an important distinction is between continuous-time models and discrete-event models: in the former case, models are referred to as fluid models, in which all the quantities are represented by means of continuous variables; in the latter case, the whole system dynamics is driven by the occurring of asynchronous events which usually change the values of system state variables. Many continuous-time models have been developed for representing production systems. Among them, an important research stream consists in the determination of analytical solutions for production systems that have to meet a random demand [4, 5, 6]. Other works are relative to multi-inventory systems in which demand is supposed to be unknown but generally bounded and the proposed control schemes are aimed at defining appropriate inventory levels in order to meet such demand [7, 8]. On the other hand, discrete-event models are generally suited for representing real case studies with a high level of detail or for comparing different scenarios characterized by the presence of stochastic aspects, for which an analytical evaluation is too difficult. In [9] a discrete-event simulation model is used for an integrated product supply chain system, where different decision makers act and real-time information is provided. In [10] a simulation-based optimization framework involving simultaneous perturbation stochastic approximation is developed for supply chain systems.

The objective of the present work is the definition of a model for a single node of a supply chain and the statement and solution of a relative optimization problem. The ultimate objective of our research work consists in studying the behaviour of a supply chain system, by defining some coordination mechanisms among the different nodes of the network; as a consequence, each node needs to be analytically represented at a quite aggregate level of detail. Moreover, in our model, we want to focus on the interactions among the different nodes; this means that for each production node the process of raw material arrivals and finite product departures must be represented in detail. For all these reasons, the model that we have defined integrates some aspects of continuous-time models with other aspects of discrete-event approaches. On one hand, fluid models generally allow to determine closed-loop solutions (that is, solutions expressed as functions of the current system state) for planning/control problems, while they normally do not represent in detail the interactions with external entities (such as arrivals and departures of parts). On the other hand, discrete-event models can describe a supply chain system with a high level of detail, managing stochastic aspects as well, but they do not allow the definition of analytical solutions. Since our objective is that of defining an analytical approach for a single production node interacting with other nodes in a supply chain, the resulting model we propose is a *hybrid model* combining a continuous dynamics (corresponding to the production process) with discrete-event processes (representing the arrivals of raw materials and departures of finite products).

In this work, we propose an optimization procedure based on this hybrid model, acting at a tactical/operational decision level and being relevant to the minimization of order costs, inventory costs and costs due to deviations from the external demand; decisions concern the process of raw material arrivals, the process of finite product deliveries and the production effort. The resulting optimization problem is hard to be solved and, then,it is decomposed into two subproblems. The former subproblem refers to the determination of the optimal production effort and the optimal product departure process, whereas the latter subproblem corresponds to the determination of the optimal replenishment policy. The overall optimization problem and its decomposition have already been presented in [11, 12, 13, 14], where only the first subproblem has been fully described and dealt with. One of the main novelties of the present work is a detailed analysis of the second subproblem (which has been defined in [15]) and, thus, the definition of a solution procedure for the overall optimization problem.

As a further innovative aspect, in this work some basic multi-site structures are considered, which exploit the solution procedure determined for the single nodes of a supply chain. Generally speaking, if a supply chain is represented as a decentralized structure, several decision agents are considered, each one provided with its own information set, giving rise to either a cooperative or a competitive environment. In the case of cooperation among different agents, the main aspect to consider is the way in which the coordination of the various decisional entities may be achieved. On the contrary, in the case of competition, the decisional agents are provided not only with different information sets, but also with individual performance objectives. In this work, two simple schemes are presented: a competitive environment of two parallel producers which compete for serving a customer and a cooperative framework of two producers belonging to subsequent stages in the supply chain.

The paper is organized as follows. In Sect. 2 the proposed model and optimization problem for a production node in a supply chain system is defined. The decomposition of the optimization problem into two subproblems is also discussed in the same section. Sections 3 and 4 are devoted to the statement and solution of the first and second subproblem, respectively. Some simple multi-site structures are discussed in Sect. 5, where the described optimization procedure is adopted and exploited by different decision agents in the network.

## 2 The Optimization Problem for a Single Node in a Supply Chain

A single node of a supply chain is here modeled as a production center, where raw parts arriving from suppliers (or from previous nodes in the chain) are manufactured and immediately transformed into final products, without considering any assembly operation. More specifically, each part entering the single node is supposed to be processed by a single operation and, then, it is transformed into one product. A schematization of this model is represented in Fig. 1, where the two inventories refer to raw materials and final products, respectively; the corresponding inventory levels at time $t$ are indicated with variables $\xi(t)$ and $x(t)$.

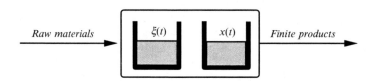

**Fig. 1.** Schematization of the production node

As previously specified in the Introduction, the proposed model is a hybrid model, which combines a continuous-time dynamics, related to the production process, together with a discrete-event dynamics, associated with arrivals of raw materials and departures of finite products. The production process is represented by a production effort $k(t)$, which models the portion of the overall work-capacity $K$ assigned to the production at time $t$. On the other hand, arrivals of raw materials and departures of finite products are modelled as discrete-event processes, in which the events (arrivals and departures) are not equally spaced in time. In particular, in the arrival process, $\Gamma$ is the number of raw material arrivals, within the considered time horizon, $\delta_i$, $i = 1, \ldots, \Gamma$, is the time instant at which the $i$-th arrival takes place, and $\Theta_i$, $i = 1, \ldots, \Gamma$, is the amount of raw materials entering the node at time instant $\delta_i$, that is the $i$-th ordered quantity (Fig. 2). In an analogous way, the flow of products delivered to clients is represented as a finite sequence of departures (Fig. 3), characterized by the following quantities: $N$, that is the number of finite products requests, within the considered time horizon, $t_i$, $i = 1, \ldots, N$, that is the time instant at which the $i$-th departure of finite products occurs, $Q_i$, $i = 1, \ldots, N$, that is the amount of finite products leaving the system at time instant $t_i$. Moreover, the external demand is characterized by the due-date of the $i$-th departure of finite products (i.e., the $i$-th lot), namely $t_i^*$, $i = 1, \ldots, N$. It is assumed that the required quantities in the external demand are satisfied and then they correspond to $Q_i$.

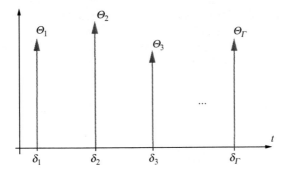

**Fig. 2.** The arrival of raw materials

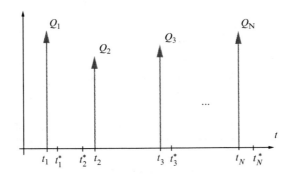

**Fig. 3.** The delivery of finite products

The system *state variables* are inventory levels $\xi(t)$ and $x(t)$, whereas *decisions* concern the following issues:

- The arrival process, i.e., how many orders to place ($\Gamma$), the ordering time instants ($\delta_i$, $i = 1, \ldots, \Gamma$), and the quantities to order ($\Theta_i$, $i = 1, \ldots, \Gamma$);
- The production process, corresponding to the function $k(t)$, $0 \leq t \leq t_N$;
- The departure process, i.e., the delivering time instants for each lot of finite products ($t_i$, $i = 1, \ldots, N$).

Taking into account the asynchronous time instants which characterize the arrival and the departure processes, the state equations of the proposed single node model can be written, for the raw material inventory and for the finite product inventory, respectively, as

$$\xi(\delta_{i+1}) = \xi(\delta_i) - \int_{\delta_i}^{\delta_{i+1}} k(t)\,dt + \Theta_{i+1} \qquad i = 0, \ldots, \Gamma - 1 \tag{1}$$

$$x(t_{i+1}) = x(t_i) + \int_{t_i}^{t_{i+1}} k(t)\,dt - Q_{i+1} \qquad i = 0, \ldots, N - 1 \tag{2}$$

where $\delta_0 = 0$, $t_0 = 0$, and $\xi(0)$ and $x(0)$ are given initial inventory levels.

The proposed optimization takes into account the order cost $C^O$ due to the acquisition of raw materials from suppliers, the cost $C^I$ relevant to the inventory occupancy, and the cost $C^T$ relevant to the deviations from the due-dates, stated, respectively, as in the following

$$C^O = \sum_{i=1}^{\Gamma} \left(c_f + c_v \Theta_i\right) \tag{3}$$

$$C^I = H_\xi \int_0^{\delta_\Gamma} \xi(t)\,dt + H_x \int_0^{t_N} x(t)\,dt \tag{4}$$

$$C^T = \alpha \sum_{i=1}^{N} (t_i - t_i^*)^2 \tag{5}$$

where $c_f$ and $c_v$ are the fixed and variable unitary order costs, respectively, $H_\xi$ and $H_x$ are the unitary inventory costs for raw materials and finite products, respectively, while $\alpha$ is a suitable parameter, weighing the deviation of the delivering times from the corresponding due-dates. As regards the expression of cost term $C^T$, note that early and late deliveries are equally penalized. Every deviation from the corresponding due-date yields a decrease of the node service level, which results to be a crucial performance indicator in a supply chain system. This is the reason why a quadratic form of the cost has been chosen, aiming at strongly penalizing deviations from due-dates.

The overall optimization problem can then be stated as follows.

**Problem 1.** Given the initial conditions $\delta_0 = 0$, $t_0 = 0$, $\xi(0) \geq 0$, and $x(0) \geq 0$, find

$$\min_{\substack{\Gamma,\, \delta_i, \Theta_i,\, i=1,\dots,\Gamma \\ t_i,\, i=1,\dots,N \\ k(t),\, 0 \leq t \leq t_N}} C_1 = C^O + C^I + C^T$$

subject to (1), (2), and

$$0 \leq k(t) \leq K \qquad 0 \leq t \leq t_N \tag{6}$$

$$\delta_{i+1} > \delta_i \qquad i = 0, \dots, \Gamma - 1 \tag{7}$$

$$t_{i+1} > t_i \qquad i = 0, \dots, N - 1 \tag{8}$$

$$\xi(t) \geq 0 \qquad 0 < t \leq \delta_\Gamma \tag{9}$$

$$x(t) \geq 0 \qquad 0 < t \leq t_N \tag{10}$$

$$\Theta_i > 0 \qquad i = 1, \dots, \Gamma \tag{11}$$

$\square$

Problem 1 is a *nonlinear functional optimization problem*, thus a simplified version of this problem is needed. The heuristic approach proposed here is that of decomposing it into two subproblems (of course, this decomposition approach leads to a sub-optimal solution of Problem 1):

1. The first subproblem consists in minimizing the inventory cost for final products and the deviations from due-dates; the decision variables regard the production effort $(k(t), 0 \leq t \leq t_N)$ and the departure process $(t_i,\ i = 1, \ldots, N)$, assuming to have *unlimited available raw materials*;
2. The second subproblem is relevant to the minimization of the inventory cost for raw materials and order costs; the decision variables are associated with the arrival process $(\Gamma, \delta_i$ and $\Theta_i, i = 1, \ldots, \Gamma)$, with the *fixed production effort* coming as a solution of the first subproblem.

By defining

$$C_2 = H_x \int_0^{t_N} x(t)dt + \alpha \sum_{i=1}^N (t_i - t_i^*)^2 \tag{12}$$

$$C_3 = \sum_{i=1}^{\Gamma} \left( c_f + c_v \Theta_i \right) + H_\xi \int_0^{\delta_\Gamma} \xi(t)dt \tag{13}$$

the first and the second subproblem can be stated, respectively, as follows.

**Problem 2.** Given the initial conditions $t_0 = 0$ and $x(0) \geq 0$, find

$$\min_{\substack{t_i, i=1,\ldots,N \\ k(t), 0 \leq t \leq t_N}} C_2$$

subject to (2), (6), (8) and (10).

□

**Problem 3.** Given the initial conditions $\delta_0 = 0$ and $\xi(0) \geq 0$, and given $k(t) = k^o(t)$ solution of Problem 2, find

$$\min_{\Gamma, \delta_i, \Theta_i, i=1,\ldots,\Gamma} C_3$$

subject to (1), (7), (9), and (11).

□

Note that Problem 2 still is a functional optimization problem, while Problem 3 is a parametric optimization problem. In the following, these two problems will be studied, in order to find some properties of their optimal solutions and, thus, to define solution procedures for the two problems. Problem 2 will be analysed in Sect. 3, whereas Problem 3 will be studied in Sect. 4.

## 3 Solution of Problem 2

In this section, Problem 2 is analysed and some fundamental results about the optimal solution of the problem will be reported. The proofs of the theorems are only sketched in the present section, whereas their complete version can be found in [14].

As previously pointed out, Problem 2 is a *functional optimization problem*, as a function $k(t)$ has to be determined over the considered optimization interval. A first simple result can be provided, in order to convert Problem 2 into a more tractable (parametric) optimization problem. Note that, in the following, $k^\circ(t)$ refer to the optimal pattern of the decision variable $k(t)$.

**Proposition 1.** *In the optimal solution of Problem 2, the function $k(t)$ is such that in each time interval between two subsequent delivery instants, e.g., $(t_i, t_{i+1}]$, the following conditions hold*

$$k^\circ(t) \equiv 0, \quad t_i < t < t_i + \tau_i, \quad i = 0, \ldots, N - 1 \tag{14}$$

$$k^\circ(t) \equiv K, \quad t_i + \tau_i < t \leq t_{i+1}, \quad i = 0, \ldots, N - 1 \tag{15}$$

*for some $\tau_i$ such that $0 \leq \tau_i \leq t_{i+1} - t_i$. Note that the value $k^\circ(t_i + \tau_i)$ is irrelevant, and thus it is not necessary to precise it in the statement of the proposition; it is possible to set either $k^\circ(t_i + \tau_i) = 0$ or $k^\circ(t_i + \tau_i) = K$, indifferently.*

$\square$

*Proof.* Suppose, *ab absurdo*, that an optimal solution of Problem 2 exists that does not satisfy conditions (14), (15). This implies that a time interval $\tau_i$ such that $0 \leq \tau_i \leq t_{i+1} - t_i$ exists, where one or both of the following conditions hold:

(a) there is an interval $(t_i + \tau_i, t_{i+1})$ of nonzero length in which the value of $k(t)$ is not constantly its maximum value, that is,

$$k(t) \not\equiv K, \quad t_i + \tau_i < t \leq t_{i+1}$$

with $\tau_i < t_{i+1} - t_i$
(b) there are some intervals (at least one) of nonzero length, preceding time instant $t_i + \tau_i$, in which the value of $k(t)$ is not identically zero, that is,

$$k(t) \not\equiv 0, \quad \bar{t}_i < t < \bar{\bar{t}}_i$$

with $t_i \leq \bar{t}_i < \bar{\bar{t}}_i \leq t_i + \tau_i$.

Then, in case condition (a) occurs, it is immediate to understand that a new solution can be obtained by "reducing" the length of the interval $(t_i + \tau_i, t_{i+1})$, and imposing that in such interval the production effort is at its maximum value, i.e.,

$$k(t) \equiv K, \quad t_i + \tau_i' < t \le t_{i+1}$$

being $\tau_i' > \tau_i$, and $\int_{t_i + \tau_i}^{t_{i+1}} k(t) \, dt = \int_{t_i + \tau_i'}^{t_{i+1}} k(t) \, dt$.

Evidently, this new solution is characterized by the same value of cost $C^T$, but it has a lower value of cost $C^I$, then the original solution cannot be optimal.

Similar considerations apply in connection with condition b), or even with reference to the combination of the two conditions.  ∎

The result provided by Proposition 1 allows to convert the functional optimization problem into a parametric one, by restricting the search for optimal solutions of Problem 2 to those characterized by functions $k(t)$ satisfying conditions (14), (15). Thus, it is possible to define

- $\tau_i$, $i = 0, \ldots, N - 1$, that represents the (nonnegative) *idle time* between $t_i$ and $t_{i+1}$;
- $T_i$, $i = 0, \ldots, N - 1$, that is the (nonnegative) *production time* between $t_i$ and $t_{i+1}$.

Thanks to Proposition 1, Problem 2 can also be stated as a *multistage optimal control problem*. First of all, the following assumptions will be made concerning the parameters characterizing Problem 1:

- The initial inventory level is all consumed to satisfy the first order, that is

$$x(0) < Q_1 \tag{16}$$

This assumption is not restrictive since, if this condition is not verified, in the optimal solution of Problem 2 no production is realized ($T_i = 0$) for a certain number of orders starting from the first one; then, the beginning of the sequence of orders can be simply shifted onward till meeting condition (16);

- The two terms of cost function $C_2$ are "well balanced", that is the inventory cost has not a prevailing effect with respect to the deviation cost from due-dates; this corresponds to suppose that

$$H_x K < \alpha \tag{17}$$

If such a condition were not fulfilled, then the optimization problem would be of poor interest.

Taking into account Proposition 1 and on the basis of the above considerations, cost function $C_2$ can now be written as

$$C_4 = H_x \sum_{i=0}^{N-1} \left( x(t_i)(\tau_i + T_i) + \frac{K T_i^2}{2} \right) + \alpha \sum_{i=0}^{N-1} \left( \tau_i + T_i + t_i - t_{i+1}^* \right)^2 \tag{18}$$

Furthermore, the state equation (2) and the time instants $t_i$, $i = 1, \ldots, N$ can be expressed, respectively, as

$$x(t_i) = x(t_{i-1}) + K T_{i-1} - Q_i \quad i = 1, \ldots, N \tag{19}$$

$$t_i = t_{i-1} + \tau_{i-1} + T_{i-1} \quad i = 1, \ldots, N \tag{20}$$

On these bases, Problem 2 can be re-stated as follows.

**Problem 4.** Given the initial conditions $t_0 = 0$ and $x(0) \geq 0$, find

$$\min_{\tau_i, T_i, i=0,\ldots,N-1} C_4$$

where $C_4$ is given by (18), subject to (19), (20) and

$$T_i \geq 0 \qquad i = 0, \ldots, N-1 \tag{21}$$

$$\tau_i \geq 0 \qquad i = 0, \ldots, N-1 \tag{22}$$

$$x(t_i) \geq 0 \qquad i = 1, \ldots, N \tag{23}$$

$\square$

It is apparent that Problem 4 is structured into $N$ stages, being $x(t_i)$ and $t_i$ the state variables at stage $i$, $\tau_{i-1}$ and $T_{i-1}$ the control variables acting at the same stage. Obviously, Problem 4 can be also viewed as a *mathematical programming problem* with non linear objective and non linear constraints. It can be solved by mathematical programming solvers, yielding the optimal control law in an open-loop form for a specific value of the initial conditions. In this work, instead, we are interested in adopting optimal control strategies defined as functions of the system state, that is, solutions typically denoted as closed-loop ones. Then, we will find the solution of Problem 4 as a set of optimal *feedback control strategies*. In order to do that, it is first of all necessary to discuss some significant properties of the optimal solution of Problem 4. Note that in the following propositions the values $T_i^\circ$ and $\tau_i^\circ$, $i = 0, \ldots, N-1$, refer to the optimal values of the decision variables $T_i$ and $\tau_i$, $i = 0, \ldots, N-1$, respectively.

**Proposition 2.** *In the optimal solution of Problem 4, the decision variable $T_i$ is always positive, that is*

$$T_i^\circ > 0 \quad i = 0, \ldots, N-1 \tag{24}$$

$\square$

*Proof.* The value of the inventory level just after the generic delivery time instant $t_i$ is given by

$$x(t_i) = x(t_{i-1}) + KT_{i-1} - Q_i \geq 0, \qquad i = 1, \ldots, N \tag{25}$$

Note that, in an optimal solution of Problem 4, the inequality $x(t_{i-1}) < Q_i$ must hold. In fact, if $x(t_{i-1})$ were greater or equal to $Q_i$, this would imply that the *whole* quantity of products required at the delivery time $t_i$ has been manufactured during the time intervals preceding $t_{i-1}$ (remember also that condition (16) prevents the initial inventory contents from being still partially available after the first delivery time instant). But, due to the structure of the cost function (including the inventory cost), this is never convenient since it makes the cost function value increase without providing any advantage. Then, some part of the required quantity $Q_i$ has to be produced just during time interval $(t_{i-1}, t_i)$, that is, $x(t_{i-1}) < Q_i$, as above claimed. Thus, by taking into account (25), condition $T^\circ_{i-1} > 0$, $i = 1, \ldots, N$, is proved.

∎

**Proposition 3.** *In the optimal solution of Problem 4, if $\tau_i > 0$ for some $i \in \{1, \ldots, N - 1\}$, that is, if the i-th time interval $(t_i, t_{i+1})$ includes a nonzero idle time, then the inventory level $x(t_i)$ (that is, just after the delivery at time instant $t_i$) is zero.*

□

*Proof.* Suppose, *ab absurdo*, $x(t_i) = \bar{x} > 0$ and $\tau_i > 0$, in an optimal solution. Products corresponding to this inventory level $x(t_i)$ have been realised in time interval $(t_{i-1}, t_i)$ or in previous time intervals, and they cannot be part of the initial inventory that is all consumed, by assumption, to satisfy the first order. These products are not necessary for order deliveries realised in $t_i$ or before and certainly they are not all necessary for order deliveries after $t_i$, owing to condition $\tau_i > 0$. This implies that a solution with $\tau_i > 0$ and $x(t_i) = \bar{\bar{x}}$ such that $0 \le \bar{\bar{x}} < \bar{x}$ is feasible and guarantees the same deliveries in the same time instants (thus yielding the same value of cost $C^T$), but with a lower value of the inventory cost $C^I$. This proves that any solution with $\tau_i > 0$ and $x(t_i) > 0$ cannot be optimal.

∎

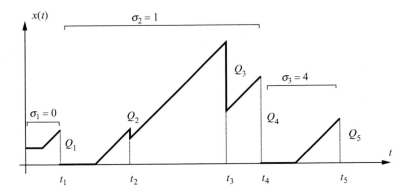

**Fig. 4.** Behaviour of the state variable $x(t)$ in the optimal solution

On the basis of the above propositions, it turns out that the behaviour of the state variable $x(t)$ in the optimal solution is of the type represented in Fig. 4. Note that the overall sequence of time intervals can be decomposed into a set of $S$ ($1 \le S \le N$) subsequences of time intervals in which the resource works at the maximum production effort, except in a portion of the first time interval of the subsequence for which $k(t) = 0$, $t_i \leq t \leq t_i + \tau_i$, being $\tau_i > 0$. Note also that, save the first time interval $(t_0, t_1)$ (which may include, for the possible presence of a nonzero initial inventory level, a time interval in which $k(t) = 0$ with a nonzero inventory level), for all subsequent idle intervals the inventory level must be equal to zero, as represented in Fig. 4.

For notational purposes, still referring to an optimal solution of Problem 4, a set $\varXi$ is defined, gathering the indexes corresponding to the beginning of the different subsequences, namely:

$$\varXi = \{\sigma_j, \ j = 1, \ldots, S\} \tag{26}$$

where $\sigma_1 = 0$ and obviously $\tau_{\sigma_j} > 0$, $j = 2, \ldots, S$, by definition of the set $\varXi$.

The $S$ subsequences correspond to $\{\sigma_1, \ldots, \sigma_2 - 1\}, \{\sigma_2, \ldots, \sigma_3 - 1\}, \ldots, \{\sigma_S, \ldots, \sigma_{S+1} - 1\}$. Moreover, for the sake of notational convenience, define $\sigma_{S+1} = N$. For instance, in the example reported in Fig. 4, it turns out $S = 3$, $\varXi = \{0, 1, 4\}$ and the three subsequences are defined as $\{0\}$, $\{1, 2, 3\}$, $\{4\}$.

Note that the decomposition of the overall optimization horizon into $S$ subsequences of time intervals allow to decompose the problem of finding the optimal control strategies at each stage $i$, $i = 0, \ldots, N - 1$, into $S$ independent control problems, each conditioned by the initial values, i.e., $t_{\sigma_j}$, $x(t_{\sigma_j})$, $j = 1, \ldots, S$. In order to find the optimal solution of Problem 4 it is then necessary to first solve the problem by mathematical programming to find the positive idle times, hence the set $\varXi$. Once known this set $\varXi$, optimal strategies for each subsequence are derived, as explained in the following.

For the sake of simplicity, from now on, the weights appearing in the cost function will be expressed in a more compact form, introducing the term $\gamma = \alpha \backslash H_x$. Condition (17) then yields:

$$K < \gamma \tag{27}$$

**Theorem 1.** *Consider a generic j-th subsequence composed of o > 1 orders (with $o = \sigma_{j+1} - \sigma_j$), the optimal solution at each stage of the subsequence is the following.*

*Case 1) For the first stage of the subsequence, that is stage $(i - 1) = \sigma_j$:*

- *if the inventory level at the end of stage $(i - 1)$ is positive:*

$$T_{i-1}^{\circ} = t_i^* - \frac{1}{v+1} C_{i+1} - \frac{(v+2)K - 2(v+1)\gamma}{2(v+1)\gamma K} Q_i + \frac{K - 2(v+1)\gamma}{2(v+1)\gamma K} x(t_{i-1}) \tag{28}$$

$$\tau_{i-1}^{\circ} = \frac{1}{v+1} C_{i+1} - t_{i-1} + \frac{K - 2(v+1)\gamma}{2(v+1)\gamma K} Q_i - \frac{K - 2(v+1)\gamma}{2(v+1)\gamma K} x(t_{i-1}) \tag{29}$$

- *if the inventory level at the end of stage $(i - 1)$ is zero:*

$$T_{i-1}^{\circ} = \frac{Q_i - x(t_{i-1})}{K} \tag{30}$$

$$\tau_{i-1}^{\circ} = \frac{1}{v+1} t_i^* + \frac{1}{v+1} C_{i+1} - t_{i-1} - \frac{Q_i}{K} - \frac{K - 2(v+1)\gamma}{2(v+1)\gamma K} x(t_{i-1}) \tag{31}$$

*where v is the total number of stages in the subsequence whose inventory level goes
to zero, and the generic constant $C_{i+1}$ is calculated iteratively depending on data of
the whole subsequence, following these steps:*

- *for $i = \sigma_{j+1}$:*

$$C_i = t_i^* - \frac{Q_i}{K} \tag{32}$$

- *from $i = \sigma_{j+1} - 1$ backward to $i = \sigma_j + 3$:*
  - *if the inventory level at the end of the stage i is positive:*

$$C_i = C_{i+1} - \frac{v_i + 1}{K} Q_i + \frac{Q_i}{2\gamma} \tag{33}$$

  - *if the inventory level at the end of the stage i is zero:*

$$C_i = C_{i+1} + t_i^* - \frac{v_i + 1}{K} Q_i \tag{34}$$

*where $v_i$ is the number of stages whose inventory is equal to zero, from the stage
i (included) to the last stage of the subsequence.*

*Case 2) For the intermediate stage of the subsequence, that is stage $(i - 1) = \sigma_j +
1, \ldots, \sigma_{j+1} - 2$, when $o > 2$:*

- *if the inventory level at the end of stage $(i - 1)$ is positive:*

$$T_{i-1}^\circ = t_i^* - t_{i-1} - \frac{Q_i}{2\gamma} \qquad \tau_{i-1}^\circ = 0 \tag{35}$$

- *if the inventory level at the end of stage $(i - 1)$ is zero:*

$$T_{i-1}^\circ = \frac{Q_i - x(t_{i-1})}{K} \qquad \tau_{i-1}^\circ = 0 \tag{36}$$

*Case 3) For the last stage of the subsequence, that is stage $(i - 1) = \sigma_{j+1} - 1$:*

$$T_{i-1}^\circ = \frac{Q_i - x(t_{i-1})}{K} \qquad \tau_{i-1}^\circ = 0 \tag{37}$$

□

*Sketch of the proof.* The results stated in the Theorem have been found applying
dynamic programming techniques, starting from the last interval (stage) of a subsequence and proceeding backwards up to the first interval (stage). First of all, the last
stage of a subsequence, that is stage $(n - 1)$, is considered and the relative problem is
stated, as a function of the initial conditions $t_{n-1}$, $x(t_{n-1})$, and of the decision variables
$\tau_{n-1}$ and $T_{n-1}$. In this case, the optimal values of the decision variables are found, as
stated in the Theorem, and the optimal cost-to-go at stage $(n - 1)$ is obtained.

Then, the intermediate stage is considered, but different cases must be analysed. Actually, the intermediate stage can be stage $(n - 2)$, or $(n - 3)$, or previous stages; thus, each of these stages has been considered, starting from stage $(n - 2)$. In this case, the problem is stated adding the cost-to-go at stage $(n - 1)$, already computed, and the optimal solutions are found considering first-order Kuhn–Tucker conditions; two cases arise, depending on whether the inventory level at the end of stage $(n - 2)$, i.e., $x(t_{n-1})$, is positive or null. Therefore, two optimal solutions are found and two costs-to-go are determined, corresponding to the two cases. When considering the intermediate stage $(n - 3)$, two different problems must be stated according to the behaviour of the optimal solution at stage $(n - 2)$ (either $x(t_{n-1}) > 0$ or $x(t_{n-1}) = 0$), which correspond to two different costs-to-go. For each of the two problems stated for stage $(n-3)$ two different solutions are found, corresponding to whether $x(t_{n-2})$ is positive or null (this means that four different cases are developed for stage $(n - 3)$). Generalising this approach to a generic intermediate stage of a subsequence, the result stated in the Theorem is proved.

The optimal values of the decision variables for the first stage of a subsequence are computed following the same reasoning line already described for the intermediate stage, since the first stage of the subsequence can be either stage $(n-2)$, or $(n-3)$ or previous stages, depending on the length of the subsequence. As a matter of fact, the problems to be solved at each stage are the same for the intermediate and the first stage, while the solutions are different because they correspond to different cases of Kuhn–Tucker conditions.

■

**Theorem 2.** *Consider a generic $j$-th subsequence, composed of $o = 1$ orders (with $o = \sigma_{j+1} - \sigma_j$), the optimal solution at stage $(i - 1) = \sigma_j$ is:*

$$T^\circ_{i-1} = \frac{Q_i - x(t_{i-1})}{K} \tag{38}$$

$$\tau^\circ_{i-1} = t^*_i - t_{i-1} - \frac{Q_i - x(t_{i-1})}{K} - \frac{x(t_{i-1})}{2\gamma} \tag{39}$$

□

*Sketch of the proof.* The previous statement is proved considering the last stage of the subsequence, i.e., stage $(n - 1)$, which is both the first and the last stage (since the considered subsequence is composed of 1 order). First-order Kuhn–Tucker conditions are developed and the optimal values of the decision variables are found.

■

# 4 Solution of Problem 3

Once Problem 4 has been solved, in particular finding the feedback control law, Problem 3 still needs to be studied. As previously described and as it is clear in the statement of Problem 3, the optimal solution of Problem 4 provides some data characterizing Problem 3, related to $k(t)$. A major feature of the optimal solution of Problem 4 is the optimal behaviour of the production effort $k(t)$ that turns out to be either null or equal to its maximum value $K$, as depicted in Fig. 5. The set of subsequences defined by (26) and identified when solving Problem 4 corresponds to a set of $S$ time intervals during which production is active (and, correspondingly, $S$ idle periods). For the sake of simplicity, quantities $\mu_j$ and $\rho_j$ are introduced and defined as follows:

$$\mu_j = t_{\sigma_j} + \tau^\circ_{\sigma_j} \qquad j = 1, \ldots, S \tag{40}$$

$$\rho_j = t_{\sigma_{j+1}} \qquad j = 1, \ldots, S \tag{41}$$

It is straightforward that

$$k(t) = \begin{cases} 0 & t \in (\rho_{j-1}, \mu_j] \\ K & t \in (\mu_j, \rho_j] \end{cases} \qquad j = 1, \ldots, S \tag{42}$$

where $\rho_0 = 0$. The time intervals $(\mu_j, \rho_j]$, $j = 1, \ldots, S$, will be in the following denoted as *production intervals*.

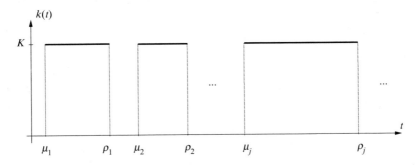

**Fig. 5.** Plot of $k(t)$ in the optimal solution of Problem 2

With the above assumptions and considerations, some important properties of the optimal solution of Problem 3 can now be derived, as provided in the following propositions.

**Proposition 4.** *In the optimal solution of Problem 3, no material arrival occurs in the intervals in which $k(t) = 0$, i.e.,*

$$\delta^\circ_i \notin (\rho_{j-1}, \mu_j] \qquad i = 1, \ldots, I \qquad j = 1, \ldots, S \tag{43}$$

□

*Proof.* Suppose, ab absurdo, that $\delta_i \in (\rho_{j-1}, \mu_j]$ in the optimal solution of Problem 3. It is straightforward that a solution with $\delta_i = \mu_j$ would be feasible (since no production is realized in $(\delta_i, \ \mu_j])$ and it would be also characterized by the same order cost and by a lower inventory cost. This proves that any solution with $\delta_i \in (\rho_{j-1}, \ \mu_j]$ cannot be optimal.

∎

The above proposition states that it is never convenient to have material arrivals in the time intervals in which production is not active. This fact is actually quite intuitive since it can be easily argued that placing orders in such time intervals would only yield an increased inventory cost. This is actually the same reasoning line that leads to prove the following result.

**Proposition 5.** *In the optimal solution of Problem 3, the raw materials arrived at time $\delta_{i-1}$ are all consumed in the production process before the following arrival occurs at $\delta_i$. This means that*

$$\xi^\circ(\delta_i^-) = 0 \qquad i = 1, \ldots, \Gamma \tag{44}$$

$$\xi^\circ(\delta_i^+) = \Theta_i \qquad i = 1, \ldots, \Gamma \tag{45}$$

□

*Proof.* Suppose, ab absurdo, that in the optimal solution of Problem 3, the inventory level in the time instant $\delta_i$ is positive. It is easy to verify that a solution in which the $i$-th order is placed in the time instant $\bar{\delta}_i > \delta_i$ (with $\bar{\Theta}_i = \Theta_i$) and such that the inventory level in $\bar{\delta}_i$ is equal to zero would be feasible and would guarantee a lower inventory cost (with the same order cost). This proves that any solution in which a generic order is placed when the inventory level is still positive cannot be optimal.

∎

Proposition 4 implies that the inventory level $\xi(t)$ keeps constant during the time intervals $(\rho_{j-1}, \ \mu_j]$, $j = 1, \ldots, S$, thus yielding

$$\xi(\mu_j) = \xi(\rho_{j-1}) \qquad j = 1 \ldots, S \tag{46}$$

Moreover, thanks to Proposition 5, it is possible to state that in the time instant in which the $i$-th material arrival occurs, the inventory level changes instantaneously from 0 to the ordered quantity $\Theta_i$. A possible behaviour of $\xi(t)$ is then depicted in Fig. 6.

In the following, some results will be provided concerning a generic production interval $j$, for which a cost $C_j$ is considered, made of the fixed order cost and the inventory cost. Actually, the variable order cost (which has been considered in the

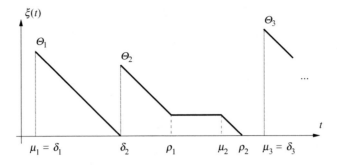

**Fig. 6.** A possible pattern of $\xi(t)$

definition of Problem 3) can be neglected when searching for some properties of the optimal solution of Problem 3. This is motivated by the fact that the variable order cost (obtained as the sum of the ordered quantities $\Theta_i$, $i = 1, \ldots, \Gamma$, times a cost term $c_v$) does not depend on the problem decision variables. This fact is due to the following consideration. The decision variables $\Theta_i$, $i = 1, \ldots, \Gamma$, can be derived if the order time instants $\delta_i$, $i = 1, \ldots, \Gamma$, are known. The quantity $\Theta_i$ to order at each time instant $\delta_i$ must assure that the inventory level $\xi(t)$ never becomes negative and that it becomes equal to 0 exactly in $\delta_{i+1}$, as stated in Proposition 5. Thus, supposing to have a null initial inventory level, it must be

$$\sum_{j=1}^{S} K(\rho_j - \mu_j) = \sum_{i=1}^{\Gamma} \Theta_i \tag{47}$$

The variable order cost

$$\sum_{i=1}^{\Gamma} c_v \Theta_i \tag{48}$$

can be written as

$$c_v \sum_{i=1}^{\Gamma} \Theta_i = c_v K \sum_{j=1}^{S} (\rho_j - \mu_j) \tag{49}$$

thus, it depends only on problem data. For this reason, this cost term will be from now on neglected.

Consider a generic $j$-th production interval, with an initial inventory $\xi(\mu_j)$ and a final inventory $\xi(\rho_j)$, as depicted in Fig. 7. From now on, we define as $n_j$ the number of orders to be placed in the $j$-th interval and $\delta_{j,i}$, $i = 1, \ldots, n_j$, the time instant in which the $i$-th order is placed in the $j$-th production interval. Of course, it must be

$$\sum_{j=1}^{S} n_j = \Gamma \tag{50}$$

Analogously, the corresponding ordered quantities will be referred to as $\Theta_{j,i}$, $i = 1, \ldots, n_j$. With these considerations, in a production interval $j$, there are $n_j + 1$ reorder

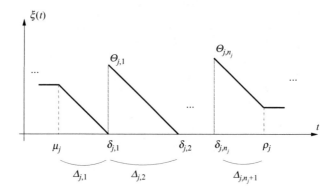

**Fig. 7.** $\xi(t)$ in a generic production interval $j$

periods, whose length is denoted as $\Delta_{j,i}$, $i = 1, \ldots, n_j + 1$. In the following proposition, a property relevant to the values of these quantities in the optimal solution is reported.

**Proposition 6.** *In the optimal solution of Problem 3, if $n_j$ orders ($n_j > 1$) are placed within the $j$-th production interval, then the time intervals between two subsequent order deliveries inside the interval (except the first and the last one) have the same length. The reorder periods are obtained as a function of the initial and final inventory level, $\xi(\mu_j)$ and $\xi(\rho_j)$, and of the number of orders $n_j$.*

□

*Proof.* The generic cost $C_j$ associated with the interval $j$ can be written as

$$C_j = n_j c_f + H_\xi \left[ \frac{1}{2} \Delta_{j,1} \xi(\mu_j) + \sum_{i=2}^{n_j} \frac{1}{2} K \Delta_{j,i}^2 + \frac{1}{2} K \Delta_{j,n_j+1}^2 + \Delta_{j,n_j+1} \xi(\rho_j) \right] \qquad (51)$$

where, thanks to Proposition 5, $\Delta_{j,1}$ is given by

$$\Delta_{j,1} = \frac{1}{K} \xi(\mu_j) \qquad (52)$$

Moreover, variables $\Delta_{j,i}$ are related by the following expression

$$\sum_{i=1}^{n_j+1} \Delta_{j,i} = \rho_j - \mu_j \qquad (53)$$

which, considering (52), becomes

$$\frac{1}{K}\xi(\mu_j) + \frac{1}{K}\sum_{i=2}^{n_j+1} \Delta_{j,i} = \rho_j - \mu_j \tag{54}$$

that can be written as

$$h_j = \frac{1}{K}\xi(\mu_j) + \frac{1}{K}\sum_{i=2}^{n_j+1} \Delta_{j,i} - \rho_j + \mu_j = 0 \tag{55}$$

The problem concerning interval $j$ can be stated as follows, being $\Delta_{j,i}$, $i = 2,\dots,n_j+1$ the decision variables, while $\xi(\rho_j)$, $\xi(\mu_j)$ and $n_j$ are considered as known data

$$\min_{\Delta_{j,i},\, i=2,\dots,n_j+1} C_j$$

where $C_j$ is given by (51), subject to

$$h_j = 0 \tag{56}$$

$$\Delta_{j,i} > 0 \qquad j = 2,\dots,n_j+1 \tag{57}$$

with $h_j$ defined as in (55).

This problem is a quadratic programming problem, characterized by a quadratic objective function, $n_j$ decision variables and one linear equality constraint. Considering Kuhn–Tucker conditions for this problem, one Lagrangian multiplier, i.e., $\lambda$, must be considered, leading to the following equations:

$$\frac{\partial C_j}{\Delta_{j,i}} + \lambda\frac{\partial h_j}{\Delta_{j,i}} = 0 \qquad i = 2,\dots,n_j+1 \tag{58}$$

which become

$$K\Delta_{j,i} + \lambda = 0 \qquad i = 2,\dots,n_j \tag{59}$$

$$K\Delta_{j,n_j+1} + \xi(\rho_j) + \lambda = 0 \tag{60}$$

These equations, together with the relation (54), let us obtain $\Delta_{j,i}$, $i = 2,\dots,n_j+1$ as

$$\Delta_{j,i} = \Delta_j = \frac{1}{Kn_j}\left[K(\rho_j - \mu_j) + \xi(\rho_j) - \xi(\mu_j)\right] \qquad i = 2,\dots,n_j \tag{61}$$

$$\Delta_{j,n_j+1} = \Delta_j - \frac{1}{K}\xi(\rho_j) \tag{62}$$

■

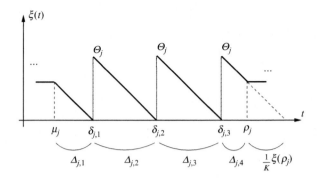

**Fig. 8.** The optimal behaviour of $\xi(t)$ for the case $n_j = 3$

The result proposed in Proposition 6 implies that, in a given production interval $j$, the reorder periods are equal, except the first and the last one. In the following, such intermediate reorder periods will be referred to as $\Delta_j, j = 1, \ldots, S$ (without using $\Delta_{j,i}$). Moreover, the ordered quantities are the same (an example is provided in Fig. 8 for the case of $n_j = 3$). Therefore, also the ordered quantities can be simply indexed by $j$ (instead of using the pair $j, i$), and will be in the following referred to as $\Theta_j$ such that:

$$\Theta_j = \frac{1}{n_j} \left[ K(\rho_j - \mu_j) + \xi(\rho_j) - \xi(\mu_j) \right] \tag{63}$$

The statement of Proposition 6 does not consider the case in which $n_j = 1$, that is when only one order is placed in the $j$-th production interval. In this case, the reorder periods can be simply derived by remembering the result of Proposition 5 (implying that the inventory level goes to zero before reordering again). As also depicted in Fig. 9, the following relations hold:

$$\Delta_{j,1} = \frac{\xi(\mu_j)}{K} \tag{64}$$

$$\Delta_{j,2} = \rho_j - \mu_j - \frac{\xi(\mu_j)}{K} \tag{65}$$

$$\Theta_j = K(\rho_j - \mu_j) + \xi(\rho_j) - \xi(\mu_j) \tag{66}$$

The propositions previously reported define some significant properties of the optimal solution of Problem 3. Such properties help in rewriting Problem 3 in a simplified way. For doing this, it is necessary to introduce a binary variable indicating whether in a given production interval any orders are placed or not; to this end, we define $\omega_j \in \{0, 1\}$ as follows:

$$\omega_j = \begin{cases} 0 & \text{if} \quad n_j = 0 \\ 1 & \text{if} \quad n_j > 0 \end{cases} \qquad j = 1, \ldots, S \tag{67}$$

Before rewriting Problem 3, it is still necessary to express the inventory cost. Note that such a cost is the area defined by $\xi(t)$ multiplied by the unitary cost $H_\xi$.

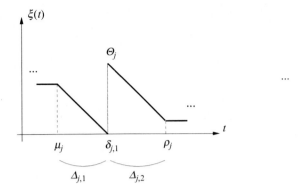

**Fig. 9.** The optimal behaviour of $\xi(t)$ for the case $n_j = 1$

Such an area is computed by exploiting the geometric properties of $\xi(t)$ coming from the previous propositions. Two different expressions of the inventory cost in the $j$-th production interval can be written, depending on whether any orders are placed in a production interval or not. In the former case (corresponding to $\omega_j = 1$), the area defined by $\xi(t)$ in the production interval $j$ is:

$$A_j^1 = \frac{1}{2K}\left[\xi(\mu_j)\right]^2 + \frac{1}{2}Kn_j\Delta_j^2 - \frac{1}{2K}\left[\xi(\rho_j)\right]^2 \tag{68}$$

On the contrary, if $\omega_j = 0$, the area defined by $\xi(t)$ in the production interval $j$ is:

$$A_j^0 = \frac{1}{2}(\rho_j - \mu_j)\left[\xi(\mu_j) + \xi(\rho_j)\right] \tag{69}$$

Moreover, for completing the expression of the inventory cost, it is still necessary to add the term corresponding to a positive inventory level during idle intervals. This is the last term included in the cost function of the following problem which is the new version of Problem 3.

**Problem 5.** Given the initial conditions $\rho_0 = 0$ and $\xi(0) = 0$, find

$$\min_{\substack{\omega_j, n_j, \Delta_j, \xi(\rho_j), \xi(\mu_j) \\ j=1,\ldots,S}} c_f \sum_{j=1}^{S} n_j + H_\xi \sum_{j=1}^{S}\left[\omega_j \cdot A_j^1 + (1-\omega_j)\cdot A_j^0\right] + H_\xi \sum_{j=1}^{S-1}\left[\xi(\rho_j)(\mu_{j+1}-\rho_j)\right]$$

where $A_j^1$ is given by (68), $A_j^0$ is given by (69), subject to

$$\xi(\mu_j) = \xi(\rho_{j-1}) \qquad j = 1,\ldots,S \tag{70}$$

$$\xi(\rho_j) = \xi(\mu_j) + Kn_j\Delta_j - K(\rho_j - \mu_j) \qquad j = 1,\ldots,S \tag{71}$$

$$\xi(\rho_j) \geq 0 \qquad j = 1,\ldots,S \tag{72}$$

$$\xi(\mu_j) \geq 0 \qquad j = 1, \ldots, S \tag{73}$$

$$\Delta_j \leq \rho_j - \mu_j \qquad j = 1, \ldots, S \tag{74}$$

$$\Delta_j \geq 0 \qquad j = 1, \ldots, S \tag{75}$$

$$M\omega_j - n_j \geq 0 \qquad j = 1, \ldots, S \tag{76}$$

$$\omega_j - n_j \leq 0 \qquad j = 1, \ldots, S \tag{77}$$

$$\omega_j \in \{0, 1\} \qquad j = 1, \ldots, S \tag{78}$$

where $M$ is a positive and sufficiently large number.

□

In constraints (71), the values of the inventory level at the end of each production interval are defined. Note that such constraints hold both for the case in which orders are placed within the production interval and for the opposite case (in which, of course, $n_j = 0$).

Problem 5 is a *nonlinear mixed-integer mathematical programming problem*, which can be solved by nonlinear solvers included in mathematical programming software tools. The solution of Problem 5 provides decisions about how many orders must be forwarded to suppliers, the time instants corresponding to order deliveries and the ordered quantities. Actually, the provided solution is typically a local optimum.

Note that Problem 5, which is here considered as a subproblem within the optimization procedure defined for a node of a supply chain, is actually a general replenishment problem that significantly extends the classical Wagner-Within model. More specifically, in the model proposed here, demand is a time-varying deterministic quantity (as in Wagner-Within model), but the time instants in which orders are placed are asynchronous (and no time discretization has been realized). Moreover, in the present case, the inventory serves a production process characterized by a piecewise constant production effort (which can be also null in specified time intervals).

## 5 The Multi-Site Case

The optimization approach regarding a single node is extended to consider the interactions among different stages of the supply chain structure. In this section, two simple multi-site schemes will be analyzed, the former being related to a competitive case, the latter considering a cooperative structure.

The first multi-site structure we want to analyze is a simple case in which different production nodes can serve a customer (the simple case of two competitive nodes is reported in Fig. 10). In particular, the considered problem regards the definition of an optimization procedure in order to define the best production centre able to serve a given demand. It is assumed that each producer has to optimally realize a predefined

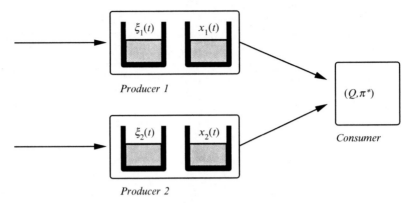

**Fig. 10.** A simple competitive multi-site structure

job sequence, over a certain time horizon, and that, at a given time instant, a new request comes by a generic consumer, in terms of a required quantity and a required delivery time instant.

In order to formalize the optimization problem concerning production and delivery of finite products to a consumer in a supply chain, some assumptions have been considered. First of all, the requests coming from consumers are processed one at a time and they are sequenced depending on a chronological order; moreover, each consumer request cannot be split, this means that it must be satisfied by only one producer. Furthermore, transportation operations are modelled in a very simplified way, thus each request corresponds to one transportation operation and transportation times between each producer and each consumer are known and constant.

The decision process through which a consumer chooses the best producer for satisfying its demand can be schematized as a sequence of some decision steps. First of all, the consumer makes a request $(Q, \pi^*)$, where $\pi^*$ is the due date of the request and $Q$ is the required quantity of products. Then, each producer $p = 1, \ldots, P$ (in the case shown in Fig. 10, it is $P = 2$) answers to the request following these steps:

- Producer $p$ determines the required time instant for ending the production as $t_p^* = \pi^* - t_p$, where $t_p$ is the transportation time between the production site $p$ and the consumer;
- Producer $p$ inserts $(t_p^*, Q)$ in its job sequence;
- Producer $p$ applies the optimization procedure defined for single production nodes (Problem 2) and it obtains its optimal solution, determining its optimal delivering time $dt_p$ and the production cost $CP_p$ yielded by the insertion in the job sequence of the considered request;
- Producer $p$ computes its transportation cost $CT_p$;
- Producer $p$ can thus determine the characteristics of the product delivery to the consumer and, specifically, the total cost $C_p = CT_p + CP_p$ and the delivering time to the consumer $tc_p$, where $tc_p = dt_p + t_p$.

Once received this information by all the producers, the consumer chooses the best producer to which the required production can be assigned; its choice is realized by adequately weighing the price and the supply characteristics, in terms of delivery time.

A second example of multi-site structure consists in the connection of two production nodes belonging to two subsequent stages of the supply chain system. Note that, in this case, the production nodes operate in a cooperative framework. A sketch of this two-node framework is given in Fig. 11.

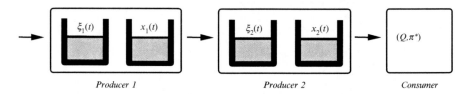

$$\text{Producer 1} \qquad\qquad \text{Producer 2} \qquad\qquad \text{Consumer}$$

**Fig. 11.** A simple cooperative multi-site structure

First of all, finite horizon time periods are defined and correspond to optimization horizons for both the nodes; such intervals are driven by the external demand which reaches Producer 2 in the form of the already defined finite sequence of asynchronous requests $(Q_i, \pi_i^*)$, $i = 1, \ldots, N$. As previously described, considering the transportation time, $\pi_i^*$ are transformed in $t_i^*$, as in the definition of Problem 2.

In particular, by solving Problem 2, Producer 2 finds the optimal closed-loop production policy, determining the optimal production effort $k^o(t)$, the optimal values of the idle times $\tau_i^o$, $i = 0, \ldots, N - 1$, and the optimal values of the production times $T_i^o$, $i = 0, \ldots, N - 1$. Then, by solving Problem 3, Producer 2 calculates the optimal raw material replenishment policy, computing the number of orders to be placed at each production interval, the reorder periods at each production interval, and the corresponding ordered quantities. This also leads to the determination of the time instants at which orders are placed. The replenishment policy is thus defined for Producer 2 and it corresponds to a finite sequence of asynchronous requests for Producer 1.

The way in which Producer 1 deals with the received requests is based on its internal optimization procedure (solution of Problem 2 and Problem 3) and on the following considerations. An early delivery by Producer 1 to Producer 2 would involve an increased inventory cost for Producer 2, whereas a tardy delivery by Producer 1 to Producer 2 would cause a higher cost relevant to the deviations form due-dates of the external demand. If the unitary costs defined by Producer 2 are known by Producer 1, and as far as a cooperative framework is considered, the cost relevant to deviations from due-dates is no more symmetric for Producer 1 and it can be computed on the basis of the unitary costs of Producer 2. More in detail, we suppose that the unitary raw material inventory cost and the unitary cost for deviations from due-dates are $H_{\xi,2}$ and $\alpha_2$ for Producer 2; then, two different unitary costs for deviations from due-dates are defined for Producer 1, respectively, for earliness and tardiness, defined as $\alpha_{E,1}$ and $\alpha_{T,1}$. Thus, it will be $\alpha_{E,1} = H_{\xi,2}$ and $\alpha_{T,1} = \alpha_2$.

In this case as well, transportation operations are very simply modelled as fixed delays. Moreover, it would be desirable that decisions taken by Producer 1 did not "perturb" too much the production policy of Producer 2. More specifically, it would be interesting to maintain the set of subsequences found by Producer 2 unchanged. As a matter of fact, by preserving the defined subsequences, Producer 2 could also admit early or tardy deliveries by Producer 1, since it is provided with control strategies, function of the system state and in particular of the current inventory level. If Producer 1 does not succeed in fulfilling the requests of Producer 2 by preserving the set of subsequences, Producer 2 needs to re-run the overall optimization procedure.

## 6 Conclusions

A hybrid model for a production node of a supply chain system is proposed in the paper. The node is represented by defining the dynamics of two inventories referred to raw materials and final products, respectively. On the basis of such a model, an optimization problem relevant to the minimization of order costs, inventory costs and costs due to deviations from the external demand has been stated. The decision variables refer to the process of raw material arrivals, the process of finite product deliveries and the production effort. The resulting optimization problem is a nonlinear functional optimization problem. A solution procedure based on the decomposition of the problem into two subproblems is proposed. The former subproblem refers to the determination of the optimal production effort and the optimal product departure process, whereas the latter subproblem corresponds to the determination of the optimal replenishment policy. Finally, the exploitation of the proposed solution procedure in two very simple multi-site structures is described.

Present and future research is devoted to the definition of more complex decentralized decisional structures in which production nodes equipped with the described optimization algorithm interact both in cooperative and in competitive frameworks.

## References

[1] Chopra S and Meindl P (2001) *Supply chain management: strategy, planning, and operation*. Prentice-Hall.
[2] Min H, Zhou G (2002) Supply chain modeling: past, present and future. *Computers and Industrial Engineering* 43:231–249.
[3] Tan KC (2001) A framework of supply chain management literature. *European Journal of Purchasing and Supply Management* 7:39–48.
[4] Gershwin SB (1994) *Manufacturing Systems Engineering*. Prentice-Hall.
[5] Tan B, Gershwin SB (2004) Production and Subcontracting Strategies for Manufacturers with Limited Capacity and Volatile Demand. *Annals of Operations Research* 125: 205–232.

[6] Hu JQ, Vakili P, Huang L (2004) Capacity and Production Managment in a Single Product Manufacturing System. *Annals of Operations Research* 125: 191–204.

[7] Blanchini F, Miani S, Ukovich W (2000) Control of production-distribution systems with unknown inputs and system failures. *IEEE Transactions on Automatic Control* 45: 1072–1081.

[8] Bauso D, Blanchini F, Pesenti R (2006) Robust control strategies for multi-inventory systems with average flowconstraints. *Automatica* 42: 1255–1266.

[9] Xu J, Hancock KL, (2004) Enterprise-Wide Freight Simulation in an Integrated Logistics and Transportation System. *IEEE Transactions on Intelligent transportation Systems* 5: 342–346.

[10] Schwartz JD, Wang W, Rivera DE, (2006) Simulation-based optimization of process control policies for inventory management in supply chains. *Automatica* 42: 1311–1320.

[11] Giglio G, Minciardi R, Sacone S, Siri S (2005) A hybrid model for optimal control of single nodes in supply chains. Proceedings of 16th IFAC World Congress.

[12] Giglio G, Minciardi R, Sacone S, Siri S (2006) Supply chain management and optimization. Proceedings of LT'06, International Workshop on Logistics and Transportation.

[13] Siri S (2006) Modelling, optimization and control of logistic systems. PhD Thesis, University of Genova, Italy.

[14] Giglio G, Minciardi R, Sacone S, Siri S (2007) Optimal control of single nodes in supply chains by a hybrid model. DIST Technical Report - June 2007.

[15] Giglio G, Minciardi R, Sacone S, Siri S (2007) Optimal replenishment policies in production nodes of supply chain models. Proceedings of the European Control Conference.

# Approximate Optimal Order Batch Sizes
# in a Parallel-aisle Warehouse

Yeming Gong[1] and René de Koster[2]

[1] RSM Erasmus University, The Netherlands ygong@rsm.nl
[2] RSM Erasmus University, The Netherlands rkoster@rsm.nl

**Summary.** The past warehousing literature dealing with order picking and batching assumes batch sizes are given. However, selecting a suitable batch size can significantly enhance the system performance. This paper is one of the earliest to search optimal batch sizes in a general parallel-aisle warehouse with stochastic order arrivals. We employ a sample path optimization and perturbation analysis algorithm to search the optimal batch size for a warehousing service provider facing a stochastic demand, and a central finite difference algorithm to search the optimal batch sizes from the perspectives of customers and total systems. We show the existence of optimal batch sizes, and find past researches underestimate the optimal batch size.

**Key words:** Stochastic optimization, Sample path optimization, Order picking, Batch size

## 1 Introduction

Order picking – the process of retrieving products from storage (or buffer areas) in response to a specific customer request – is the most labor-intensive operation in warehouses with manual systems, and a very capital-intensive operation in warehouses with automated systems (Goetchalckx and Ashayeri, 1989; Tompkins et al., 2003). Managing order picking systems effectively and efficiently is a challenging process in many warehouses. Order picking efficiency can often be improved by order batching (Gademann and Van De Velde, 2005), which is a method to group a set of orders into a number of sub-sets, each of which can then be retrieved by a single picking tour (De Koster et al., 2007).

The earlier papers dealing with order batching problem usually assume the batch size is directly given. A natural question is: are these given batch sizes suitable? Considering setup time and unit service time for order picking, the total service time and batch size are not related linearly: the setup time will take a bigger proportion in the total service time for a small batch, while the unit service time will take a bigger

L. Bertazzi et al. (eds.), *Innovations in Distribution Logistics*, Lecture Notes
in Economics and Mathematical Systems 619, DOI: 10.1007/978-3-540-92944-4,
© 2009 Springer-Verlag Berlin Heidelberg

proportion in the total service time for larger batches. Therefore it is an interesting question to explore the optimal batch size when orders arrive according to a stochastic process. A following research question is: how to find an optimal batch size if it exists? Most research involved in optimizing batch sizes, with the objective to minimize total service times assumes the order set is given. Gademann and Van de Velde (2005) have pointed out that the deterministic version of the batch problem with optimal routing is $\mathcal{NP}$- hard when the batch size is larger than 2, in a parallel-aisle layout. It is tough or even infeasible to determine optimal batch sizes for the stochastic version of the problem. Few papers explore optimum batch sizes in a stochastic context. Chew and Tang (1999) assume orders arrive according to a Poisson process and approximate the travel time in a rectangular warehouse and use this approximate expression to minimize the total throughput time of the first order in a batch. They compare their results with simulation. Le-Duc and De Koster (2007) extend these results by determining the optimal batch size minimizing the throughput time of a random order in a two-block warehouse. All methods use approximation methods and do not directly optimize the batch size, as this is very cumbersome. In this paper we opt for a different approach, by efficient simulation optimization. Simulation optimization can help the search for an improved policy while allowing for complex features that are typically outside of the scope of analytical models. In this paper, we will employ SPO (sample path optimization), a simulation optimization technique with the advantage of high efficiency and convenience. However, SPO requires a technique to estimate the gradient of the objective function with respect to the batch size.

A large number of gradient estimation techniques exist, such as Infinitesimal Perturbation Analysis (IPA), Likelihood Ratios, Symmetric Difference, and Simultaneous Perturbation (Fu, 2002). IPA is mainly used to calculate a sample path derivative with respect to an input parameter in a discrete event simulation (Heidelberger et al., 1998). We will employ this technique since it is an "efficient gradient estimation technique" (Ho et al., 1979), which can "expedite the process of performing experiments on discrete event simulation models" (Johnson and Jackman, 1989). The implicit assumption of IPA is that the average of the change which results from the perturbation equals the change in expectation, and it yields an unbiased estimator. Convergence is an important issue for the implementation of IPA. Heidelberger et al. (1988) have studied the convergence properties of IPA sample path derivative, and derived the necessary and sufficient condition for the convergence. Applications of perturbation analysis have been reported in simulations of Markov chains (Glasserman, 1992), inventory models (Fu, 1994), supply chain problems (Gong and Yücesan, 2006), manufacturing systems (Glasserman, 1994), finance (Fu and Hu, 1997), and statistical process control (Fu and Hu, 1999). In some formulations in this paper, we will face complicated objective functions. In order to obtain optimal batch sizes, we need to compute the gradient of these objective functions. However, either their gradients are not available in explicit form or they are given by complicated expressions. We therefore resort to a finite difference method, which makes it possible to use arithmetic operations to determine the gradient.

In this paper, we consider the optimal order batch size problem with stochastic demand in a parallel-aisle warehouse (see Fig. 1), with cross aisles at the front and back of the aisles. The warehouse faces a demand with a given distribution. An order picker travels at a constant velocity with a S-shape routing policy, one of most common routing policies in practice. In order to improve picking efficiency, orders are batched.

The research objective in this paper is to minimize the operational costs by optimizing the order batch size, defined as *the set of orders that are picked by one order picker in one route, and batch size q is the number of items in the batch, with constraint $q^{LB} \leq q \leq q^{UB}$, where the upper bound $q^{UB}$ is determined by the capacity of pick devices (pallets or bins) and the lower bound $q^{LB}$ is specified by an additional condition like system stability.* To achieve this research objective, we consider three major research questions and build corresponding models as follows.

First of all, we examine the operational cost from the perspective of a warehousing service provider and build the corresponding Model-1. This model focuses primarily on an internal objective by minimizing the average total service time, which is the sum of setup time and travel time. We exclude picking time as this is not influenced by the batching policy. Orders are picked in a FIFO sequence. Model-1 emphasizes the impact of order batching on performance of a warehousing service provider. Secondly,we examine the cost for customers and build a corresponding Model-2, which is taken from Chew and Tang (1999). The contribution in Model-2 is to provide an efficient finite difference algorithm. While using straightforward simulation takes much time to obtain a solution by enumeration, our method takes on average 6 seconds to get one solution. Finally,we consider the total cost for both the warehousing service provider and the customers by combining Model-1 and Model-2 into a new Model-3. The contribution of this research is twofold. First we show SPO and perturbation analysis algorithms are efficient in deriving optimal values. Second we combine the perspectives of both customers and a warehousing provider in one model and show it can also be solved by perturbation analysis and an SPO algorithm.

The remainder of the paper is organized as follows: in the following section, we search optimal batch sizes for warehousing service providers in a general stochastic parallel-aisle warehouse by sample path optimization and infinitesimal perturbation analysis techniques. Section 3 is devoted to an efficient finite difference algorithm to search optimal batch sizes for customers. In Sect. 4, we present a model with the objective of minimizing the total cost, and provide an efficient finite difference algorithm to search the optimal batch size. We conclude with final discussion, contribution summary and further research in Sect. 5.

# 2 Optimal Order Batch Sizes for Model-1

## 2.1 Model

Model-1's objective is to minimize the total expected operation time of a warehousing service provider $E[T_p(q, D)]$, where $q$ is the decision variable, the subscript $p$ indicates a warehousing service **p**rovider, and $D$ is the demand generated from a given distribution $f(D)$. $E[T_p(q, D)]$ is the product of the expected number of batches $E[D/q]$ and the expected operational time of one batch. Following [4], we do not consider picking time since the batching policy does not influence the total picking time for a given demand. But the batch size does influence the total setup time and the total expected travel time. Therefore, in our model the expected operational time of one batch is the sum of a setup time $\beta$ and an expected travel time $E[L(q)/v]$, where $E[L(q)]$ is the expected travel distance and $v$ is a constant travel velocity. $E[L(q)]$ depends on the warehouse layout and the routing method. We assume a rectangular, parallel-aisle layout, as sketched in Fig. 1 and an S-shape routing method (see De Koster et al., 2007). For this environment, Chew and Tang (1999) have found a closed form approximate expression for $E[L(q)]$. We have

$$Model - 1 : \min_{D \sim f(D), 0 \leq q \leq D} E[T_p(q, D)]$$

$$s.t. \quad E[T_p(q, D)] = E[L(q)/v + \beta]E[D/q]$$

## 2.2 Algorithm

This section demonstrates how to obtain the optimal batch size quantities in a parallel-aisle warehouse with stochastic demand. Our scheme is to use the simulation optimization algorithm by combining sample path optimization and perturbation analysis to examine optimal order batch sizes.

### Algorithm Description

To compute the optimal batch size values, we adopt a sample path optimization technique as main algorithm, where we use IPA (Infinitesimal Perturbation Analysis) to calculate the gradient value. We start with an arbitrary batch size $q^1$. After randomly generating an instance of the demand, we construct and solve Model-1 in a deterministic fashion. Then, we compute gradient values by the decision tree from the perturbation analysis. The procedure is summarized in a pseudo-code format in the following procedure, where $K$ denotes the total number of steps taken in a search path of the main algorithm, $U$ represents the total number of steps in one inner cycle which is to provide a gradient estimation at one step of the main algorithm, $\alpha_k$ represents the step size at the each iteration $k$, and $q^k$ represents the batch size at the $k^{th}$ step. The choice of step size is important to guarantee convergence of the batch size. A proper choice will be explained further in Theorem 4.

*Algorithm 1*
(I) *Initialization.*
    (I.1) *Initialize K.*
    (I.2) *Initialize U.*
    (I.3) *Initialize $q^1$.*
    (I.4) *Initialize $\alpha_1 = \alpha/1$ for a constant $\alpha$.*
(II) *Set $k \leftarrow 1$.*
*Repeat.*
    *Set $u \leftarrow 1$.*
*Repeat.*
    (II.1) A. *Generate the demand $d_u^k$ from $f(D)$.*
    (II.1) B. *Compute the objective value of Model-1 in a deterministic fashion.*
    (II.1) C. *Compute and accumulate gradients $dL_u^k$,*
    *also record generated demand $d_u^k$ and realized travel distance $L_u^k$.*
    *$u \leftarrow u + 1$,*
*Until u=U.*
    (II.2) *Compute the desired gradients $\frac{\partial E[L]}{\partial q}|_{q=q^k} = \frac{1}{U} \sum_{u=1}^{U} dL_u^k$,*
    *$E[D]^k = \frac{1}{U} \sum_{u=1}^{U} d_u^k$, and $E[L]^k = \frac{1}{U} \sum_{u=1}^{U} L_u^k$ at the k step.*
    (II.3) *Calculate the desired gradients $dT_q^k = (\frac{1}{v} \frac{\partial E[L]}{\partial q}|_{q=q^k} + \beta)\frac{E[D]^k}{q^k}$*
    *$+(\frac{1}{v}E[L]^k + \beta)(-\frac{E[D]^k}{(q^k)^2})$.*
    (II.4) *Update the batch size by $q^{k+1} \leftarrow \lfloor q^k - \alpha_k dT_q^k \rfloor$, where $\alpha_k = \alpha/k$.*
    *$k \leftarrow k + 1$,*
*Until $k = K$.*
(III) *Return the $\{q^k\}_{k=1}^K$ and the objective function value.*

We explain the procedure as follows:

(I) Initialization. The algorithm starts with an arbitrary value for the batch size $q^1$. $K$ and $U$ are given and can be determined by a pilot study to solve the following trade-off: while a small $K$ cannot provide sufficient data, and output will have a big variance, a too large $K$ is inefficient to improve the optimal value.

(II) The main loop in step (II) is an outer loop with $K$ steps. Each step includes a $U$-step inner loop computation in step (II.1), IPA analysis in step (II.2), the desired gradient calculation in step (II.3), and the updating of batch sizes in step (II.4).

We first run an inner loop with $U$ steps. At each step of the inner loop, we generate the demand from distribution $f(D)$, solve the problem of Model-1 in a deterministic fashion once the demand is observed, and calculate the perturbation value $dL_u^k$. Secondly, we conduct critical computation $\frac{\partial E[L]}{\partial q}|_{q=q^k} = \frac{1}{U} \sum_{u=1}^{U} dL_u^k$, which is just the IPA technique. Thirdly, we compute the gradient of the expected travel time with respect to the batch size by $dT_q^k = (\frac{1}{v} \frac{\partial E[L]}{\partial q}|_{q=q^k} + \beta)\frac{E[D]^k}{q^k} + (\frac{1}{v}E[L]^k + \beta)(-\frac{E[D]^k}{(q^k)^2})$. Finally, we update batch sizes $q^{k+1}$ by $q^{k+1} \leftarrow \lfloor q^k - \alpha_k dT_q^k \rfloor$ at the $k^{th}$ step. Also note that since the algorithm stops at $k = K$, we do not need an extra stopping rule here.

(III) Return the $q^k$ and objective function value at each step. Then we can conduct the output analysis.

**Algorithm Justification**

If an algorithm can converge and the objective function subject to minimization is convex, the algorithm can provide a global optimal value. In order to justify Algorithm 1, we build four theorems. Theorem 1 is to justify the convexity of our objective function. Theorem 2, 3, and 4 will show the convergence of Algorithm 1.

In order to examine the convexity of objective function $E[T_p(q, D)]$. We first establish Theorem 1 as follows.

**Theorem 1.** *The objective function $E[T_p(q, D)] = E[L(q)/v + \beta]E[D/q]$ is a convex function of $q$.*

Proof: For item locations with uniform distribution, Chew and Tang [4] have given approximate distance estimation for the S-shape routing policy. Based on their result, we have

$$E[\frac{L(q)}{v} + \beta] = \frac{M}{v}H[1 - (1 - \frac{1}{M})^q] + 2\frac{\omega}{v}[M - \sum_{j=1}^{M-1}(\frac{j}{m})^q] + \frac{H}{2v} + \beta. \quad (1)$$

For a constant $\theta$ with $0 < \theta < 1$, $f(q) = -\theta^q$ is a concave function of $q$. So $E[L(q)/v + \beta]$ here is a concave function. $E[D]/q$ is nonincreasing convex function of $q$. From Boyd and Vandenberghe (2004), the product $E[L(q)/v + \beta]E[D/q]$ of a concave function $E[L(q)/v + \beta]$ and a nonincreasing convex function $E[D]/q$ is convex. $\qquad\square$

In this algorithm, it is critical to find an efficient gradient estimator. We use perturbation analysis to compute this gradient. Perturbation analysis is a powerful technique for the efficient performance analysis of dynamic systems. Its fundamental approach is to keep track of information along a perturbed path. The main principle behind perturbation analysis is that if a decision variable of a system is perturbed by a small amount, the sensitivity of the response of the system to that variable can be estimated by "tracing its pattern of propagation through the system" (Carson and Maria, 1997). This will be a function of "the fraction of the propagations that die before having a significant effect on the response of interest" (Carson and Maria, 1997). The fact that all derivatives can be derived from the same simulation run represents a significant advantage to IPA in terms of the efficiency. With the support of this technique, we have the Theorem 2.

**Theorem 2.** *The gradient of expected travel time with respect to batch size can be computed by $dT_p = (\frac{1}{v}\frac{\partial E[L]}{\partial q} + \beta)\frac{E[D]}{q} + (\frac{E[L]}{v} + \beta)(-\frac{E[D]}{(q^k)^2})$, where $\frac{\partial E[L]}{\partial q}$ can be calculated by the perturbation analysis and decision tree method.*

Proof: From Model-1, we have

$$\frac{\partial E[T_p(q)]}{\partial q}\bigg|_{q=q^k} = (\frac{1}{v}\frac{\partial E[L]}{\partial q}\bigg|_{q=q^k} + \beta)\frac{E[D]}{q^k} + (\frac{E[L]}{v} + \beta)(-\frac{E[D]}{(q^k)^2}). \quad (2)$$

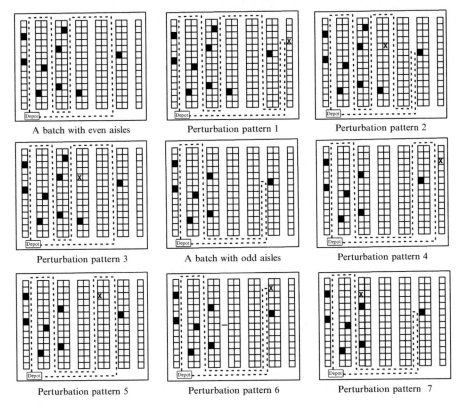

**Fig. 1.** Perturbation analysis for a batch in a parallel-aisle warehouse with S-shape routing policy

In this formula, the critical issue is to compute the gradient of the travel distance with respect to batch sizes. Gong and Yücesan [10] have provided an implementation framework and theoretical justification of SPO and IPA, and they compute the gradient by an analytical duality method. Different from them, we conduct direct perturbation analysis, and then derive a decision tree from perturbation patterns.

We conduct a perturbation analysis for a single batch with S-shape routing policy here. For a batch with batch size $q$, we give the system a perturbation, i.e., let the batch size increase by 1. In Fig. 1 the item with a cross "X" is the perturbed item. By comparing the distance before and after perturbation, we can compute the perturbation of distance. When the number of visited aisles is even, there are three perturbation patterns. (see perturbation patterns 1, 2, 3 in Fig. 1). When the number of visited aisles is odd, there are four perturbation patterns. (see perturbation patterns 4, 5, 6, 7 in Fig. 1).

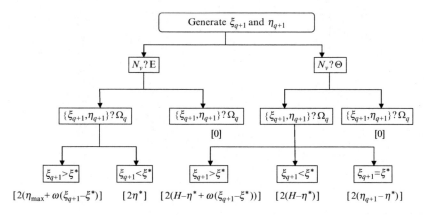

**Fig. 2.** Decision tree from perturbation analysis

In the following research,we use the notation below:

$M$ = the number of aisles;

$H$ = the length of aisles. We assume a bin containing one kind of item has one unit length;

$\omega$ = the cross distance between two consecutive aisles;

$\overline{D}$ = the mean of demand;

$N_v$ = the number of visited aisles;

$E$ = the even number set;

$\Theta$ = the odd number set;

$\Psi = \{1, 2, ..., q\}$ the indices set of $q$ items;

$\xi_i$ = the aisle position of item $i$, $\xi_i = 1, ..., M$;

$\eta_i$ = the location position of item $i$, $\eta_i = 1, ..., H$;

$\Omega_q$ = the position set covered by the routing when batch size is $q$. The position of an item $i \in \Psi$ is indicated by $(\xi_i, \eta_i)$;

$\xi^* = max(\xi_i, i \in \Psi)$ be the farthest aisle visited;

$\eta^* = max(\eta_i, \forall i, s.t.\xi_i = \xi^*)$ be the farthest position at the farthest aisle visited.

By tracing its pattern of propagation through the system, we can build a decision tree for the gradient computation in Fig. 2. Theorem 2 follows from IPA analysis in Fig. 1 and the decision tree in Fig. 2.                                              □

**Theorem 3.** *If demand D has a density on $(0, \infty)$ and $E[D] < \infty$, batch size $q \in \mathcal{R}^+$ and $q < \infty$, the gradients obtained by Theorem 2 are bounded with probability 1.*

Proof: Generate the aisle position $\xi_{q+1}$ and location position $\eta_{q+1}$ of a perturbed item. For S-shape routing policy, from the decision tree in Fig. 2, the gradient can be computed as follows:

$$\frac{\partial L}{\partial q} = \begin{cases} 2[\eta_{q+1} + \omega(\xi_{q+1} - \xi^*)], & \xi_{q+1} > \xi^* \text{ and } N_v \in E \\ 2\eta^*, & \xi_{q+1} < \xi^*, (\xi_{q+1}, \eta_{q+1}) \notin \Omega_q \text{ and } N_v \in E \\ 0, & \xi_{q+1} = \xi^*, (\xi_{q+1}, \eta_{q+1}) \notin \Omega_q \text{ and } N_v \in E \\ 2[H - \eta^* + \omega(\xi_{q+1} - \xi^*)], & \xi_{q+1} > \xi^* \text{ and } N_v \in \Theta \\ 2(H - \eta^*), & \xi_{q+1} < \xi^*, (\xi_{q+1}, \eta_{q+1}) \notin \Omega_q \text{ and } N_v \in \Theta \\ 2(\eta_{q+1} - \eta^*), & \xi_{q+1} = \xi^*, \eta_{q+1} > \eta^* \text{ and } N_v \in \Theta \\ 0, & (\xi_{q+1}, \eta_{q+1}) \in \Omega_q \text{ and } N_v \in \Theta. \end{cases} \tag{3}$$

From (3), we have

$$\frac{\partial L}{\partial q} \le max\{2[\eta_{q+1} + \omega(\xi_{q+1} - \xi^*)], 2\eta^*, 2[H - \eta^* + \omega(\xi_{q+1} - \xi^*)],$$

$$2(H - \eta^*), 2(\eta_{q+1} - \eta^*)\} \le 2H + 2\omega(M - 1). \tag{4}$$

The boundedness of gradient follows from (4).                    □

**Theorem 4.** *By the sample path optimization in Algorithm 1 for a proper choice of step size, the batch size $\{q^k\}_{k=1}^{\infty}$ converges with probability 1.*

Proof: In order to ensure the convergence, a key issue is the selection of a suitable step size $\alpha_k$, where we have

Condition (1): A criterion for choosing $\alpha_k$ is to let step size go to zero fast enough so that the algorithm can converge, but not so fast that it will induce a wrong value. One condition to meet that criterion is $\sum_{k=1}^{\infty} \alpha_k = \infty$ and $\sum_{k=1}^{\infty} \alpha_k^2 < \infty$.

For instance, $\alpha_k = \alpha/k$ for some fixed $\alpha > 0$ satisfies Condition (1). The first part of this condition facilitates convergence by ensuring that the steps do not become too small too quickly. However, if the algorithm is to converge, the step sizes must eventually become small, as ensured by the second part of the condition.

For a convex objective function $E[T_p(q)]$, a bounded gradient (see Theorem 2 and Theorem 3), and a step size $\alpha_k$ which satisfies the condition (1), according to Robbins and Monro (1951), we have a limit point of $\{q^k\}_{k=1}^{\infty}$, which is stationary with probability 1.                    □

## 2.3 Results

We implement Algorithm 1 in Matlab. Experiments are conducted on a computer with 1.73GHz CPU and 516MB RAM. After acquiring characteristic information like warehouse size and generating the demand by a normal distribution $N(\mu, \sigma^2)$ to specify the problem, the distance computation program can return batch sizes and objective values at each step to the main program. Then by the gradient computing algorithm, the main program can update the batch size until it converges. Here we adopt an initial step size $\alpha_1 = 0.5$ by a pretest experiment. Since the objective function is convex, this convergence will lead to a global optimum.

Without loss of generality, the position of our depot is the first aisle and the first location. The probability to visit an aisle is equal for all aisles and uniformly distributed. We have aisle number $\xi_i \sim U(1, M), \forall i \in \Psi$. The probability to visit a location in a visited aisle is also equal, i.e., location position $\eta_i \sim U(1, H), \forall i \in \Psi$. In order to verify the result from the simulation optimization, we compare it with the result of enumeration, where we enumerate all the possible batch size values from $q^{LB}$ to $q^{UB}$. In combination with Monte Carlo simulation, the enumeration is conducted as follows. First, we generate a very large order number, and then generate the position of each item by given distributions $\xi_i \sim U(1, M), \forall i \in \Psi$ and $\eta_i \sim U(1, H), \forall i \in \Psi$. Then, for every value of $q$, we determine the batches of size $q$ in an FCFS sequence. Third we compute the routing length and corresponding warehouse operation time of each batch by the S-shape policy. For every batch size we compute the expected warehousing operation time and hence finally find the optimal batch size.

We present experiments in Table 1. We have conducted two groups of experiments: varying the aisle number $M$ (experiments 1 to 5) and varying the aisle length $H$ (experiments 6 to 10). The computation results include the items below:

$q^E =$ the optimal batch size obtained by enumeration;

$\widehat{q} =$ the statistical estimation of batch size by the stochastic simulation algorithms, which includes the mean batch size $\overline{q}$ and half width ($HW$) of the 95% confidence interval ($CI$);

$R(\overline{q}) =$ the rounded integer value of the estimated batch size;

$\Delta_1 = |\overline{q} - q^E|/q^E$, the direct bias of statistical estimation;

$\Delta_2 = |R(\overline{q}) - q^E|/q^E$, the indirect bias of rounded statistical estimation;

We compute the average direct bias $\overline{\Delta_1} = 1/N \sum_n |\overline{q} - q_n^E|/q_n^E = 0.255\%$ and the average indirect bias $\overline{\Delta_2} = 1/N \sum_n |R(\overline{q}) - q_n^E|/q_n^E = 0\%$. The average direct bias of statistical estimation is less than 1%, and the average indirect bias of rounded statistical estimation is negligible.

**Table 1.** Experiment result for Model-1

| No. | $M$ | $H$ | $\omega$ | $q^{UB}$ | $q^{LB}$ | $q^E$ | $\widehat{q_1} = \overline{q} \pm HW$ | $R(\overline{q})$ | $\Delta_1$ | $\Delta_2$ |
|-----|-----|-----|----------|----------|----------|-------|----------------------------------------|-------------------|-----------|-----------|
| 1 | 25 | 20 | 3 | 50 | 1 | 50 | 49.8694±0.0876 | 50 | 0.26% | 0 |
| 2 | 30 | 20 | 3 | 50 | 1 | 50 | 49.7793±0.0773 | 50 | 0.44% | 0 |
| 3 | 35 | 20 | 3 | 50 | 1 | 50 | 49.8895±0.0271 | 50 | 0.22% | 0 |
| 4 | 40 | 20 | 3 | 50 | 1 | 50 | 49.9391±0.0763 | 50 | 0.12% | 0 |
| 5 | 45 | 20 | 3 | 50 | 1 | 50 | 49.8394±0.0745 | 50 | 0.32% | 0 |
| 6 | 40 | 25 | 3 | 60 | 1 | 60 | 59.7696±0.0876 | 60 | 0.38% | 0 |
| 7 | 40 | 27 | 3 | 60 | 1 | 60 | 59.9196±0.0773 | 60 | 0.13% | 0 |
| 8 | 40 | 28 | 3 | 60 | 1 | 60 | 59.8195±0.0272 | 60 | 0.30% | 0 |
| 9 | 40 | 30 | 3 | 60 | 1 | 60 | 59.9093±0.0765 | 60 | 0.15% | 0 |
| 10 | 40 | 32 | 3 | 60 | 1 | 60 | 59.8597±0.0743 | 60 | 0.23% | 0 |

From the experiment, we observe that

*the optimal batch size for Model-1 equals to the upper bound $q^* = q^{UB}$,*

which is robust in both groups of simulation experiments. We can understand this result from the limiting system behavior. By increasing the batch size, the travel and setup time per batch will converge to a constant.

$$\lim_{q \to \infty} E[\frac{L(q)}{v} + \beta] = \lim_{q \to \infty} \frac{M}{v} H[1 - (1 - \frac{1}{M})^q] + 2\frac{\omega}{v}[M - \sum_{j=1}^{M-1}(\frac{j}{m})^q] + \frac{H}{2v} + \beta$$

$$= \frac{2MH + 4M\omega + H}{2v} + \beta. \tag{5}$$

However, with an increasing batch size $q$, the number of batches $E[D]/q$ will continue to decrease, and therefore the total operation time, which is the product of the two items, will also decrease. That is the reason why the optimal batch size will converge to its upper bound.

# 3 Optimal Order Batch Sizes for Model-2

## 3.1 Model

Model-1 in Sect. 2 considers the main operation time from the perspective of a warehousing service provider. It does not measure the time of customers and the service level. It is also necessary to examine the time spent by customers in a warehousing system. We therefore adopt the turnover time of a consumer's order, which is from [4], as the objective to build the Model-2 as follows.

$$Model - 2 : Min \quad T_{TO}(q)_{q^{LB} \leq q \leq q^{UB}}$$

$$s.t. \quad T_{TO}(q) = W_1(q) + W_2(q) + E[S]$$

Chew and Tang [4] focus on the first order in a batch. However, their results can be generalized to a random order in a batch. In Model-2, the objective $T_{TO}(q)$ is the turnover time of a customer's order, i.e., the duration an order stays in the system, when batch size is $q$ with $q^{LB} \leq q \leq q^{UB}$. $T_{TO}(q)$ consists of three parts: expected batch time $W_1(q)$, expected waiting time $W_2(q)$ and expected service time $E[S]$. Let the order arrival rate be $\lambda$. The expected batch time $W_1(q)$ is given by Chew and Tang (1999) as $W_1(q) = (q - 1)/\lambda$. Expected waiting time $W_2(q)$ is approximately computed by the linear combination of expected waiting times of $M/M/1$, $M/D/1$, and $D/M/1$. Expected service time $E[S]$ consists of travel time, picking and sorting time.

## 3.2 Algorithm

The essential problem in searching for the optimal batch size is the choice of the computation method. From Chew and Tang (1999), our objective function is a specially complicated function of $q$, and an analytical gradient computation method is infeasible. Perturbation analysis is also highly complicated in Model-2, especially for the perturbation analysis of $W_2$. Even if we had obtained the decision tree by perturbation analysis, its computation will not be efficient. Therefore we use the finite differences (FD) for our objective function. There are two FD approximations: forward FD and central FD. The forward difference derivative approximations consume less computer time, but they are usually not as precise as central difference method. Therefore we mainly use central FD as our gradient computation method.

We use finite difference optimization algorithm to examine order batch problem in a parallel-aisle warehouse, and demonstrates how to obtain the optimal batch size quantities. The procedure is summarized in a pseudo-code format in Algorithm 2, where we start with a batch size $q^1$, usually $q^1 = q^{UB}$, $K$ denotes the total number of steps taken in a search path, $\alpha_k$ represents the step size at the each iteration $k$, and $q^k$ represents the batch size at the $k^{th}$ step.

*Algorithm 2*
(I) *Initialization.*
    (I.1) *Initialize $K$.*
    (I.2) *Initialize $q^1$ to $q^{UB}$.*
(II) *Set $k \leftarrow 1$.*
*Repeat.*
    (II.1) *Compute $T_{TO}^k(q^k + h)$ and $T_{TO}^k(q^k - h)$, when $k > 1$.*
    (II.2) *Compute the desired gradients.*
    $dT_{TO}^k = \frac{T_{TO}^k(q^k+h)-T_{TO}^k(q^k)}{h}$, *when $k = 1$.*
    $dT_{TO}^k = \frac{T_{TO}^k(q^k+h)-T_{TO}^k(q^k-h)}{2h}$, *when $k > 1$.*
    (II.3) *Update the batch size,$q^{k+1} \leftarrow \lfloor q^k - \alpha_k dT_{TO}^k \rfloor$.*
    $k \leftarrow k + 1$,
*Until $k = K$.*
(III) *Return the $\{q^k\}_{k=1}^K$ and the objective function value.*

**Theorem 5.** *By the finite difference algorithm 2 and a step size $\alpha_k$ which can satisfy Condition (1), the batch sizes $\{q^k\}_{k=1}^\infty$ in Model-2 can converge.*

Proof: From Bertsekas (1999), for the finite difference algorithm with a step size which can satisfy Condition (1), if the objective function in Model-2 is convex, the batch size can converge. Chew and Tang (1999) have showed that $T_{TO}(q) = W_1(q) + W_2(q) + E[S]$ is a convex function of $q$ for $q^{LB} \le q \le q^{UB}$, where $q^{LB}$ is determined by the system equilibrium condition since if the arrival rate $\lambda$ is too high the system will become unstable and $q^{UB}$ is specified by facility capability limitation.    □

## 3.3 Results

Based on Chew and Tang (1999) formulation and our optimization algorithm, we implement the optimization procedure in Matlab. One run of simulation takes only 6 seconds on average. Let $q^1 = q^{UB}$, for the number of aisle ranging from 25 to 45, we obtain the search paths indicated in Fig. 3. All the experiments converge in the last 500 steps. We compute the statistical estimation by the transient deletion technique.

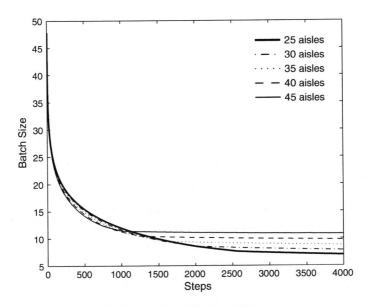

**Fig. 3.** Search path by Algorithm 2

In order to verify the result from optimization algorithm 2, we compare the result with that by enumeration, where we traverse all possible batch size values from $q^{LB}$ to $q^{UB}$. For all possible values of batch size $q$, we compute the expected objective values from Chew and Tang [4] and find the optimal batch size. We present the result in Fig. 4.

We present experiments in Table 2. The left part of Table 2 is the indices of experiments. The middle part of Table 2 is the experiment setting: $M$, $H$, $\omega$, $q^{LB}$ and $q^{UB}$. We have conducted two groups of experiments: varying the aisle number $M$ (experiments 1 to 5) and varying the aisle length $H$ (experiments 6 to 10). The right part presents the computation results, which includes the items below:

$q^E$ = the optimal batch size obtained by enumeration;

$\widehat{q}$ = the statistical estimation by Algorithm 2, which includes the mean batch size $\overline{q}$ and $HW$ of the 95%$CI$;

$R(\overline{q})$ = the rounded integer value of the estimated batch size;

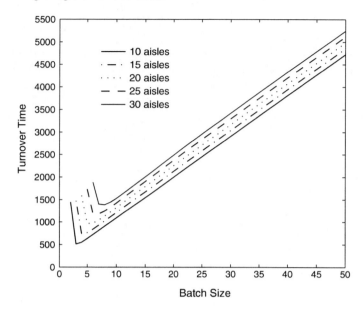

**Fig. 4.** Turnover time versus batch sizes in Model-2

$\Delta_1 = |\overline{q} - q^E|/q^E$, the direct bias of statistical estimation;
$\Delta_2 = |R(\overline{q}) - q^E|/q^E$, the indirect bias of rounded statistical estimation;
We compute the average direct bias $\overline{\Delta_1} = 1/N \sum_n |\overline{q} - q_n^E|/q_n^E = 2.785\%$ and the average indirect bias $\overline{\Delta_2} = 1/N \sum_n |R(\overline{q}) - q_n^E|/q_n^E = 0.769\%$. The average direct bias of statistical estimation is less than 3%, and the average indirect bias of rounded statistical estimation is less than 1%.

**Table 2.** Experiment result for Model-2

| No. | M | H | $\omega$ | $q^{UB}$ | $q^{LB}$ | $q^E$ | $\widehat{q_1} = \overline{q} \pm HW$ | $R(\overline{q})$ | $\Delta_1$ | $\Delta_2$ |
|---|---|---|---|---|---|---|---|---|---|---|
| 1 | 25 | 20 | 3 | 50 | 6 | 7 | 6.9129±0.0061 | 7 | 1.24% | 0 |
| 2 | 30 | 20 | 3 | 50 | 7 | 8 | 7.7847±0.0046 | 8 | 3.08% | 0 |
| 3 | 35 | 20 | 3 | 50 | 8 | 9 | 8.7647±0.0034 | 9 | 2.61% | 0 |
| 4 | 40 | 20 | 3 | 50 | 9 | 10 | 9.8098±0.0025 | 10 | 1.90% | 0 |
| 5 | 45 | 20 | 3 | 50 | 10 | 11 | 10.8911±0.0018 | 11 | 0.99% | 0 |
| 6 | 40 | 25 | 3 | 50 | 10 | 12 | 12.4991±0.00005 | 12 | 4.16% | 0 |
| 7 | 40 | 27 | 3 | 50 | 10 | 13 | 13.5003±0.00006 | 14 | 3.85% | 7.69% |
| 8 | 40 | 28 | 3 | 50 | 10 | 14 | 14.4994±0.00006 | 14 | 3.57% | 0 |
| 9 | 40 | 30 | 3 | 50 | 10 | 15 | 15.4996±0.00007 | 15 | 3.33% | 0 |
| 10 | 40 | 32 | 3 | 50 | 10 | 16 | 16.4998±0.00005 | 16 | 3.12% | 0 |

From the result we can observe that, the optimal batch size for customers is close to its lower bound and less than its upper bound $q^{LB} < q_2^* < q^{UB}$, which is robust in both groups of simulation experiments. This result is similar to results of Chew and Tang (1999).

# 4 Optimal Order Batch Sizes for the Total System

The objective functions of both Model-1 and Model-2 are unilateral. The result from Model-2 is similar to Chew and Tang (1999) and Le-Duc (2005), and this result possibly underestimates the positive effect of batch procedure. The result from Model-1 also possibly overestimates the positive effect of batch procedure. While a large batch size brings short-run minimal cost to warehouse service providers, it will also cause long throughput times for the customers, and may therefore harm the long-run interest of warehouse service providers. Considering both sides, we therefore build Model-3 and measure the total system cost.

## 4.1 Model

The objective in Model-2 is the turnover time for a single customer's order while the objective in Model-1 is the total service time for the total customers. So we need to transform the data in Model-1, and compute the time spent by service provider on a single customer, that is $\overline{T_P(q)} = \frac{E[L(q)]/v+\beta}{q}$. Without loss of generality, we assume a single customer corresponds to a single order. Let $c_1$ be the operation cost per unit time for service provider, and $c_2$ be the waiting cost per unit time for customer. Then $c_1\overline{T_P(q)} + c_2 T_{TO}(q)$ is the total system cost $C(q)$ for one customer. We have:

$$Model - 3 : Min \quad C(q)_{q^{LB}\le q\le q^{UB}}$$

$$s.t. \quad C(q) = c_1\overline{T_P(q)} + c_2 T_{TO}(q)$$

$$\overline{T_P(q)} = \frac{E[L(q)]/v + \beta}{q}$$

$$T_{TO}(q) = W_1(q) + W_2(q) + E[S]$$

The ratio of $c_1$ and $c_2$ in Model-3 is used to measure the weight of both sides in the system. We define: the unit cost ratio $\gamma = \frac{c_1}{c_2}$.

## 4.2 Algorithm

We mainly use central FD as gradient computation method since the objective function in Model-3 is a specially complicated function of $q$. The procedure is summarized in a pseudo-code format in Algorithm 3, where we start with an initial batch size $q^1$, for example $q^1 = q^{UB}$, $K$ denotes the total number of steps taken in a searching path, $\alpha_k$ represents the step size at the each iteration $k$, and $q^k$ represents the batch size at the $k^{th}$ step.

*Algorithm 3*
(I) *Initialization.*
   (I.1) *Initialize $K$.*
   (I.2) *Initialize $q^1$ to $q^{UB}$.*
(II) *Set $k \leftarrow 1$.*
*Repeat.*
   (II.1) *Compute $C^k(q^k + h)$ and $C^k(q^k - h)$, when $k > 1$.*
   (II.2) *Compute the desired gradients.*
   $dC^k = \frac{C^k(q^k+h)-C^k(q^k)}{h}$, *when $k = 1$.*
   $dC^k = \frac{C^k(q^k+h)-C^k(q^k-h)}{2h}$, *when $k > 1$.*
   (II.3) *Update the batch size, $q^{k+1} \leftarrow \lfloor q^k - \alpha_k dC^k \rfloor$.*
   $k \leftarrow k + 1$,
*Until $k = K$.*
(III) *Return the $\{q^k\}_{k=1}^K$ and the objective function value.*

**Theorem 6.** *By the finite difference algorithm 3 and a step size $\alpha_k$ which can satisfy Condition (1), the batch sizes $\{q^k\}_{k=1}^{\infty}$ in Model-3 can converge.*

Proof: From [1], for the finite difference algorithm with a step size which can satisfy Condition (1), if the objective function in Model-3 is convex, the batch size can converge. In Sect. 2, we have proven $T_P(q)$ and therefore $c_1\overline{T_P(q)}$ are convex function. Chew and Tang [4] have showed that the objective function $T_{TO}(q)$ is a convex function of $q$. So $C(q) = c_1\overline{T_P(q)} + c_2 T_{TO}(q)$ is a convex function of $q$. The convexity ensures the algorithm will converge to a global optimum.     $\square$

### 4.3 Results

Based on the formulation in Model-3, we implement the optimization algorithm 3 in Matlab. The running time ranges from 13 seconds to 19 seconds. For the coefficient $\gamma$ ranging from 30 to 70, we obtain the search paths in the Fig. 5. For all the experiments we conducted, we observe the search paths converge in the last 500 steps. We use the transient deletion technique to conduct the statistical estimation.

In order to verify the result from the finite difference optimization, we compare it with the result of enumeration, where we traverse all the batch size values from $q^{LB}$ to $q^{UB}$. We compute the expected objective values for all possible values of $q$ and find the optimal batch size.

For the aisle numbers ranging from 25 to 45, we respectively compute their "total cost", "part 1 cost" which is the cost of warehousing service providers, and "part 2 cost" which is the cost for the customers. The result is presented in Fig. 6 and

**Table 3.** Experiment result for Model-3

| No. | $M$ | $H$ | $\omega$ | $q^{UB}$ | $q^{LB}$ | $\gamma$ | $q^E$ | $\widehat{q}_1 = \overline{q} \pm HW$ | $R(\overline{q})$ | $\Delta_1$ | $\Delta_2$ |
|---|---|---|---|---|---|---|---|---|---|---|---|
| 1 | 25 | 20 | 3 | 50 | 6 | 20 | 12 | 12.2801±3.5136e-004 | 12 | 2.33% | 0 |
| 2 | 30 | 20 | 3 | 50 | 7 | 20 | 12 | 12.4999±1.1592e-005 | 12 | 4.17% | 0 |
| 3 | 35 | 20 | 3 | 50 | 8 | 20 | 12 | 12.5000±1.3900e-005 | 13 | 4.17% | 8.33% |
| 4 | 40 | 20 | 3 | 50 | 9 | 20 | 13 | 12.7900±9.0225e-004 | 13 | 1.62% | 0 |
| 5 | 45 | 20 | 3 | 50 | 10 | 20 | 13 | 13.4995±9.3120e-005 | 13 | 3.84% | 0 |
| 6 | 40 | 25 | 3 | 50 | 10 | 20 | 14 | 14.9999±1.8239e-005 | 14 | 3.57% | 0 |
| 7 | 40 | 27 | 3 | 50 | 10 | 20 | 15 | 15.4998±1.7737e-005 | 15 | 3.33% | 0 |
| 8 | 40 | 28 | 3 | 50 | 10 | 20 | 16 | 15.5001±2.3669e-005 | 16 | 3.12% | 0 |
| 9 | 40 | 30 | 3 | 50 | 10 | 20 | 16 | 16.9980±2.4165e-005 | 16 | 3.12% | 0 |
| 10 | 40 | 32 | 3 | 50 | 10 | 20 | 17 | 17.5000±2.5194e-006 | 18 | 2.94% | 5.88% |
| 11 | 40 | 20 | 3 | 50 | 10 | 30 | 15 | 15.4992±9.7966e-006 | 15 | 3.33% | 0 |
| 12 | 40 | 20 | 3 | 50 | 10 | 40 | 18 | 18.1962±6.0757e-004 | 18 | 1.09% | 0 |
| 13 | 40 | 20 | 3 | 50 | 10 | 50 | 20 | 20.4999±6.3132e-006 | 20 | 2.50% | 0 |
| 14 | 40 | 20 | 3 | 50 | 10 | 60 | 23 | 22.9373±1.3000e-003 | 23 | 0.27% | 0 |
| 15 | 40 | 20 | 3 | 50 | 10 | 70 | 25 | 25.2757±7.0235e-004 | 25 | 1.10% | 0 |

summarized in Table 3. The first column of Table 3 is the experiment index. The second part is the experiment setting: we have conducted three groups of experiments: varying the aisle number $M$ (experiments 1 to 5), the aisle length $H$ (experiments 6 to 10) and the cost ratio $\gamma$ (experiments 11 to 15). The third part of Table 2 is the computational results by Algorithm 3 and enumeration, containing $q^E$, $\widehat{q}$, $R(\overline{q})$, $\Delta_1$, and $\Delta_2$.

We compute the average direct bias $\overline{\Delta_1} = 1/N \sum_n |\overline{q} - q_n^E|/q_n^E = 2.700\%$ and the average indirect bias $\overline{\Delta_2} = 1/N \sum_n |R(\overline{q}) - q_n^E|/q_n^E = 0.947\%$ . The average direct bias of statistical estimation is less than 3%, and the average indirect bias of rounded statistical estimation is less than 1%.

From the results, we can observe that the optimal batch size for the total system is less than the optimal batch size $q_1^*$ in Model-1 and larger than $q_2^*$ in Model-2. We have $q_2^* \leq q_3^* \leq q_1^*$, which is robust in all the experiments with a different number of aisles, different aisle length, and different cost ratios $\gamma$. The result also shows that existing research underestimates the optimal batch size. This serious underestimation is due to the unilateral objective function, and it leads to an inferior performance of warehousing service providers. Our problem is a basic economic equilibrium problem with two sides of agents: a warehousing service supplier and consumers. From the perspectives of different agents, the optimal batch sizes are different.

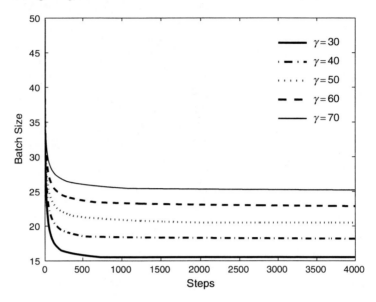

**Fig. 5.** Search path by Algorithm 3

## 5 Concluding Remarks

This paper studies the optimal order batch size problem in a parallel-aisle warehouse with stochastic order arrivals. The contribution of this paper is twofold in both application and methodology.

While existing literature directly assumes the batch size value, this paper shows that an optimal batch size exists. A too large batch size will harm the throughput time of consumers, and a too small batch size will bring a negative impact to warehousing costs. Existing literature focusing on the customer perspective only claims a suitable batch size will be close to its lower bound. Our research shows an optimal batch size will be larger than its lower bound when the costs of warehousing service providers are considered.

Past literature has not provided an efficient method to search optimal batch sizes. This paper provides an IPA and SPO stochastic approximate optimal implementation scheme to search the batch sizes for the warehousing service providers in a general parallel-aisle warehouse setting. This paper also presents an efficient FD algorithm to search the optimal batch sizes for customers and the total system. The estimation biases of the proposed algorithms are satisfactory.

A further topic for research could be to investigate the optimal batch size with the different routing policies in a general parallel-aisle warehouse with stochastic order arrivals. This paper employs an S-shape routing method. It is also possible to research the optimal batch size with other heuristic routing policies like the midpoint routing policy and the largest gap routing policy, or the optimal routing policy.

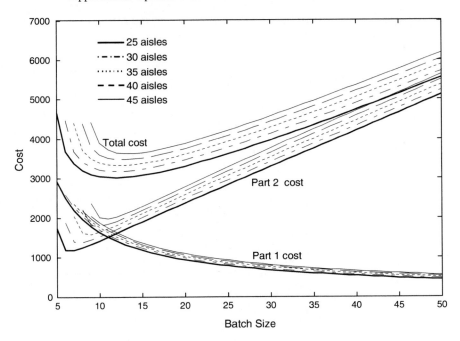

**Fig. 6.** Cost vs. batch sizes in Model-3

The latter will probably be hard to solve.

**Acknowledgements** The authors would like to thank the referees for their constructive comments that have led to many improvements in the exposition of this paper.

# References

[1] Bertsekas D (1999) Nonlinear programming, Athena Scientific, Belmont, Massachusetts, USA

[2] Boyd S, Vandenberghe L (2004) Convex optimization. Cambridge University Press, UK

[3] Carson Y, Maria A (1997) Simulation optimization: Methods and applications. In: Andradótir S, Healy KJ, Withers DH, and Nelson BL (eds) Proceedings of the 1997 Winter Simulation Conference. Atlanta, Georgia

[4] Chew EP, Tang LC (1999) Travel time analysis for general item location assignment in a rectangular warehouse. European Journal of Operational Research 112:582–597

[5] De Koster R, Le-Duc T, Roodbergen KJ (2007) Design and control of warehouse order picking: A literature review. European Journal of Operational Research 182:481–501

[6] De Koster R, Van der Poort, ES and Wolters, M (1999) Efficient orderbatching methods in warehouses. International Journal of Production Research 37(7):1479–1504

[7] Gademann N, Van De Velde S (2005) Order batching to minimize total travel time in a parallel-aisle warehouse. IIE Transactions 37(1):63–75

[8] Glasserman, P (1992) Derivative Estimates from Simulation of Continuous-Time Markov Chains. Operations Research 40:292–308

[9] Goetschalckx M and Ashayeri J (1989) Classification and design of order picking systems. Logistics World, June, 99–106

[10] Gong Y, Yücesan E (2006) The Multi-Location Transshipment Problem with Positive Replenishment Lead Times. ERIM working paper

[11] Heidelberger P, Cao X-R, Zazanis MA, Suri R (1988) Convergence properties of infinitesimal perturbation analysis estimates. Management Science 34(11):1281–1302

[12] Ho YC, Eyler MA, and Chien TT (1979) A Gradient Technique for General Buffer Storage Design in a Serial Production Line. International Journal of Production Research 17:557–580

[13] Fu MC (1994) Sample Path Derivatives for (s, S) Inventory Systems. Operations Research 42:351–364

[14] Fu MC (2002) Optimization for Simulation: Theory vs Practice. INFORMS Journal on Computing 14(3):192–215

[15] Fu MC, Hu JQ (1997) Conditional Monte Carlo: Gradient Estimation and Optimization Applications. Kluwer, Boston

[16] Fu MC, Hu JQ (1999) Efficient Design and Sensitivity Analysis of Control Charts using Monte Carlo Simulation. Management Science 45:395–413

[17] Johnson ME, Jackman J (1989) Infinitesimal perturbation analysis: a tool for simulation. Journal of the Operational Research Society 40(3):243–254

[18] Robbins H, Monro S (1951) A stochastic approximation method. Annals of Mathematical Statistics 22:400–407

[19] Tompkins JA, White JA, Bozer YA, Frazelle EH and Tanchoco JMA (2003) Facilities planning, Wiley, NJ x

[20] Le-Duc T, de Koster R (2007) Travel time estimation and order batching in a 2-block warehouse. European Journal of Operational Research 176(1):374–388

# Supply Chain Optimization for the Liquefied Natural Gas Business

Roar Grønhaug[1] and Marielle Christiansen[2]

[1] Department of Industrial Economics and Technology Management, Norwegian University of Science and Technology, Norway `roar.gronhaug@iot.ntnu.no`

[2] Department of Industrial Economics and Technology Management, Norwegian University of Science and Technology, Norway `marielle.christiansen@iot.ntnu.no`

**Summary.** The importance of natural gas as an energy source is increasing. Natural gas has traditionally been transported in pipelines, but ships are more efficient for transportation over long distances. When the gas is cooled down to liquid state it is called *liquefied natural gas* (LNG). The LNG supply chain consists of exploration, extraction, liquefaction, transportation, storage and regasification. Maritime transportation is a vital part of the LNG supply chain, and LNG is transported in special designed ships, LNG tankers. The demand for LNG tankers has increased considerably as the entire LNG industry continues to see strong growth. Hence, there is a great potential and need for optimization based decision support to manage the LNG fleet, liquefaction plants, and regasification terminals in this business.

Here, we are studying the LNG supply chain in close cooperation with a worldwide actor within the LNG business. This actor is responsible for the LNG supply chain management except the exploration and extraction.

We describe the real planning problem and present both an arc-flow and a path-flow model of the problem. Both models are tested and compared on instances motivated from the real-world problem. It is a very complex problem, so only small instances can be solved to optimality by these solution approaches.

**Key words:** Maritime transportation, Inventory routing

## 1 Introduction

Worldwide, there are large reserves of natural gas. Several existing gas producers are increasing their production capacity and new sources are explored. However, in some of these areas there are no significant markets (for instance North Africa, West Africa, South America, The Caribbean, The Middle East, Indonesia, Malaysia and

L. Bertazzi et al. (eds.), *Innovations in Distribution Logistics,* Lecture Notes in Economics and Mathematical Systems 619, DOI: 10.1007/978-3-540-92944-4,
© 2009 Springer-Verlag Berlin Heidelberg

Northwestern Australia). Some of the natural gas is liquefied at these locations for shipping to areas far away where usage of natural gas exceeds indigenous production. Such markets include Japan, Taiwan, Korea, Europe and the U.S. The transformation process from gas to liquefied natural gas (LNG) is done by cooling down the gas at atmospheric pressure at a temperature of $-260°F$ ( $-162°C$) before loading it into special designed tank ships, LNG tankers. By liquefying the natural gas into LNG the volume is reduced by a factor of 610 [8]. The reduction in volume makes transportation and storage more efficient. In addition, LNG offers greater trade flexibility than pipeline transport, allowing cargoes of natural gas to be delivered where the need is greatest and the commercial terms are most competitive.

Natural gas as an energy source is of increasing importance as the world's demand for natural gas is expected to increase by 70% between 2002 and 2025 [9]. Hence, the demand for LNG tankers is increasing. In 2007 there were 220 LNG tankers in operation, and 35 LNG tankers were scheduled for delivery in 2007. Furthermore, by 2015 the number of LNG tankers in operation will almost double to 400 [10]. As a consequence of the increasing market for LNG, the supply chain management has become more complex and the need for decision support has become even more evident. We consider a real tactical supply chain optimization problem for LNG including the production volumes, liquefaction, transportation, storage, regasification and sale volumes. Suez Energy International (SEI) is a global energy actor and is facing such a planning problem. The company is involved within most of the LNG supply chain except exploration and extraction, and is using a number of liquefaction plants and regasification terminals throughout the world. For the company's activity, the LNG can be considered a single product. The natural gas is cooled down at the liquefaction plants, stored at given pick-up ports, and transported at sea by LNG tankers to inventories at delivery ports before regasification. Inventory storage capacities are given at all ports. The production and consumption volumes are variable at all terminals. The transportation at sea is carried out with SEI's own heterogeneous fleet of LNG tankers. The hold at the LNG tanker is separated into several cargo tanks. It is assumed that an LNG tanker is always fully loaded when it leaves the pick-up port, but it is possible to unload a variable number of cargo tanks at each regasification terminal. In fact, the LNG is at boiling state in the cargo tanks. Thus, some of the LNG evaporates during a voyage. Hence the term boil-off. This gas is used as fuel. The planning problem is to maximize the profit by designing routes and schedules for the fleet, including determining the production and consumption volumes at all terminals, without exceeding the ship capacities and the inventory limits of the storages. We call this problem the LNG inventory routing problem (LNG-IRP).

Maritime transport optimization is a well established field of research within transportation planning with reviews in [12, 13, 5, 4]. Though the attention to maritime transportation has been limited compared to other modes of transportation, we have witnessed an accelerating amount of research in the literature during the last decade and the interest in these types of problems is increasing.

In maritime transportation, usually large quantities are loaded and unloaded at each port call (ship visit at port). Both the (un)loading and transportation are time consuming and expensive. Thus, the potential is great if the planning of the transportation and the inventory management at each end of a sailing leg is integrated. In practice, we can find several maritime supply chains where one of the actors has the responsibility for both the transportation and the inventory management. For instance, [3] studies such a problem for a company producing and consuming ammonia. Here the company both produces and consumes the product and is controlling the fleet of ships. Furthermore, [1] consider a maritime inventory routing problem with multiple chemical and oil products. These products have to be transported in separated compartments on board the ship and stored in separate storages at the ports. Moreover, [11] study a planning problem that integrates both the shipment planning of petroleum products from refineries to depots, and the production scheduling at the refineries. More maritime inventory routing problems are referred in [6].

However, no research on LNG-IRP is reported in the literature as far as we know. With increased focus on this type of problems in the industry, we expect several contributions in near future.

The purpose of this paper is to introduce a new type of problem within maritime transportation and provide two types of formulations for the same problem. Moreover, it will contribute to increased knowledge about the LNG supply chain from an OR point of view.

The rest of the paper is organized as follows: Sect. 2 gives some insights into the LNG industry and describes the real planning problem considered. The problem is formulated as an arc-flow model in Sect. 3, while Sect. 4 is devoted to the path-flow model. Computational results on small instances of the real planning problem are reported in Sect. 5. Finally, concluding remarks and future research follow in Sect. 6.

## 2 Description of a Real LNG Supply Chain Planning Problem

Suez Energy International (SEI) is a global energy actor. The company is a subsidiary of the international conglomerate Suez and together with its sister company, Suez Energy Europe (SEE) has the responsibilities of maintaining Suez' energy operations. SEI are involved within most of the liquefied natural gas (LNG) supply chain except exploration, extraction, and transportation to end-customers. Hence, the company is involved in liquefaction, transportation, storage and regasification of LNG. In addition, the company can influence the amount produced at the liquefaction plants and sold at the regasification terminals. Figure 1 shows the LNG supply chain and highlights the considered parts of the chain.

SEI is engaged in LNG supply chain planning at all levels, ranging from strategic decisions as determining the fleet size and mix, acquisition of plants and terminals, long-term contracts, to operational planning like determining the speed of each LNG tanker. In this paper, we consider the tactical supply chain planning problem and the typical planning horizon spans two to four months.

At the liquefaction plants the natural gas is cooled down to liquid state. Then, from ports located close to the liquefaction plants, the LNG is transported in special

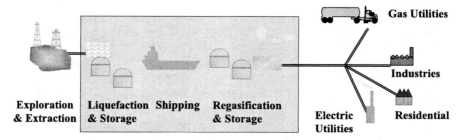

**Fig. 1.** The LNG supply chain

purpose vessels to ports close to storages and regasification terminals. Here, the LNG is converted from the liquefied state to the gaseous state, ready to be moved to the final destination through the natural gas pipeline system.

SEI controls two regasification terminals located in Zeebrugge, Belgium and Boston, USA. It has also 10% equity participation in a liquefaction plant located in Trinidad and Tobago. In addition, the company uses third-party facilities for pick-up and deliveries in other parts of the world. SEI currently purchases and distributes approximately 8 million tons of LNG per year from Algeria, Quatar, Trinidad and Tobago. The LNG operations are continuously increasing. For instance, a sales and purchase contract for 2.5 million tons of LNG per year was signed with Yemen LNG in August 2005. This contractual supply is expected to begin in 2009, and has 20 years duration.

Due to the expected increase in activity, SEI sees the need for an advanced decision support tool to coordinate and manage the fleet, and the inventories at both liquefaction and regasification terminals. For that reason, inventory management considerations are included at all ports. A reality consisting of 10 liquefaction plants and 10 regasification terminals is not impossible to imagine in the future. The number of spot cargoes in the LNG business is still limited, but we will see an increase in this activity in the future. In order to limit the size of the model and the level of details, we have disregarded possible spot trade in this paper.

The production of LNG at the liquefaction plants is normally at a maximum level. However, it is possible to regulate the production within certain limits. At each plant there exists given capacities of the storages. Production costs are dependent on volume and plant.

At the other end of the supply chain, the gas is unloaded from the LNG tankers and stored in storages with specified capacities. SEI's customers are governments, industrial corporations, the service industry, and residential users throughout the world. The sales contracts include fixed contracts where the agreed volume cannot be violated, contracts with lower and upper limits on quantities to deliver, and short term contracts which should be satisfied only if profitable. From this contract structure, we assume that for each port, we can specify upper and lower limits of demand for gas and an associated revenue per day. In reality, the consumption rate varies from day to day.

Extraction of natural gas, and hence the production of LNG, takes place all over the world. Natural gas is transported by either pipe lines or by ship to the customers. The main flows of natural gas and LNG in the world are shown in Fig. 2. Hence, the ports associated to the LNG liquefaction plants and regasification terminals are placed all over the world.

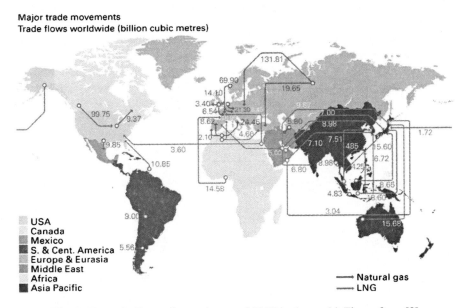

**Fig. 2.** The main flows of natural gas and LNG in the world. Figure from [2]

The LNG is transported by a fleet of LNG tankers controlled by SEI or by the associated company Suez Energy Europe. This fleet consists currently of 6 LNG tankers which they either have ownership over, or have chartered on long-term agreements. However, this number of LNG tankers will increase with increased activity. The LNG tankers have different cost structure, load capacity and specific ship characteristics. The hold of an LNG tanker is separated into several cargo tanks. Since the LNG is at boiling state in the cargo tanks, some of the cargo evaporates each day. This is called boil-off. Each day, the amount of boil-off in each cargo tank is a constant rate of the cargo capacity in the tank. Usually, the boil-off is used as fuel. It is the cargo itself that keeps the tanks cool, so if a cargo tank runs empty, the temperature will gradually increase. It is costly and time consuming to recool the cargo tanks before loading. Thus, there should always be some LNG left in the cargo tanks to keep them cool until (re)loading starts. Then, only a safety level should be left in the cargo tanks. No boil-off is assumed for the active tanks during loading and unloading in a port, while boil-off is considered for the tanks not affected at delivery ports. The loading and unloading of a ship are assumed to take one time period (one day) independent of the quantity loaded.

Successive calls at liquefaction plants are not relevant to consider. Furthermore, we can disregard the safety level from the calculations if we reduce the tank capacities with an appropriate safety level. Hence, when we speak about an 'empty cargo tank', there is a safety level of cargo left in the tank. In an optimal plan, an LNG tanker always arrives at a pick-up port and starts loading in the moment all the cargo tanks are empty and departs from the port fully loaded. However, at the delivery ports it is possible to unload partially. This means that several regasification terminals might be called in sequence, and a maximum number of successive delivery ports is given. In practice, this number is two. Due to sloshing problems for some types of LNG tankers, it is assumed that it is impossible to unload partial cargo tanks. Thus, a number of full cargo tanks adjusted for future boil-off until the next call to a liquefaction plant, must be unloaded at each delivery port. Figure 3 shows four snapshots of a voyage for an LNG tanker containing four cargo tanks. In Fig. 3a), the LNG tanker leaves the liquefaction plant fully loaded and the storage there is in one of its extreme situation; empty. The LNG tanker sails to a regasification terminal, and it has to arrive this terminal before the storage is empty. In Fig. 3b), we see that some of the gas has evaporated while sailing. The LNG tanker can then unload one or several of its cargo tanks. In this example, all tanks have been unloaded in one regasification terminal. Then, the LNG tanker returns in Fig. 3d) to the same liquefaction plant and the LNG tanker is just empty when it arrives the port.

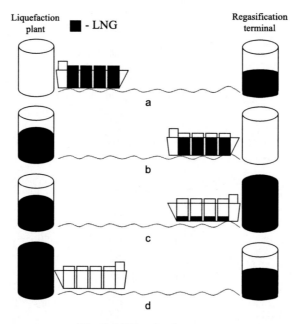

**Fig. 3.** LNG tanker inventory

The sailing time from one port to another is calculated based on the speed of the LNG tanker and the distance, but does not depend on the load aboard. There might also be several paths between two ports with different time consumptions and costs. For instance, this gives the possibility to use the Suez Canal or to sail around Africa.

The berth capacity of the ports is limited. Thus, a maximum number of LNG tankers can visit each port in each time period. However, it is possible to wait outside a port before loading and unloading, and the maximum number of waiting days outside each port is given. Normal boil-off is assumed during such waiting days. In contrast, no boil-off is assumed for the time periods from the last port call in a ship route until the end of the planning horizon. There is no natural depot for the LNG tankers. The initial position of an LNG tanker may be at a port or a point at sea. Furthermore, the LNG taker might be empty or loaded, and there is a set of first port call candidates in its route. Since there is no depot for the LNG tankers, there is no requirement for a specified position for any ship at the end of the planning horizon. In fact, the LNG tanker will end their route in one of the ports in the planning problem.

The LNG tankers are very specialized tank ships without any other area of application. In the short-term, there is no option to change the fleet size. The ship costs consist of several components. The fixed costs are the time charter rates which exist for all ships, while the variable costs consist of port and canal fees, and bunker oil costs.

In contrast to pickup and delivery vehicle routing problems [7], the number of calls to a port is not known, the quantity loaded or unloaded at each call is unknown and finally, there exist no pickup and delivery pairs. The LNG-inventory routing problem (LNG-IRP) aims at maximizing the profit by designing ship routes and schedules for the fleet in the planning period. Furthermore, the problem consists of deciding the production volumes of LNG, and determining the level of demand fulfillment. Finally, feasible inventory levels at both port types and load aboard the LNG tankers regarding the ship capacity and boil-off must be ensured.

# 3 Arc-Flow Formulation

This section describes the arc-flow formulation of the LNG-IRP. First, in Sect. 3.1 we introduce the network and describe the ship routing and scheduling constraints for the problem. Then, in Sect. 3.2 we present the constraints representing the ship inventory management. Section 3.3 is devoted to the activities at the ports, including the port inventory management. Finally, the objective function is addressed in Sect. 3.4.

The notation is based on the use of lower-case letters to represent decision variables and indices, while capital letters represent sets, constants and any constant superscripts.

## 3.1 Ship Routing and Scheduling

In the mathematical description of the problem, let $N$ be the set of physical ports indexed by $i$. This set consists of pick-up ports $N^P$ and delivery ports $N^D$. Further, let $V$, indexed by $v$, represent the heterogeneous fleet of ships (LNG tankers) available for routing and scheduling. Then, the set $N_v^{PD}$ denotes all ports feasible for ship $v$ (except its origin and destination node). Furthermore, $(N_v, \mathcal{A}_v)$ is the total network associated with a specific ship $v$. Here, $N_v = N_v^{PD} \cup \{o(v), d(v)\}$ is the set of ports that ship $v$ can visit, and $o(v)$ and $d(v)$ are the (artificial) origin node and (artificial) destination node, respectively. The set $\mathcal{A}_v$ contains all feasible arcs for ship v, which is a subset of $\{i \in N_v\} \times \{i \in N_v\}$. This set will be calculated based on capacity, time and inventory constraints, and other restrictions such as those based on precedence of pick-up and delivery nodes. From these calculations, we can extract the sets $N_v^P = N^P \cap N_v$ and $N_v^D = N^D \cap N_v$ consisting of pick-up and delivery nodes that ship $v$ may call, respectively.

The length of the planning horizon is given by the parameter $T^{MX}$. Moreover, $\mathcal{T}$ denotes the set of time periods, $\mathcal{T} = \{1, 2, \ldots, T^{MX}\}$, which is indexed by $t$. Let the parameter $T_{ijv}$ represent the sailing time on arc $(i, j)$ for ship $v$. Sailing on arc $(i, i)$ is considered waiting outside port $i$, $T_{iiv} = 1$. The maximum number of time periods a ship can wait outside a port before loading or unloading is denoted $T^W$. Cargo handling, i.e. loading and unloading, is assumed to take one time period. To ease the representation, the cargo handling at port $i$ is assumed to take place during the first time period on the sailing on arc $(i, j), i \neq j$. Each ship has a number of cargo tanks, $W_v^{MX}$, where the set of cargo tanks on each ship is given by $W_v$ and $w$ is the corresponding index.

The binary flow variable $x_{ijvt}$, $(i, j) \in \mathcal{A}_v$, $v \in V$, $t \in \mathcal{T}$ serves two purposes; sailing between two ports and waiting outside a port. If the variable equals 1 and $i = j$, ship $v$ waits one time period outside port $i$. On the other side, when $i \neq j$ and $x_{ijvt} = 1$, ship $v$ either loads or unloads at port $i$ in time period $t$ before it immediately starts sailing toward port $j$. The decision to load or unload a cargo tank is handled by the binary variable $z_{iwvt}$, $i \in N_v^{PD}$, $w \in W_v$, $v \in V$, $t \in \mathcal{T}$, which equals 1 if ship $v$ decides to load or unload cargo tank $w$ in port $i$ during time period $t$. Furthermore, the binary variable $u_{ivt}$, $i \in N_v^{PD}$, $v \in V$, $t \in \mathcal{T}$ equals 1 if ship $v$ loads or unloads any cargo tank in port $i$ during time period $t$.

In order to increase the readability of the arc-flow model, we eliminate the possibility of several paths between two nodes. However, this can easily be included in the model by introducing an additional index for the paths on the flow variable.

Then, the routing and scheduling part of the arc-flow LNG-IRP formulation is as follows:

$$\sum_{i \in \mathcal{N}_v^{PD}} x_{jivt} - \sum_{i \in \mathcal{N}_v^{PD}|t > T_{ijv}} x_{ijv(t-T_{ijv})} = 0, \qquad \forall j \in \mathcal{N}_v^{PD}, v \in \mathcal{V}, t \in \mathcal{T}, \qquad (1)$$

$$\sum_{j \in \mathcal{N}_v} x_{o(v)jvt} = 1, \qquad \forall v \in \mathcal{V}, t = 1, \qquad (2)$$

$$\sum_{t \in \mathcal{T}} \sum_{i \in \mathcal{N}_v} x_{id(v)vt} = 1, \qquad \forall v \in \mathcal{V}. \qquad (3)$$

Constraints (1)-(3) describe the flow on the route used by ship $v$. The first sailing from ship $v$'s origin node, $o(v)$, is handled by constraints (2), while constraints (3) give the end conditions for ship $v$, i.e. the ship must end its route in the destination node $d(v)$.

$$\sum_{j \in \mathcal{N}_v|j \neq i} x_{ijvt} - u_{ivt} = 0, \qquad \forall i \in \mathcal{N}_v^{PD}, v \in \mathcal{V}, t \in \mathcal{T}, \qquad (4)$$

$$x_{iivt} u_{ivt} = 0, \qquad \forall i \in \mathcal{N}_v^{PD}, v \in \mathcal{V}, t \in \mathcal{T}, \qquad (5)$$

$$\sum_{\tau = t|\tau \leq T^{MX} - T^W - 2\min_j\{T_{ijv}\}}^{t+T^W+2\min_j\{T_{ijv}\}} x_{iiv\tau} \leq T^W, \qquad \forall i \in \mathcal{N}_v^{PD}, v \in \mathcal{V}, t \in \mathcal{T}, \qquad (6)$$

$$x_{ijvt}\left(\sum_{w \in \mathcal{W}_v} z_{iwvt} + \sum_{\tau = t|\tau \leq T^{MX} - T_{ijv}}^{t+T^W} \sum_{w \in \mathcal{W}_v} z_{jwv(\tau+T_{ijv})} - W_v^{MX}\right) = 0,$$

$$\forall i \neq j, i \in \mathcal{N}_v^D, j \in \mathcal{N}_v^D, (i, j) \in \mathcal{A}_v, v \in \mathcal{V}, t \in \mathcal{T}. \qquad (7)$$

In constraints (4) we describe the connection between the cargo handling and the sailing. If a ship starts sailing between two ports, it must either load or unload depending on the type of port. Moreover, constraints (5) state that a ship cannot wait outside a port, i.e. traverses on an arc $(i, i)$, in the same time period it loads or unloads. Constraints (6) limit the number of waiting days outside port $i$ for ship $v$. Furthermore, constraints (7) assure that if a ship calls two consecutive delivery ports, all cargo tanks must unload at these ports.

$$z_{iwvt} \in \{0, 1\}, \qquad \forall i \in \mathcal{N}_v^{PD}, w \in \mathcal{W}_v, v \in V, t \in \mathcal{T}, \qquad (8)$$

$$u_{ivt} \in \{0, 1\}, \qquad \forall i \in \mathcal{N}_v^{PD}, v \in \mathcal{V}, t \in \mathcal{T}, \qquad (9)$$

$$x_{ijvt} \in \{0, 1\}, \qquad \forall (i, j) \in \mathcal{A}_v, v \in V, t \in \mathcal{T}. \qquad (10)$$

Finally, the formulation involves binary requirements (8)-(10).

A possible route through the network is illustrated in Fig. 4. The ship starts from its origin node in time period 1. The ship is either at sea and spends one time period to reach pick-up port $i$ or it waits outside port $i$ in this time period. After the arrival at port $i$ in time period 2 the ship loads its cargo tanks in this time period, before it sails to delivery port $j$. The sailing time between $i$ and $j$ is three time periods. When the ship has unloaded some of its cargo tanks in port $j$ in time period 6, the ship starts sailing towards delivery port $k$ in time period 7. Here, the ship unloads the rest of the cargo tanks and starts sailing towards port $i$. The ship arrives port $i$ in time period 14, but waits one day outside the port, before the ship loads all its cargo tanks and sails towards the destination node.

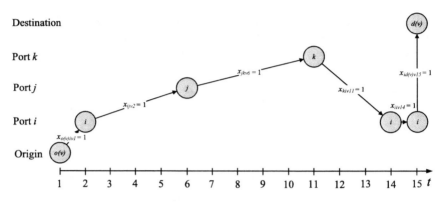

**Fig. 4.** Illustration of a possible route through the network for ship $v$

## 3.2 Ship Inventory Management

In order to describe the ship inventory management part of the LNG-IRP, we need the following additional notation.

The capacity of cargo tank $w$ on ship $v$ is given by $L_{wv}$. There is no loading or unloading at the origin node $o(v)$, but all cargo tanks have an initial load at the beginning of the planning horizon, $L_{wv}^O$. Furthermore, the parameter $I_i$, equals $-1$ if port $i$ is a pick-up port, and 1 if the port is an delivery port. The boil-off parameter $B_{wv}^F$, states the amount of cargo evaporating in each time period. A *duty* is a journey which starts when a ship either loads all its cargo tanks in a pick-up port or leaves the origin node, and the duty ends immediately before the ship starts loading at the next call at a pick-up port or when the ship reaches the destination node. We can for ship $v$ calculate upper and lower bounds on the total sailing time including waiting for the duties with visiting delivery port $i$. These upper and lower bounds are denoted $T_{iv}^{DMN}$ and $T_{iv}^{DMX}$, respectively.

The load in cargo tank $w$ at ship $v$ at the end of time period $t$ is measured by the continuous variable $l_{wvt}$, $w \in \mathcal{W}_v, v \in \mathcal{V}, t \in \mathcal{T}$. Note the initial condition $l_{wv0} = L_{wv}^O$. Finally, the continuous variable $q_{iwvt}$, $i \in \mathcal{N}_v^{PD}, w \in \mathcal{W}_v, v \in \mathcal{V}, t \in \mathcal{T}$ measures the amount of cargo loaded into or unloaded from cargo tank $w$ at ship $v$ in time period $t$.

$$- l_{wvt} + l_{wv(t-1)} + B_{wv}^F \left( \sum_{i \in \mathcal{N}_v^{PD}} z_{iwvt} + \sum_{i \in \mathcal{N}_v} \sum_{\tau=1}^{t-1} x_{id(v)v\tau} \right)$$

$$- \sum_{i \in \mathcal{N}_v^{PD}} I_i q_{iwvt} = B_{wv}^F, \forall w \in \mathcal{W}_v, v \in \mathcal{V}, t \in \mathcal{T}, \quad (11)$$

$$W_v^{MX} u_{ivt} - \sum_{w \in \mathcal{W}_v} z_{iwvt} = 0, \qquad \forall i \in \mathcal{N}_v^P, v \in \mathcal{V}, t \in \mathcal{T}, \quad (12)$$

$$q_{iwvt} - L_{wv} z_{iwvt} = 0, \qquad \forall i \in \mathcal{N}_v^P, w \in \mathcal{W}_v, v \in \mathcal{V}, t \in T, \quad (13)$$

$$u_{ivt} - z_{iwvt} \geq 0, \qquad \forall i \in \mathcal{N}_v^D, w \in \mathcal{W}_v, v \in \mathcal{V}, t \in T, \quad (14)$$

$$u_{ivt} - \sum_{w \in \mathcal{W}_v} z_{iwvt} \leq 0, \qquad \forall i \in \mathcal{N}_v^D, v \in \mathcal{V}, t \in \mathcal{T}, \quad (15)$$

$$q_{iwvt} - \left( L_{wv} - B_{wv}^F T_{iv}^{DMX} \right) z_{iwvt} \geq 0, \qquad \forall i \in \mathcal{N}_v^D, w \in \mathcal{W}_v, v \in \mathcal{V}, t \in \mathcal{T}, \quad (16)$$

$$q_{iwvt} - \left( L_{wv} - B_{wv}^F T_{iv}^{DMN} \right) z_{iwvt} \leq 0, \qquad \forall i \in \mathcal{N}_v^D, w \in \mathcal{W}_v, v \in \mathcal{V}, t \in \mathcal{T}, \quad (17)$$

$$0 \leq l_{wvt} \leq L_{wv}, \qquad \forall w \in W_v, v \in V, t \in \mathcal{T}. \quad (18)$$

The inventory balance on the ships are handled by constraints (11), which calculate the volume of LNG in each cargo tank on each ship in every time period. The amount of LNG in a cargo tank on a ship decreases at sea at an amount $B_{wv}^F$ in each time period, although there is no boil-off from a cargo tank that is being loaded or unloaded. In addition, there is no boil-off from the cargo tanks on a ship at the destination node. An illustration on how the variables and parameters affect the ship inventory is given in Fig. 5. Constraints (12)-(13) ensure that all cargo tanks are fully loaded at pick-up ports, when a loading starts. On the other hand, constraints (14)-(15) ensure that if a ship is unloading at least one cargo tank is unloaded, and vice versa. Moreover, constraints (16) give lower limits to the amount of LNG unloaded at the delivery ports, where the amount of cargo not delivered must be less or equal to the accumulated boil-off during the longest possible trip between two pick-up ports. The boil-off rate is usually a small percentage. Hence, constraints (16) allow only one unloading of a cargo tank before it is reloaded. Furthermore, constraints (16) in combination with constraints (7) limit the ships to sail to maximum two consecutive delivery ports. The upper limits on the unloading volumes at the delivery ports are given by (17). Finally, bounds on the ship inventory variable are described in constraints (18).

If a ship ends its route in an delivery port, the cargo tanks that have been unloaded should have sufficient LNG left aboard to reach a pick-up port. This amount is represented by the parameter $L_{wv}^E$. The constraints required for handling the end conditions for the delivery ports are as follows:

$$x_{id(v)vt} z_{iwvt} \left( l_{wvt} - L_{wv}^E \right) = 0, \qquad \forall i \in \mathcal{N}_v^D, w \in \mathcal{W}_v, v \in \mathcal{V}, t \in \mathcal{T}, \quad (19)$$

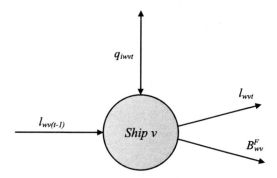

**Fig. 5.** Inventory balance for the ships

$$\sum_{\tau=t}^{t+T^W} x_{id(v)v\tau} \sum_{j\in N_v^D} x_{jiv(t-T_{ijv})} \left(l_{wvt} - L_{wv}^E\right) = 0,$$

$$\forall i \in \mathcal{N}_v^D, w \in \mathcal{W}_v, v \in \mathcal{V}, t \in \mathcal{T}. \quad (20)$$

Constraints (19) ensure that if there is an unloading immediately before the destination node, the cargo left in these cargo tanks at the unloading node should equal the parameter $L_{wv}^E$. Constraints (20) have similar purpose when a ship calls two consecutive delivery ports before the destination node. If a ship calls two consecutive delivery ports, all cargo tanks must have been unloaded when leaving the second delivery port. Thus, there is no need for including $z_{iwvt}$ or $z_{jwvt}$ in constraints (20).

**Linearization of Constraints**

Constraints (19)–(20) are nonlinear when we relax the binary requirements of the variables. In this section we linearize those constraints.

$$l_{wvt} - L_{wv}^E x_{id(v)vt} \geq 0, \qquad \forall i \in \mathcal{N}_v^D, w \in \mathcal{W}_v, v \in \mathcal{V}, t \in \mathcal{T}, \quad (21)$$

$$l_{wvt} + L_{wv} \left(x_{id(v)vt} + z_{iwvt}\right) \leq 2L_{wv} + L_{wv}^E, \quad \forall i \in \mathcal{N}_v^D, w \in \mathcal{W}_v, v \in \mathcal{V}, t \in \mathcal{T}. \quad (22)$$

Constraints (21) assure that all ships should have at least $L_{wv}^E$ of LNG aboard each cargo tank when they reach the destination node. Furthermore, constraints (22) are bounding when $x_{id(v)vt}z_{iwvt} = 1$. When $x_{id(v)vt}z_{iwvt} = 1$, i.e. when a ship unloads a cargo tank before sailing to the destination node, constraints (22) limits the cargo aboard that cargo tank to be less or equal than $L_{wv}$. Hence, constraints (21)-(22), can be regarded as linearized reformulations of constraints (19).

$$l_{wv\left(t+T_{id(v)v}\right)} + L_{wv} \left(\sum_{\tau=t}^{t+T^W} x_{id(v)v\tau} + \sum_{j\in N_v^D} x_{jivt}\right) \leq 2L_{wv} + L_{wv}^E,$$

$$\forall i \in \mathcal{N}_v^D, w \in \mathcal{W}_v, v \in \mathcal{V}, t \in \mathcal{T}. \quad (23)$$

When a ship unloads at two consecutive delivery ports before sailing to the destination node, there should be exact $L^E_{wv}$ of LNG left in the cargo tanks. Constraints (23) are bounding when a ship calls consecutive delivery ports before sailing to the destination node. Then, the cargo left in each cargo tank should be equal or less than $L^E_{wv}$. Hence, we can use constraints (23) in combination with constraints (21) to linearize constraints (20).

### 3.3 Port Operations and Inventory Management

Here, the constraints handling both the port operations and the inventory management at the ports are presented. We need to introduce the following additional parameters and variables:

The daily production and sales of LNG at the ports have to be within a given interval $\left[\underline{Y}_{it}, \overline{Y}_{it}\right]$, which can change from one time period to another. Furthermore, the inventory levels at the ports should be within the upper and lower limits $\left[\underline{S}_i, \overline{S}_i\right]$. The maximum number of ships at a port in a time period is given by the parameter $N^{CAP}_i$.

The sales and production of LNG are given by the continuous variable $y_{it}$, $i \in \mathcal{N}$, $t \in \mathcal{T}$, while the continuous variable $s_{it}$, $i \in \mathcal{N}$, $t \in \mathcal{T}$ represents the inventory level at the liquefaction plants and regasification terminals in the different time periods. Note that $s_{i0}$ represents the initial inventory.

$$s_{it} - s_{i(t-1)} - \sum_{v \in \mathcal{V}} \sum_{w \in \mathcal{W}_v} I_i q_{iwvt} + I_i y_{it} = 0, \qquad \forall i \in \mathcal{N}, t \in \mathcal{T}, \qquad (24)$$

$$\sum_{v \in \mathcal{V}} u_{ivt} \leq N^{CAP}_i, \qquad \forall i \in \mathcal{N}, t \in \mathcal{T}, \qquad (25)$$

$$\underline{S}_i \leq s_{it} \leq \overline{S}_i, \qquad \forall i \in \mathcal{N}, t \in \mathcal{T}, \qquad (26)$$

$$\underline{Y}_{it} \leq y_{it} \leq \overline{Y}_{it}, \qquad \forall i \in \mathcal{N}, t \in \mathcal{T}. \qquad (27)$$

The port inventory balances are given by constraints (24), and are further illustrated in Fig. 6. Constraints (25) ensure that the port capacity in the number of ships in each time period is not exceeded. The upper and lower bounds for the variables are given in constraints (26)-(27).

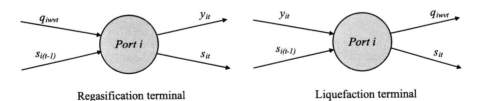

Regasification terminal               Liquefaction terminal

**Fig. 6.** Inventory balance at regasification and liquefaction terminals

### 3.4 Objective Function

Finally, we can present the objective function for the arc-flow LNG-IRP formulation. We need to introduce the following revenue and cost parameters.

The parameter $REV_{it}$ represents the unit revenue in each time period for selling LNG to the customers at the delivery ports, while $COST_{it}$ is the unit cost from producing LNG at the pick-up ports. Finally, $C_{ijv}$ is the transportation cost, i.e. the cost of traversing arc $(i, j)$ for ship $v$. The transportation cost parameter is a compound cost parameter, consisting of daily operating costs for ship $v$, port fees at port $i$, and any canal fees.

$$\max \sum_{i \in \mathcal{N}^D} \sum_{t \in \mathcal{T}} REV_{it} y_{it} - \sum_{i \in \mathcal{N}^P} \sum_{t \in \mathcal{T}} COST_{it} y_{it} - \sum_{(i,j) \in \mathcal{A}_v} \sum_{v \in \mathcal{V}} \sum_{t \in \mathcal{T}} C_{ijv} x_{ijvt} \qquad (28)$$

The objective function (28) maximizes total profit of selling LNG to end-customers while minimizing the costs of production and transportation.

# 4 Path-Flow Formulation

In the arc-flow formulation, the routes including arrivals times and load quantities are constructed based on the values on the variables, while these routes are enumerated a priori and feed into the path-flow formulation. The model is presented in Sect. 4.1, while the algorithm for enumerating the paths is described in Sect. 4.2.

### 4.1 The Model

In the path-flow formulation of the LNG-IRP, a route $r \in \mathcal{R}_v$ contains the geographical route and the schedule with information about the arrival times for all port calls for ship $v$ in the planning horizon. In addition, the route contains information about the quantities loaded and unloaded at the liquefaction plants and regasification terminals, respectively.

The parameter $Z_{ivtr}$ equals 1 if ship $v$ calls port $i$ in time period $t$ on route $r$ and 0 otherwise, while the corresponding (un)loading volume is given by the parameter $Q_{ivtr}$. The cost of sailing route $r$ for ship $v$ is given by the parameter $C_{vr}$. The cost parameters are composed of ship operation cost, port fees, and canal fees.

The binary variable $\lambda_{vr}$, $v \in \mathcal{V}$, $r \in \mathcal{R}_v$ is the ship route variable, and equals 1 if ship $v$ chooses to sail route $r$, and 0 otherwise.

Then, the path-flow formulation of the LNG-IRP can be modeled as follows:

$$\max \sum_{i \in \mathcal{N}^D} \sum_{t \in \mathcal{T}} REV_{it} y_{it} - \sum_{i \in \mathcal{N}^P} \sum_{t \in \mathcal{T}} COST_{it} y_{it} - \sum_{v \in \mathcal{V}} \sum_{r \in \mathcal{R}_v} C_{vr} \lambda_{vr}, \qquad (29)$$

subject to

$$s_{it} - s_{i(t-1)} - \sum_{v \in V} \sum_{r \in \mathcal{R}_v} I_i Q_{ivtr} \lambda_{vr} + I_i y_{it} = 0, \qquad \forall i \in \mathcal{N}, t \in \mathcal{T}, \qquad (30)$$

$$\sum_{v \in V} \sum_{r \in \mathcal{R}_v} Z_{ivtr} \lambda_{vr} \leq N_i^{CAP}, \qquad \forall i \in \mathcal{N}, t \in \mathcal{T}, \qquad (31)$$

$$\underline{S}_i \leq s_{it} \leq \overline{S}_i, \qquad \forall i \in \mathcal{N}, t \in \mathcal{T}, \qquad (32)$$

$$\underline{Y}_{it} \leq y_{it} \leq \overline{Y}_{it}, \qquad \forall i \in \mathcal{N}, t \in \mathcal{T}, \qquad (33)$$

$$\sum_{r \in \mathcal{R}_v} \lambda_{vr} = 1, \qquad \forall v \in \mathcal{V}, \qquad (34)$$

$$\lambda_{vr} \in \{0, 1\}, \qquad \forall v \in \mathcal{V}, r \in \mathcal{R}_v. \qquad (35)$$

The objective function (29) maximizes the profit from the LNG activities. The inventory balances are given in constraints (30), while the port capacity constraints are given by (31). The bounds in constraints (32)-(33) are identical to constraints (26)-(27). Constraints (34) are the convexity constraints, limiting the ships to sail exactly one route. Finally, the formulation involves binary requirements (35) on the ship route variables $\lambda_{vr}$.

## 4.2 Path Enumeration Algorithm

Here, we present an algorithm for complete enumeration of all possible paths, called routes, for the path-flow LNG-IRP formulation. By use of a recursive algorithm we can identify all possible routes with information regarding the geographical route, the arrival times and the number of cargo tanks loaded and unloaded at all ports. When a route has been found, the algorithm must identify all the duties within the routes. These duties are in fact distinct subpaths in the route. The algorithm must calculate the aggregated boil-off from each cargo tank during the duties in order to calculate the exact amount of cargo unloaded at the delivery ports. This can be expressed mathematically if we introduce some new notation. Let $\Delta_{vr}$, indexed by $d$, denote the set of duties for ship $v$ on route $r$. The aggregated boil-off in cargo tank $w$ on ship $v$ on duty $d$ on route $r$ is given by $B_{wvrd}^{FA}$. Furthermore, $X_{ijvtrd} = 1$ if arc $(i, j)$ is traversed by ship $v$ starting in time period $t$ on duty $d$ on route $r$, and 0 otherwise. Moreover, $Z_{iwvtrd}^{\Delta} = 1$ if ship $v$ loads or unloads cargo tank $w$ in time period $t$ on duty $d$ on route $r$, and 0 otherwise.

$$B^{FA}_{wvrd} = B^F_{wv} \left( \sum_{(i,j) \in \mathcal{A}_v} \sum_{t \in \mathcal{T}} X_{ijvtrd} T_{ijv} - \sum_{i \in \mathcal{N}_v^{PD}} \sum_{t \in \mathcal{T}} Z^{\Delta}_{iwvtrd} \right),$$

$$\forall w \in \mathcal{W}_v, v \in \mathcal{V}, r \in \mathcal{R}_v, d \in \Delta_{vr}, \tag{36}$$

$$Q_{ivtr} = \begin{cases} \sum_{w \in \mathcal{W}_v} \sum_{d \in \Delta_{vr}} Z^{\Delta}_{iwvtrd} L_{wv}, & \forall i \in \mathcal{N}_v^P, v \in \mathcal{V}, t \in \mathcal{T}, r \in \mathcal{R}_v, \\ \sum_{w \in \mathcal{W}_v} \sum_{d \in \Delta_{vr}} Z^{\Delta}_{iwvtrd} \left( L_{wv} - B^{FA}_{wvrd} \right), & \forall i \in \mathcal{N}_v^D, v \in \mathcal{V}, t \in \mathcal{T}, r \in \mathcal{R}_v, \end{cases} \tag{37}$$

$$Z_{ivtr} \geq Z^{\Delta}_{iwvtrd}, \forall i \in \mathcal{N}_v, w \in \mathcal{W}_v, v \in \mathcal{V}, t \in \mathcal{T}, r \in \mathcal{R}_v, d \in \Delta_{vr}. \tag{38}$$

With equations (36)–(38) we can calculate the input from the path enumeration algorithm to the path-flow model. In equations (36) we calculate the aggregated boil-off during a duty. The aggregated boil-off is used as input when calculating the amount of cargo loaded and unloaded at the ports in equations (37). Finally, equations (38) gives the relation between the cargo tank (un)loading parameter on a duty and the ship call parameter.

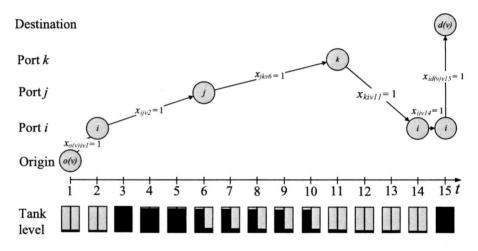

**Fig. 7.** Illustration of a possible route through the network for ship $v$ and the corresponding cargo tank levels

Figure 7 illustrates a possible route through the network for a ship with two cargo tanks. The ship loads both its cargo tanks in time period 2. Then it unloads one cargo tank in time period 6 and the other one in time period 11. Some LNG is left in the cargo tanks for the boil-off after the tanks have been unloaded. This boil-off keeps the cargo tanks cool and is also used as fuel until the cargo tanks are reloaded in time period 15.

The pseudocode for the path enumeration algorithm is described in Algorithm 1 and Algorithm 2. Here, Algorithm 1 initializes the path enumeration algorithm, while the recursive part of the algorithm which searches for all paths is given in Algorithm 2. To illustrate how the algorithm works we use the example from Fig. 7. The path enumeration algorithm will first identify the complete route, then it will identify the three duties in the network. The first duty starts in the origin node and ends in port $i$ in time period 2. The second one starts in time period 2 when the ship leaves port $i$, and ends in time period 15 when the ship once again will leave port $i$. Note that the waiting day outside port $i$ in time period 14 is considered belonging to the second duty. Finally, the third duty both starts and ends in time period 15, as it covers the sailing from port $i$ to the destination node $d(v)$.

---

**Algorithm 1** initializePath

---

**for all** $v \in \mathcal{V}$ **do**

$\quad$ $r = 0$, the route numbering starts from 0 for each ship

$\quad$ $N^{LIST} = \emptyset$, the list of nodes in the route

$\quad$ call createPath$\left(r, o\left(v\right), N^{LIST}\right)$

**end for**

---


---

**Algorithm 2** createPath$\left(r, i, N^{INLIST}\right)$

---

$N^{LIST} = N^{INLIST} \cup \{i\}$

**for all** $j|(i, j) \in \mathcal{A}_v$ **do**

$\quad$ **if** $j \neq d(v)$ **then**

$\quad\quad$ call createPath$\left(r, j, N^{LIST}\right)$

$\quad$ **else**

$\quad\quad$ $N^{LIST} = N^{LIST} \cup \{j\}$

$\quad\quad$ Identify all duties in $N^{LIST}$ and create the set $\Delta_{vr}$ and the tables $X_{ijvtrd}$, $Z_{ivtr}$, and $Z^{\Delta}_{iwvtrd}$

$\quad\quad$ Calculate the boil-off, $B^{FA}_{wvrd}$, in each duty with equations (36)

$\quad\quad$ Calculate the quantity loaded or unloaded, $Q_{ivtr}$, with equations (37)

$\quad\quad$ $r = r + 1$

$\quad$ **end if**

**end for**

---

For instance, if the cargo tanks have a capacity of 75,000 m$^3$ each and the boil-off is 115 m$^3$, we can calculate the aggregated boil-off and the amount of cargo unloaded at the ports. The aggregated boil-off for cargo tank 1 on the second duty is then calculated based on a 13 time period long duty, less one time period for loading and one for unloading: $B^{FA}_{1vr2} = 115\,(13 - 2) = 1\,265$ m$^3$. Note that the loading in time period 15 belongs to the third duty. Then, if cargo tank 1 is unloaded at port $j$ in time period 6, 73,735 m$^3$ will be unloaded there. Since the two cargo tanks are identical, the second tank will unload 73,735 m$^3$ at port $k$ in time period 11.

## 5 Computational Results

The purpose of the computational study is to evaluate the two proposed model formulations of the LNG-IRP, the arc-flow formulation presented in (1)–(28), and the path-flow formulation presented in (29)–(35). The models have been solved by use of XPRESS Optimizer v 17.1 on a computer with a 3 GHz processor and 8 GB RAM running on Rock Cluster v 4.2.1 operating system. Furthermore, the path enumeration algorithm presented in Sect. 4.2 has been programmed in C++ and solved on the same architecture.

We have created 21 instances motivated by the real planning problem faced by Suez Energy International. An overview of these instances is presented in Table 1. In this table the number of ships, cargo tanks, ports, and time periods are given for each instance. The number of pick-up and delivery ports are given in parenthesis behind the total number of ports. The number of ships ranges from 1 to 5 depending on the instance, while the number of ports is between 3 and 6. Keep in mind that each port may have ship calls several times during the planning period. Furthermore, the instances have 30, 45, or 60 time periods. To be able to test how the time horizon affect the solution time, some instances share most characteristics, but have different number of time periods. For instance, instance 1–3 have the same physical network and similar upper and lower bounds on production and sales, and 30, 45, and 60 time periods, respectively. The other instances can also be grouped similarly (4–6, 7–9, . . .).

The number of rows and columns for the instances for both formulations are presented in Table 2. Note that the number of integer columns for the path-flow formulation is the number of routes for the ships. This number gives some indication of the complexity of the problems, and how the instances escalate with the number of time periods. As we can see from the table, the path-flow formulation scales very poorly with respect to the number of time periods. Furthermore, for instances #9 and #15 we were not able to enumerate all routes as the optimizer ran out of memory during the enumeration process. For the other instances the number of routes spans from 426 for instance #10 to more than 1.5 million for instance #21.

In Table 3 the solution times for each instance are reported. For both formulations, the table reports the time to solve the linear relaxation, first integer solution, best integer solution, and the total solution time. All these solution times are measured starting from the moment the optimizer starts solving the problem, after the complete matrix has been fed into the solver. Maximum running time for the instances is 10 h. Thus, if the search is not completed within the time limit, the search is interrupted. In addition, the table reports the time needed for the path enumeration algorithm to enumerate all routes and feed them into the solver for the path-flow formulation.

The LNG-IRP is hard to solve. In general, both formulations solve the minor instances efficiently, while both formulations have problems with solving the instances with longer planning horizon to optimality. The path-flow formulation is the most efficient formulation with respect to total solution time for 7 of the instances, while the arc-flow performs better for 7 when we disregard 2 instances where we only have test results from the arc-flow formulation. The path-flow formulation suffers from its

**Table 1.** Instance overview

| # | Ships | Tanks | Ports (P,D) | Time Periods |
|---|-------|-------|-------------|--------------|
| 1 | 1 | 2 | 4(1,3) | 30 |
| 2 | 1 | 2 | 4(1,3) | 45 |
| 3 | 1 | 2 | 4(1,3) | 60 |
| 4 | 2 | 2 | 3(1,2) | 30 |
| 5 | 2 | 2 | 3(1,2) | 45 |
| 6 | 2 | 2 | 3(1,2) | 60 |
| 7 | 2 | 2 | 4(2,2) | 30 |
| 8 | 2 | 2 | 4(2,2) | 45 |
| 9 | 2 | 2 | 4(2,2) | 60 |
| 10 | 2 | 1 | 5(2,3) | 30 |
| 11 | 2 | 1 | 5(2,3) | 45 |
| 12 | 2 | 1 | 5(2,3) | 60 |
| 13 | 2 | 2 | 5(2,3) | 30 |
| 14 | 2 | 2 | 5(2,3) | 45 |
| 15 | 2 | 2 | 5(2,3) | 60 |
| 16 | 3 | 1 | 4(2,2) | 30 |
| 17 | 3 | 1 | 4(2,2) | 45 |
| 18 | 3 | 1 | 4(2,2) | 60 |
| 19 | 5 | 1 | 6(3,3) | 30 |
| 20 | 5 | 1 | 6(3,3) | 45 |
| 21 | 5 | 1 | 6(3,3) | 60 |

poor scaling capacity, which leads to large solution time even for the LP relaxation. For instance, solving the LP relaxation for instance #3 with the path-flow formulation is almost 2 h, while the corresponding solution time for the arc-flow formulation is 0 s. As a consequence of the long solution time for the path-flow formulation's LP relaxation, the time to find the first integer solution is longer than for the arc-flow formulation for 11 of the instances, and faster for only 3 of the instances.

The solution values and the MIP-gaps are presented in Table 4. The MIP-gap is defined as $|MIP^*\text{-}Bound^*|/Bound^*$, where Bound* is the best bound on the solution from the branch-and-bound procedure and MIP* is the best (mixed) integer solution. For the instances that were not solved to optimality, the MIP-gaps are in the interval [2.9%, 42.9%]. One of the reasons for the large MIP-gaps is the poor linear relaxation for the LNG-IRP. If we define the LP-gap as the percentage deviation between the best MIP solution and the LP solution, $|MIP^*\text{-}LP^*|/LP^*$, the average LP-gap is 19,4% for both model formulations. This indicates that for some of the instances where we did not manage to prove the optimal solution, the best integer solution found might be near optimal or even the optimal solution.

**Table 2.** Dimensions of the instances

|   | Arc-flow | | | Path-flow | | |
|---|---|---|---|---|---|---|
| # | Rows | Columns | Int columns | Rows | Columns | Int columns |
| 1 | 2,072 | 1,411 | 871 | 243 | 1,791 | 1,551 |
| 2 | 3,107 | 2,116 | 1,306 | 363 | 63,101 | 62,741 |
| 3 | 4,142 | 2,821 | 1,741 | 483 | 2,534,165 | 2,533,685 |
| 4 | 2,764 | 2 044 | 1,384 | 184 | 1,110 | 930 |
| 5 | 4,144 | 3,064 | 2,074 | 274 | 28 348 | 28,078 |
| 6 | 5,524 | 4,084 | 2,764 | 364 | 849,858 | 849,498 |
| 7 | 3,424 | 2,764 | 1,924 | 244 | 9,018 | 8,778 |
| 8 | 5,134 | 4 144 | 2,884 | 364 | 461,293 | 460,933 |
| 9 | 6,844 | 5,524 | 3,844 | – | – | – |
| 10 | 3,544 | 2,102 | 1,502 | 304 | 728 | 426 |
| 11 | 5,314 | 3,152 | 2,252 | 454 | 6,883 | 6,431 |
| 12 | 7,084 | 4,202 | 3,002 | 604 | 93,325 | 92,723 |
| 13 | 4,564 | 2 702 | 1,802 | 304 | 4,010 | 3,710 |
| 14 | 6,844 | 4,052 | 2,702 | 452 | 239 208 | 238,758 |
| 15 | 9,124 | 5,402 | 3,602 | – | – | – |
| 16 | 3,846 | 3,036 | 2,346 | 245 | 1,339 | 1,096 |
| 17 | 5,766 | 4,551 | 3,516 | 365 | 20,863 | 20,500 |
| 18 | 7 686 | 6,066 | 4,686 | 485 | 363,802 | 363,319 |
| 19 | 9,670 | 7,480 | 6,070 | 367 | 2 774 | 2,414 |
| 20 | 14,500 | 11,215 | 9,100 | 547 | 65,695 | 65,155 |
| 21 | 19,330 | 14,950 | 12,130 | 727 | 1,561,996 | 1,561,276 |

Looking at the quality of the first integer solutions found during branch-and-bound, we can see that for 11 of the instances the optimizer finds better solutions for the path-flow formulation than for the arc-flow formulation. Also the arc-flow formulation finds better integer solutions for 6 of the instances when we disregard instances #9 and #15. The path-flow formulation's linear relaxations are tighter than the corresponding linear relaxations for the arc-flow formulation for all instances except instances #10 and #13. We would expect that the path-flow formulation would have best linear relaxation for all instances. However, when solving these instances without presolve applied, the conclusion alters. With the optimizer's presolve turned off, the LP solutions for instances #10 and #13 are 2,433.7 for the arc-flow formulation and 2,403.1 for the path-flow formulation.

**Table 3.** Solution times. All measures in sec. LP – linear relaxation, MIP [1] - first MIP solution, MIP* – best MIP solution, Total – total solution time, Enum = time to enumerate the routes

| # | Arc-flow | | | | Path-flow | | | | |
|---|---|---|---|---|---|---|---|---|---|
| | LP | MIP[1] | MIP* | Total | LP | MIP[1] | MIP* | Total | Enum |
| 1 | 0 | 0 | 1 | 2 | 0 | 0 | 0 | 0 | 0 |
| 2 | 0 | 6 | 6 | 63 | 5 | 12 | 12 | 27 | 1 |
| 3 | 0 | 31 | 7,548 | 7,548 | 7,131 | 8,992 | 23,540 | 36,000 | 56 |
| 4 | 0 | 0 | 2 | 120 | 0 | 0 | 0 | 1 | 0 |
| 5 | 0 | 3 | 7,105 | 36 000 | 1 | 6 | 186 | 423 | 0 |
| 6 | 0 | 6 | 23,706 | 36,000 | 140 | 378 | 32,096 | 36,000 | 19 |
| 7 | 0 | 1 | 1 | 1 | 0 | 0 | 1 | 43 | 0 |
| 8 | 0 | 1 | 3 | 4 | 27 | 43 | 180 | 1,761 | 8 |
| 9 | 0 | 155 | 454 | 456 | – | – | – | – | – |
| 10 | 0 | 0 | 0 | 0 | 0 | 0 | 0 | 0 | 0 |
| 11 | 0 | 0 | 3 | 15 | 0 | 4 | 196 | 973 | 0 |
| 12 | 0 | 7 | 27 | 56 | 15 | 70 | 33,605 | 36,000 | 2 |
| 13 | 0 | 0 | 21 | 22 | 0 | 0 | 1 | 5 | 0 |
| 14 | 0 | 0 | 2,693 | 3,797 | 67 | 237 | 1,162 | 36,000 | 4 |
| 15 | 2 | 63 | 33,093 | 36,000 | – | – | – | – | – |
| 16 | 0 | 1 | 42 | 55 | 0 | 0 | 13 | 14 | 0 |
| 17 | 0 | 0 | 225 | 9,217 | 0 | 0 | 206 | 13,625 | 0 |
| 18 | 0 | 0 | 3 354 | 36,000 | 35 | 223 | 35,896 | 36,000 | 7 |
| 19 | 0 | 1 | 17 | 172 | 0 | 0 | 3 | 39 | 0 |
| 20 | 0 | 11 | 7,476 | 36 000 | 3 | 13 | 3,074 | 36,000 | 1 |
| 21 | 1 | 41 | 12,214 | 36,000 | 321 | 8 724 | 20 492 | 36,000 | 32 |

## 6 Concluding Remarks

In this paper we have introduced a new type of optimization problems, the *liquefied natural gas inventory routing problem*, LNG-IRP. This problem deals with managing the supply chain for the liquefied natural gas (LNG) business at a tactical planning level. Here, one actor controls the supply chain from liquefaction to sales, where both the production and sales levels are variable and may change from day to day. Furthermore, ship routing and scheduling of specialized ships (LNG tankers) are important parts of this supply chain. The problem is more complicated than many other maritime inventory routing problems, as it deals with variable rates of production and consumption. Moreover, the ship routing and scheduling are also more complicated, as the ships' cargo tanks should not run dry at sea, as they have to deal with a constant rate of boil-off. In addition, the ships load all their cargo tanks at the pick-up ports, and unload a discrete number of cargo tanks at the delivery ports.

**Table 4.** Solution values and MIP-gaps for the instances

| # | Arc-flow | | | | Path-flow | | | |
|---|---|---|---|---|---|---|---|---|
| | LP | MIP[1] | MIP* | MIP-gap(%) | LP | MIP[1] | MIP* | MIP-gap |
| 1 | 1,036.9 | 807.0 | 887.0 | 0.0 | 1,024.7 | 887.0 | 887.0 | 0.0 |
| 2 | 1,504.8 | 1 314.7 | 1,314.7 | 0.0 % | 1,470.4 | 1,314.7 | 1,314.7 | 0.0 |
| 3 | 1,940.0 | 1,354.7 | 1,533.0 | 0.0 | 1,896.6 | 1,133.8 | 1,331.5 | 28.8 |
| 4 | 782.7 | 490.7 | 681.0 | 0.0 | 760.9 | 537.0 | 681.0 | 0.0 |
| 5 | 1 082.1 | 507.8 | 940.8 | 2.9 | 1,044.6 | 688.5 | 940.8 | 0.0 |
| 6 | 1,375.3 | 429.2 | 1 122.6 | 15.2 | 1,326.8 | 804.2 | 1,095.3 | 16.9 |
| 7 | 1,313.6 | 992.1 | 1,018.6 | 0.0 | 1 262.5 | 905.0 | 1,018.6 | 0.0 |
| 8 | 1 838.2 | 1,142.2 | 1,492.6 | 0.0 | 1,761.3 | 1,166.2 | 1,492.6 | 0.0 % |
| 9 | 2,346.9 | 1,243.8 | 1 784.2 | 0.0 | – | – | – | – |
| 10 | 2,329.7 | 2,170.0 | 2,170.0 | 0.0 | 2 396.1 | 2,002.0 | 2,170.0 | 0.0 |
| 11 | 3 371.3 | 2,231.9 | 2,544.8 | 0.0 | 3,350.2 | 1,996.8 | 2,544.8 | 0.0 |
| 12 | 3,944.5 | 2,247.8 | 2,920.3 | 0.0 | 3 892.7 | 2,403.2 | 2,745.6 | 26.6 |
| 13 | 2 394.5 | 1,742.5 | 2,170.0 | 0.0 | 2,396.2 | 2,068.3 | 2,170.0 | 0.0 |
| 14 | 3,462.3 | 1,659.0 | 2 591.0 | 0.0 | 3,352.6 | 2,259.9 | 2,591.0 | 15.2 |
| 15 | 4,421.3 | 1,997.1 | 2,882.3 | 18.8 | – | – | – | – |
| 16 | 1,290.5 | 867.1 | 1 118.0 | 0.0 % | 1,287.3 | 1,085.7 | 1,118.0 | 0.0 |
| 17 | 1,778.4 | 1,291.3 | 1,435.4 | 0.0 | 1 771.2 | 1,291.3 | 1,435.4 | 0.0 |
| 18 | 2,244.9 | 1,243.0 | 1,579.7 | 16.6 | 2,228.2 | 953.2 | 1,579.7 | 27.6 |
| 19 | 1 910.9 | 1,530.8 | 1,697.6 | 0.0 | 1,909.4 | 1,180.8 | 1,697.6 | 0.0 |
| 20 | 2,498.4 | 947.4 | 2 032.5 | 6.0 | 2,483.1 | 1,304.0 | 2,011.3 | 16.0 |
| 21 | 3,026.6 | 910.3 | 2,062.3 | 25.1 | 2 995.7 | 1,090.4 | 1,684.5 | 42.9 |

We have proposed two formulations of the LNG-IRP; an arc-flow and a path-flow formulation. In the path-flow formulation, a path represents a possible geographical route and schedule for a ship during the entire planning horizon. In addition, the path handles the ship inventory management, the boil-off from the cargo tanks, and the amount of cargo loaded and unloaded at the pick-up and delivery ports. Moreover, we have presented an algorithm for complete enumeration of the columns in the path-flow model.

Both model formulations have been tested on instances motivated by a real planning problem. Both formulations provide good solutions to the test instances presented, although none of them were able to solve the largest instances to optimum. From the limited number of instances, it is hard to conclude which formulation is superior. The path-flow formulation was able to solve more instances faster to optimum than the arc-flow formulation, while the arc-flow formulation finds the first integer solution faster than the path-flow formulation. Furthermore, the path-flow formulation suffers from poor scaling capabilities. Hence, the optimizer ran out of memory when we tried to enumerate the columns for two of the instances.

The LNG-IRP is hard to solve. Thus, none of the proposed formulations where able to verify the optimal solutions for all instances presented. To be able to solve these, and even larger instances, more research is required. Solution approaches based on both exact methods and heuristics may be appropriate for solving larger instances of the LNG-IRP. Within exact methods, column generation seems like a particularly interesting alternative since we will be able to work with a subset of the columns for the problems. Moreover, the LP-gap for the LNG-IRP is poor. Thus, development of valid inequalities might also be valuable.

## Acknowledgments

This work was carried out with financial support from the Research Council of Norway through the INSUMAR project (Integrated supply chain and maritime transportation planning), the OPTIMAR project (Optimization in Maritime transportation and logistics) and the DOMinant project (Discrete optimization methods in maritime and road-based transportation). We would also like to thank Chief Analyst Stephane Hecq and Senior LNG Analyst Dr. Geert Stremersch from the Suez Corporation for their involvement in the project.

## References

[1] F. Al-Khayyal and S.-J. Hwang. Inventory constrained maritime routing and scheduling for multi-commodity liquid bulk, part I: Applications and model. *European Journal of Operational Research*, 176:106–130, 2007.

[2] BP. *BP Statistical Review of World Energy 2007*. British Petroleum, 2007.

[3] M. Christiansen. Decomposition of a combined inventory and time constrained ship routing problem. *Transportation Science*, 33(1):3–14, 1999.

[4] M. Christiansen, K. Fagerholt, B. Nygreen, and D. Ronen. Maritime transportation. In C. Barnhart and G. Laporte, editors, *Transportation*, volume 14 of *Handbooks in Operations Research and Management Science*, pages 189–284. Elsevier Science, 2007.

[5] M. Christiansen, K. Fagerholt, and D. Ronen. Ship routing and scheduling, status and perspectives. *Transportation Science*, 38(1):1–18, 2004.

[6] Marielle Christiansen and Kjetil Fagerholt. Maritime inventory routing problems. *Encyclopedia of Optimization*, 2007. To appear. pp. 16.

[7] G. Desaulniers, J. Desrosiers, A. Erdmann, M.M. Solomon, and F. Soumis. VRP with pickup and delivery. In P. Toth and D. Vigo, editors, *The Vehicle Routing Problem*, volume 9 of *SIAM Monographs on Discrete Mathematics and Applications*, pages 225–242. SIAM, Philadelphia, PA, 2002.

[8] EIA. *The Global Liquefied Natural Gas Market: Status and Outlook*. Energy Information Administration, U.S. Department of Energy, 2003.

 [9] EIA. *International Energy Outlook 2005*. Energy Information Administration, U.S. Department of Energy, 2005.

[10] IEA. *Natural Gas Market Review 2007: Security in a globalising market to 2015*. International Energy Agency, 2007.

[11] Jan A. Persson and Maud Göthe-Lundgren. Shipment planning at oil refineries using column generation and valid inequalities. *European Journal of Operational Research*, 163(3):631–652, 2005.

[12] David Ronen. Cargo ships routing and scheduling: Survey of models and problems. *European Journal of Operational Research*, 12(2):119–126, 1983.

[13] David Ronen. Ship scheduling: The last decade. *European Journal of Operational Research*, 71(3):325–333, 1993.

# Modeling The Pre-Auction Stage: The Truckload Case

Gianfranco Guastaroba[1], Renata Mansini[2], and M. Grazia Speranza[3]

[1] University of Brescia, Department of Quantitative Methods, Italy
   guastaro@eco.unibs.it
[2] University of Brescia, Department of Electronics for Automation, Italy
   rmansini@ing.unibs.it
[3] University of Brescia, Department of Quantitative Methods, Italy
   speranza@eco.unibs.it

**Summary.** In transportation service procurement, shipper and carriers cost functions for serving a pair of origin-destination points, usually called *lanes*, are highly dependent on the opportunity to serve neighboring lanes. Traditional single-item auctions do not allow to capture this type of preferences. On the contrary, they are perfectly modeled in combinatorial auctions where bids on bundles of items are allowed. In transportation service procurement the management of a combinatorial auction can be seen as a three-stage process. Each stage involves several complex decision making problems. All such problems have relevant practical implications but only some of them have received attention in the literature. In the present paper we focus on the pre-auction stage for transportation procurement. In particular, we analyze the problem of a shipper who has to decide between undertaking and/or outsourcing (through an auction) his transportation requests. The problem has never been analyzed before. We propose two different models for the problem in the truckload case and provide their computational comparison on randomly generated instances.

**Key words:** Electronic auctions, Truckload problem, Shipper's lane selection problem, Arc routing problems

## 1 Introduction

The worldwide use of the Internet has deeply modified the way commercial agreements in logistics are stipulated (see Cranic and Speranza [7]). As in traditional marketplaces, every agent that approaches an e-marketplace has to solve different complex problems, either in case one has to buy or to sell an item. Making the most profitable pricing decision is one of them (see Elmaghraby and Keskinocak [15]). A widely used price mechanism is the auction. Several auction protocols are known

L. Bertazzi et al. (eds.), *Innovations in Distribution Logistics,* Lecture Notes
in Economics and Mathematical Systems 619, DOI: 10.1007/978-3-540-92944-4,
© 2009 Springer-Verlag Berlin Heidelberg

in the literature and used by the practitioners. Between them, the most suitable for the procurement of goods and services is the *reverse auction*, in which a buyer demands either a Request for Quotation (RFQ) or a Request for Proposal (RFP) for the item/s needed. The price is then determined by a competition between potential sellers. Electronic auctions, conducted over the Internet, have several benefits compared to the traditional ones such as lower transaction and participation costs and the access to possibly larger markets (see [26]). Despite most of the auctions involve the sale of multiple distinct items, research in auction theory has traditionally focused on single item auctions assuming that bidders have no preferences for sets of items. On the contrary, bidders often show preferences for bundles of items expressed either as complementary or substitution effects (see de Vries and Vohra [10]). Running single-item auctions for multiple distinct items, for example in parallel or in sequence, results in inefficient allocations when bidders show complementarity or substitution preferences (see, the well-known "exposure problem" in DeMartini et al. [11]). To correctly model such preferences on items researchers have proposed the use of combinatorial auctions, where bidders are allowed to bid on any combination of the items auctioned off.

In different contexts, the potential benefits deriving from the use of combinatorial auctions are enormous. In the present paper, we consider the context of transportation procurement, so that the items auctioned off are transportation services. In this domain, buyer is a manufacturer, a distributor, a retailer and any other company that needs to move goods, while potential sellers are all the trucking companies. From now on, we will refer to the buyer as the *shipper* and to the sellers as the *carriers*. For sake of simplicity, we assume that the shipper is the auctioneer, while the carriers act as bidders of the procurement auction. Items being auctioned off are distinguishable pairs of origin-destination points, usually called *lanes*, to which the shipper associates the number of loads that have to be moved during the temporal horizon considered in the proposed agreement. Jara-Diaz [21] defines a lane as a one-way movement from an origin to a destination with the associated set of shipments for the period considered by the RFP.

E-procurement of transportation services is a typical application domain where agents participating to the procurement auction show complementarity or substitution preferences on service contracts (see Caplice [3]) and have to solve a complex decision problem. In fact, a carrier's convenience to serve a lane depends not only on the number of loads the carrier hauls on that lane (economies of scale), but also on the number of loads he might carry on other neighboring lanes (economies of scope) (see Sheffi [36]). An important factor contributing to a carrier's transportation costs is indeed the deadheading cost (also called repositioning cost) which can be defined as the cost incurred when moving an empty truck from its current position to the origin of a new lane (see Caplice and Sheffi [4] and Ergun et al. [17]). Thus, one of the main problems faced by a carrier is to find the right bid (the right bundles of lanes) and the corresponding price to submit as an auction bid. On the other side of the auction, there are shipper's decision problems. A shipper that manages a fleet of vehicles may get substantial benefits from running an auction, in the case the fleet is not sufficient to fulfill all the required transportation services and the shipper has to outsource some of them, or in case the available fleet is sizable, but the auction might provide carriers able to achieve the same requests at a significant lower cost. We identify the problem faced by the shipper in deciding which lanes he can personally serve and which ones he has to outsource as the *Shipper's Lane Selection Problem* (SLSP).

Several other complex problems can be encountered in the design and implementation of a combinatorial auction (see Abrache et al. [1] and Pekeč and Rothkopf [31]). Selecting the most suitable auction protocol (see Klemperer [23]), deciding the type of the auction (see Parkes [30] for a theoretical justification of the use of multi-round combinatorial auctions when bidders have hard valuation problems) and its bidding language (see Abrache et al.[1], Nisan [28] and [29], Sandholm [35] and Caplice and Sheffi [5] where the authors report the traditional practice in truckload transportation), managing uncertainty in shipper-carriers relationship and determining the winners of the auction (see again the chapter by Caplice and Sheffi [5]) are only some of the problems that an auctioneer has to solve when setting up a combinatorial auction. Several excellent surveys have been published on general combinatorial auctions. These include the papers by de Vries and Vohra [10], Pekeč and Rothkopf [31], Gavish [19], Kalagnanam and Parkes [22] and the volume edited by Cramton et al. [8].

The aim of this paper is twofold. We first provide a survey on the use of combinatorial auctions in transportation service procurement and detect open research areas. Secondly, we focus our analysis on one new problem, the Shipper's Lane Selection Problem, and provide two alternative mathematical formulations for it. Though such models are easily adaptable for transportation service procurement in general, we concentrate on the procurement of transportation trucking services since it is the most relevant among all the transportation modes. Indeed, the Bureau of Transportation Statistics of the U.S. Department of Transportation (BTS [2]) reports that in 2002 trucks moved more than \$6.2 trillion and 7.8 billion tons of manufactured goods and raw materials. This is about 74.3% of the value shipped and 67.2% of the weight carried. In particular, we analyze the transportation services procurement assuming the shipper is requiring truckload transportation (TL) services.

The following sections of the paper are organized as follows. In Sect. 2 we analyze the stages composing the organization of a combinatorial auction process and describe the main decision making problems encountered in each phase. In particular, when analyzing the pre-auction stage, the Shipper's Lane Selection Problem (SLSP) is introduced. In Sect. 3 the two mixed-integer linear programming formulations to model the SLSP for the truckload case are provided. Section 4 is devoted to an experimental comparison of the two models on random instances, while in Sect. 5 conclusions and future developments are drawn.

## 2 The Outline of an Auction Process

An auction process typically consists of several steps involving different actors and requiring the solution of different problems. Plummer [32] gives a description of the bidding process from the carrier's perspective. The process starts with carriers responding to Requests for Quotation (RFQ) or Requests for Proposal (RFP) moved for by the shipper. Usually a carrier decides on which RFQ (or RFP) to bid by considering his fleet size and the current commitments in his network. In Caplice and Sheffi [5] the authors describe the complete combinatorial auction framework from a more general point of view. They identify three main stages that form an auction process:

- *Pre-auction stage*: the shipper estimates his transportation service requests for the upcoming period. He then determines the carriers to be invited to the auction through a screening selection and decides what information carriers have to return back, the type of the service rate (e.g. flat rate per move, rate per mile), the service details (e.g. days of transit, capacity available) and the type of bid allowed (e.g. simple bids, combinatorial bids and possible constraints on allowed combinations).
- *Auction stage*: the shipper communicates the freight network to the carriers. Carriers conduct their own analysis and determine their bidding strategy. Every bid consists in a bundle of lanes and in the rate (the price) the carrier asks to serve it. Then, carriers submit their bids.
- *Post-auction stage*: the shipper receives bids and has to determine the winners of the auction. Once the winners have been determined, the shipper communicates to the participants the winning bids.

Each one of the described stages involves a complex decision problem in which operational research methods can represent valuable tools for decision makers. While the third stage is the most studied in the literature, very few contributions exist for the carriers' bidding problem in the second stage and no attention at all has been reserved to the pre-auction stage.

Many efforts have been made in analyzing the post-auction stage in the transportation field. The first reported use of management science tools to determine the winners of an auction can be found in Moore et al. [27]. The authors describe how the optimization and simulation tools, developed in 1988 at Reynolds Metals Company to centralize interstate truckload freight operations, have resulted in improving on-time delivery of shipments and in a decrease of the annual freight costs by over $7 million. The authors introduce a mixed integer programming model that globally minimizes transportation costs subject to several operational constraints such as upper bounds on the number of carriers to be selected and individual carriers' capacity constraints. Their model allows simple bids with volume constraints and does not allow combinatorial bids. Moreover, due to the limited computational capabilities available, the model was not fully implemented in practice. Ledyard et al. [25] describe the iterative combinatorial auction implemented by Sear Logistics Services (SLS) in 1993. The authors report that SLS has been the first procurer of trucking services that used a combinatorial auction to reduce its costs, saving about a 13% over past procurement practices. Elmaghraby and Keskinocak [15] discuss the implementation of the combinatorial auction for procuring transportation services experienced by the Home Depot. The company decided to use a single-round combinatorial auction, eventually allowing for a second round only for a limited number of lanes and a selected group of carriers. The winner determination problem was solved using an integer programming based algorithm. They point out that several other companies use the same optimization tool to procure transportation services, included Walmart Stores, Compaq Computer Co., Staples Inc., and many others. de Vries and Vohra [10] mention Logistics.com Inc.'s use of combinatorial auctions to procure transportation services for K-Mart Corporation and Ford Motor Company. Caplice and Sheffi [4] report how shippers can use optimization-based techniques for transportation services procurement. While in [5] the same authors discuss three formulations of the winner determination problem, called "carrier assignment problem", usually used in practice for the procurement of truckload transportation services. The first formulation allows only simple bids with no side constraints, the second one introduces both simple bids and combinatorial bids, and the last one is based on flexible combinatorial bids. In a typical procurement application, side constraints are often introduced to ensure additional conditions on a valid assignment are satisfied (see [9] and [16], see [5] for an illustration of the constraints most commonly used in the practice of truckload services).

In the auction stage, carriers have to determine their optimal bidding strategy, analyzing their internal costs structure determined by their own resources and by their existing commitments. Several researchers have pointed out the computational difficulties of the bid valuation and construction problem in a combinatorial auction (e.g., see Parkes [30]). Several practical cases have been described where the use of optimization tools in selecting bundles of lanes would have been greatly beneficial to carriers. See, for example, the difficulties in forming bundles experienced by bidders in the Sears Logistics Services TL combinatorial auction [25]. Caplice [3] introduces some heuristic algorithms that carriers can use to create open loop tours, closed loop tours, inbound-outbound reload packages, and short haul packages using potential savings estimates based on historical load volumes. In a simulation based study, Regan and Song [33] show that the use of combinatorial auctions is beneficial for carriers compared to traditional single-item auctions in terms of revenue. Regan and Song [34] propose a linear model in which the objective function to be minimized is the total empty movement costs subject to constraints that impose all loads for every commitment, new or pre-existing, have to be satisfied. Then, the authors suggest approximation methods to construct carrier's optimal or near optimal bids with and without pre-existing commitments. Kwon et al. [24] propose a quadratic integer programming model for the carrier's bid generation problem in which carriers employ vehicle routing-type models to identify packages of lanes based on the actual routes that a fleet of trucks will follow in practice. Some researchers have analyzed the use of advisors applied to electronic freight marketplaces (see [6] and [18]). In this context, advisors are software agents that assist carriers in making "profitable" bidding decisions, by processing the information available in the market, and realize its integration into the dynamic planning of transportation operations (see [1]).

Finally, the stage that has received less attention in the literature is the pre-auction stage. In such stage the shipper has to determine for which lanes it may be convenient to auction off the transportation service. If the shipper has a fleet of vehicles able to serve all his transportation requests, he might be interested to set up an auction to find out if there exist carriers able to offer the same transportation services at a significant lower cost. If, on the contrary, the fleet available is not enough to fulfill all the needs, the shipper has to determine which lanes to serve with his own fleet and which ones to outsource by an auction in order to serve all transportation requests at the minimum cost. We identify such problem as the Shipper's Lane Selection Problem. To the best of our knowledge, this problem has never been analyzed in the literature up to now. In the following section we provide two alternative mathematical formulations for this problem in the truckload case.

## 3 Modeling the Shipper's Lane Selection Problem: The Truckload Case

Let us consider a shipper who has a set of *transportation requests* (lanes) to be satisfied. Each lane consists in carrying a load from an origin straight to its destination (truckload (TL) case). For sake of simplicity, we suppose that every request has to be fulfilled only once in the life span of the agreement between the shipper and his customer. We assume that traveling costs are expressed as distances between two nodes and that satisfy the triangle inequality. We also assume that the shipper has a vehicle which can be used for the service. We assume, for sake of simplicity, that every carrier is allowed to submit bids only for single requests. Moreover, the shipper is able to estimate, for example by inferring from historical data, the bid price that carriers potentially participating to the auction will submit on each lane. Finally, implementing and running an electronic auction involves a fixed cost, no matter the number of carriers participating. The shipper main problem is to decide which lanes to serve directly and which ones to outsource by means of an auction. The objective is to optimize the trade-off between the travel costs of a direct service and the cost of paying carriers for the service plus the possible set-up cost of the auction. Let us formalize the problem.

Let $L$ be a set of lanes, i.e. a set of transportation requests each one identified by an origin and a destination. Given a complete bi-directed graph $G = (V, A)$ with node set $V$ (including the depot), arc set $A$ and lane set $L \subseteq A$, the problem looks for a directed cycle starting and ending at the depot covering a subset of lanes while minimizing the direct traveling costs plus the sum of costs incurred by the shipper to auction off the lanes not served by his own vehicle, i.e. the sum of the estimated costs paid to the bidders for the requests auctioned off plus a fixed cost incurred for running the auction.

Let us define as $c_{ij}$ the travel cost from node $i$ to node $j$, $(i, j) \in A$, and as $\widehat{c}_{ij}$ the estimated cheapest price asked by a potential carrier for serving the lane $(i, j) \in L$. Finally, define as $K$ the fixed cost incurred by the shipper when implementing and running the auction.

A *feasible solution* for the Shipper's Lane Selection Problem is defined by a simple directed cycle starting and ending at the depot, covering a subset of lanes

$L' \subseteq L$. Since a same node may serve either as origin or as destination of different lanes, a feasible solution is always a simple cycle but it may be non elementary.

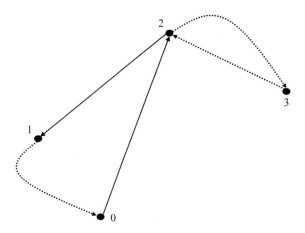

**Fig. 1.** A non elementary optimal cycle

Figure 1 shows an example where the optimal solution for a problem with four nodes (including the depot) and three lanes (set $L = \{(1,0), (2,3), (3,2)\}$) is a non elementary cycle. The travel costs are all equal to 1, whereas the estimated cheapest price requested by potential carriers is equal to $1 - \epsilon$, $0 < \epsilon < 1$, for each request and $K = 5$. The optimal solution for the shipper is to serve all the lanes with his own vehicle by a non elementary cycle traversing arc $(0,2)$, then serving lanes $(2,3)$ and $(3,2)$ and moving to node 1 to serve lane $(1,0)$, then returning to the depot. The total cost paid by the shipper is equal to 5. In this solution the selection of both arcs $(2,3)$ and $(3,2)$ forms a subtour (node 2 is visited more than once). In order to allow non elementary solutions, the mathematical formulation of the problem cannot use the generalized subtour elimination constraints (e.g. see Toth and Vigo [38]). We have to formulate constraints that allow subtours but eliminate those of them that are isolated from the depot.

Figure 2 shows an example where the optimal solution consists of a cycle which serves only a part of the available lanes while some others are auctioned off. The graph is the same as that used in Fig. 1 with the same set of lanes and the same travel costs. The estimated bid prices are the same as those used in the previous example but for that associated to lane $(1,0)$ which is set equal to $2 + \epsilon$. Finally, we set $K = 1$. Then, an optimal solution is to traverse arc $(0,1)$, to fulfill lane $(1,0)$ returning to the depot and to auction off the remaining lanes. The solution is shown in Fig. 2, where the cross-hatched lines represent the lanes auctioned off. The total cost paid by the shipper is equal to $5 - 2\epsilon$. The cheapest cost the shipper should pay by satisfying all the transportation requests with his own vehicle is equal to 5. The cost incurred by the shipper auctioning off all the requests would be equal to $5 - \epsilon$.

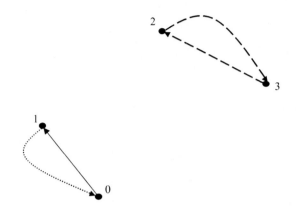

**Fig. 2.** An optimal solution with some lanes auctioned off

### 3.1 Models Formulation

As the decision maker has to determine a minimum cost cycle traversing an arc subset of a graph, the SLSP is similar to the problems belonging to the class of Arc Routing Problems (see the excellent surveys by Eiselt et al. [13, 14] and the book edited by Dror [12]). Since the objective to be minimized is a function not only of the travel cost associated to each arc but also of the costs associated to the auction, the SLSP is rather different from all the problems formerly addressed in the literature.

In the following, we present two integer programming formulations for the SLSP. The first model is new, whereas the second model derives from a well-known formulation for the capacitated arc routing problem.

Let $x_{ij}$, $(i, j) \in A$, be a binary variable that takes value 1 if arc $(i, j)$ is traversed by the shipper's vehicle, and 0 otherwise. Let $l_{ij}$, $(i, j) \in L$, be a binary variable that takes value 1 if lane $(i, j)$ is auctioned off and 0 otherwise. Finally, let $z$ be a binary variable that takes value 1 if at least one lane is auctioned off and 0 otherwise, and $u^S$, $S \subseteq V \setminus \{0\}, S \neq \emptyset$, be a binary variable which takes value 1 when at least one node in set $S$ is visited by the optimal solution and 0 otherwise. $M_1$ and $M_2$ are large constant values which can be set equal to $|L|$ and $|V|^2$, respectively.

A first formulation of the truckload Shipper's Lane Selection Problem (SLSP1) can be as follows:

$$(SLSP1) \quad \min \sum_{(i,j)\in A} c_{ij}x_{ij} + \sum_{(i,j)\in L} \widehat{c}_{ij}l_{ij} + Kz \tag{1}$$

$$x_{ij} \geq (1 - l_{ij}) \quad (i, j) \in L \tag{2}$$

$$\sum_{(i,j)\in L} l_{ij} \leq M_1 z \tag{3}$$

$$\sum_{j\in V\setminus\{i\}} x_{ji} = \sum_{j\in V\setminus\{i\}} x_{ij} \quad i \in V \tag{4}$$

$$\sum_{i\in S}\sum_{j\in S} x_{ij} \leq M_2 u^S \quad S \subseteq V\setminus\{0\}, S \neq \emptyset \tag{5}$$

$$\sum_{i\notin S}\sum_{j\in S} x_{ij} \geq u^S \quad S \subseteq V\setminus\{0\}, S \neq \emptyset \tag{6}$$

$$x_{ij} \in \{0, 1\} \quad (i, j) \in A \tag{7}$$

$$l_{ij} \in \{0, 1\} \quad (i, j) \in L \tag{8}$$

$$u^S \in \{0, 1\} \quad S \subseteq V\setminus\{0\}, S \neq \emptyset \tag{9}$$

$$z \in \{0, 1\}. \tag{10}$$

The objective function (1) minimizes the estimated total cost incurred by the shipper to serve all his lanes. The total cost is given by the sum of the transportation costs incurred by the shipper visiting customers with his own vehicle, plus the sum of the estimated cheapest prices for the requests auctioned off, plus the fixed cost paid to set up the auction.

For every request $(i, j) \in L$, constraint (2) forces variable $x_{ij}$ to take value 1 if the request is not auctioned off, i.e. if $l_{ij} = 0$. Notice that if $l_{ij} = 1$ then $x_{ij}$ is free to take any value. However, the minimization of the objective function forces variable $x_{ij}$ to take at optimum value 0 as the most convenient of the two. Constraint (3) imposes that if at least a request is auctioned off, i.e. $l_{ij} = 1$ for some $(i, j) \in L$, then the binary variable $z$ will take value 1. One may notice that if $l_{ij} = 0$ for all $(i, j) \in L$ then $z$ is forced by the objective function to take value 0 as the most convenient of the two. Constraints (4) (in-degree and out-degree constraints) establish that at optimum the total number of selected arcs entering into node $i \in V$ must be equal to the total number of selected arcs leaving it. This ensures that the number of selected arcs entering into any subset $S$, $S \subseteq V\setminus\{0\}, S \neq \emptyset$, is equal to the number of selected arcs exiting it. The set of constraints (5) along with constraints (6) eliminate isolated subtours. Indeed, the following implications are true:

$$\sum_{i\in S}\sum_{j\in S} x_{ij} > 0 \quad \Rightarrow \quad u^S = 1 \quad \Rightarrow \quad \sum_{i\notin S}\sum_{j\in S} x_{ij} \geq 1.$$

Specifically, for each subset $S \subseteq V\setminus\{0\}$, constraint (5) forces binary variable $u^S$ to take value 1 if, at optimum, at least one arc between two nodes in $S$ is selected. For each subset $S \subseteq V\setminus\{0\}$ constraint (6) imposes that, at optimum, the number of arcs entering into $S$ must be at least equal to the value taken by variable $u^S$. This imposes that if at least one arc is selected between two nodes in $S$, i.e. $u^S = 1$, then at least one arc starting from a node not in $S$ must enter into $S$. Moreover, if $\sum_{i\in S}\sum_{j\in S} x_{ij} = 0$ then $u^S$ is free to take any value, and it will take value 0 as the most convenient

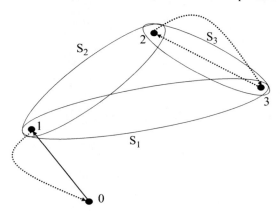

**Fig. 3.** Violated constraints. Subsets $S_1$, $S_2$ and $S_3$

one, forced by the objective function. Finally, constraints (7) – (10) define binary conditions.

In Fig. 3 we show how constraints (5) – (6) eliminate isolated subtours referring to the data of the example shown in Fig. 1. In particular, we show how the solution represented in Fig. 3 violates the constraints. Let us consider all the subsets $S \subseteq V \setminus \{0\}$ with cardinality equal to 2. We have labeled them as $S_1$, $S_2$ and $S_3$ and plotted in Fig. 3. One may notice that considering only subsets $S_1$ and $S_2$ the constraints are satisfied since there is no arc selected in the solution between the two nodes in $S_1$ and in $S_2$, respectively. Let us consider subset $S_3$. In the depicted solution $x_{23} = x_{32} = 1$, then binary variable $u^{S_3}$ is forced to take value 1 by constraint (5). By constraint (6) at least one arc must enter into subset $S_3$. Then, the current solution violates constraint (6). Finally, notice that all subsets with cardinality 1, and the unique subset with cardinality 3, do not violate any constraint.

Model (SLSP1) requires $|V| * (|V| - 1)$ binary variables $x_{ij}$, $|L|$ binary variables $l_{ij}$, $2^{|V|-1} - 1$ binary variables $u^S$ and the binary variable $z$. The total number of constraints is equal to $|L| + (2^{|V|} - 2) + |V| + 1$. Thus, even for the smallest instances, the number of variables and the number of constraints are very large and grow exponentially in the number of vertices.

An alternative mathematical formulation for the SLSP can be derived from the formulation of the Capacitated Arc Routing Problem (CARP) (see Golden and Wong [20]). The new mathematical formulation (SLSP2) is as follows:

$$\text{(SLSP2)} \qquad (1)$$

$$\sum_{i \in S} \sum_{j \in S} x_{ij} \leq |S| - 1 + M_2 u^S \quad S \subseteq V \setminus \{0\}, S \neq \emptyset \qquad (11)$$

$$(2) - (4) \text{ and } (7)-(10)$$

$$\sum_{i \in S} \sum_{j \notin S} x_{ij} \geq 1 - w^S \quad S \subseteq V \setminus \{0\}, S \neq \emptyset \tag{12}$$

$$u^S + w^S \leq 1 \quad S \subseteq V \setminus \{0\}, S \neq \emptyset \tag{13}$$

$$w^S \in \{0, 1\} \quad S \subseteq V \setminus \{0\}, S \neq \emptyset. \tag{14}$$

With respect to the (SLSP1) formulation, the model makes use of a new set of binary variables $w^S$. The meaning of all the other variables is the same as in the previous formulation.

Constraints (13) ensure that, at optimum and for every node subset $S$, only one of the two binary variables $u^S$ and $w^S$ can take value 1. Hence, if the number of arcs between nodes in a subset $S$ is strictly greater than $|S| - 1$, then the variable $u^S$ is forced to take value 1 by constraint (11). If variable $u^S$ takes value 1, then constraint (13) forces variable $w^S$ to take value 0. Finally, from constraint (12), the number of arcs leaving set $S$ must be greater than 1.

This model requires $2^{|V|-1} - 1$ additional binary variables $w^S$ and has $2^{|V|-1} - 1$ more constraints.

## 4 Experimental Analysis

In this section we compare the computational times required by the two models to optimally solve a set of random instances. The models have been solved with CPLEX 10.1. Experiments have been run on a PC with 2,992 Mhz Intel Pentium III processor and 2 Gb of RAM.

The benchmark problems have been generated starting from Solomon's instances for the VRPTW (see [37]). In detail, we have considered the first 15 customer coordinates of 5 random instances (R102, R104, R106, R108, and R110). We have changed the position of the depot assuming it is located at the origin of the axes. The latter choice has been motivated by the fact that, by maintaining Solomon's depot coordinates no optimal solution auctioned any lane off. Euclidean distances have been rounded to the nearest integer. To provide a different level of lanes density, for each Solomon's instance considered we have generated three different sets $L$, each one characterized by a different number of transportation requests. This means 15 instances altogether. Lanes have been generated randomly, assuming a predefined average number of lanes for each node, excluding the depot. The first set $L$ is composed by the minimum number of lanes needed to ensure that at least a request is incident to every node of the graph:

$$|L| = \lceil (|V| - 1)/2 \rceil.$$

The second set of lanes has been generated assuming that the ratio $|L|/(|V| - 1)$ is equal to 1. One may notice that this means to generate, on average, two lanes incident to every node. Finally, the third set of lanes has been generated assuming that the ratio $|L|/(|V| - 1)$ is equal to 2. Notice that this means to generate, on average, four lanes incident to every node. Given the number of nodes, equal to 14 without considering the depot, the cardinality of set $L$ is equal to 7, 14 and 28, respectively.

For each lane $(i, j) \in L$ the expected cheapest price $\hat{c}_{ij}$ has been uniformly generated in the interval $[\alpha * (c_{0i} + c_{ij} + c_{j0}), (c_{0i} + c_{ij} + c_{j0})]$, where $\alpha$ has been set equal to 0.2. Finally, the fixed cost $K$, paid whenever the auction is implemented, has been set equal to 70.

Table 1 provides the computational times required by the two models when solving the 15 test instances. The table is organized into five columns. The first column provides the cardinality of set $L$, i.e. the number of lanes randomly generated in the graph. The second column indicates the Solomon's problem from which the nodes coordinates have been selected. The third column provides the number of lanes (#) auctioned off in the optimal solution. Finally the last two columns provide the computational times of the models. Time is expressed in seconds.

One may notice that the SLSP1 formulation, independently of the instance, finds the optimal solution in about 7/8 mins, whereas the SLSP2 model requires computational times that are highly correlated with the type of instance. The SLSP2 model is always faster than the SLSP1 when the average number of lanes incident to every node of the graph is greater or equal than 1, i.e. for $|L| \geq 14$. Conversely, the SLSP1 model is highly faster when the number of requests is minimum ($|L| = 7$) and the optimal solution auctions some lanes off. The only exception is represented by instance R110 where the number of lanes auctioned off is zero.

**Table 1.** Computational times (in seconds)

| $|L|$ | Problem | # | SLSP1 | SLSP2 |
|---|---|---|---|---|
| 7 | R102 | 2 | 473.08 | 155829.37 |
| | R104 | 2 | 458.48 | 747.94 |
| | R106 | 2 | 481.66 | 24304.41 |
| | R108 | 2 | 463.35 | 1555.18 |
| | R110 | 0 | 451.36 | 110.39 |
| 14 | R102 | 0 | 449.97 | 54.78 |
| | R104 | 2 | 451.52 | 51.39 |
| | R106 | 3 | 451.27 | 69.09 |
| | R108 | 0 | 447.97 | 49.30 |
| | R110 | 0 | 447.09 | 53.88 |
| 28 | R102 | 6 | 449.09 | 52.94 |
| | R104 | 3 | 453.47 | 53.16 |
| | R106 | 2 | 449.32 | 54.09 |
| | R108 | 5 | 448.36 | 50.80 |
| | R110 | 3 | 453.58 | 52.63 |

# 5 Conclusions and Future Research

In this paper we have introduced the Shipper's Lane Selection Problem. We have proposed two equivalent integer linear programming formulations and compared the computational times required by the two models to solve a set of randomly generated instances. Since the problem is new, several interesting issues can be considered for further developments. From the modeling point of view, to assume a shipper's fleet composed by more than one vehicle, to consider the Less-Than-Truckload case and to allow more than one load per single lane are natural extensions. Moreover, as the problem cannot be solved in a reasonable amount of time with more than 16 customers, the design of effective and efficient heuristics seems to be a primary concern.

# References

[1] Abrache J, Crainic TG, Gendreau M (2004) Design issues for combinatorial auctions. *Quarterly Journal of the Belgian, French and Italian Operations Research Societies* 2(1):1–33.

[2] Bureau Transportation Statistics (2005) Trucks carry the most freight by weight and value. Available at: *http://www.bts .gov/press_releases/2005/bts003_05/html/bts003_05.html*.

[3] Caplice C (1996), Optimization-based bidding: a new framework for shipper-carrier relationship. PhD Thesis, Massachusetts Institute of Technology, Cambridge, MA.

[4] Caplice C, Sheffi Y (2003) Optimization-based procurement for transportation services. *Journal of Business Logistics* 24(2):109–128.

[5] Caplice C, Sheffi Y (2006) Combinatorial auctions for truckload transportation. In: Cramton P, Shoham Y, Steinberg R (eds) *Combinatorial auctions*. MIT Press, Cambridge, MA:539–571.

[6] Chang T-S, Crainic TG, Gendreau M (2002) Dynamic advisors for freight carriers for bidding in combinatorial auctions. In: Proceedings of the 5th international conference on electronic commerce research (ICECR-5), CD-ROM. Centre de Recherce sur le Transports, Montréal, Canada.

[7] Crainic TG, Speranza MG (2007) Logistics and the Internet. In: Don Taylor G (eds) *Logistics engineering handbook*. Taylor and Francis Group.

[8] Cramton P, Shoham Y, Steinberg R (2006) *Combinatorial auctions*. MIT Press, Cambridge, MA.

[9] Davenport AJ and Kalagnanam JR (2002) Price negotiations for procurement of direct inputs. In: Dietrich B, Vohra RV (eds) *Mathematics of the Internet: e-auctions and markets*. The IMA volumes in mathematics and its applications, Vol. 127. Springer, Berlin Heidelberg New York:27–44.

[10] de Vries S, Vohra RV (2003) Combinatorial auctions: a survey. *INFORMS Journal on Computing* 15(3):284–309.

[11] DeMartini C, Kwasnica AM, Ledyard JO, Porter DP (2005) A new and improved design for multi-object iterative auctions. *Management Science* 51(3):419–434.

[12] Dror M (2000) *Arc routing: theory, solutions and applications*. Kluwer Academic Publishers, Norwell, MA.

[13] Eiselt HA, Gendreau M, Laporte G (1995) Arc routing problems, part I: the chinese postman problem. *Operations Research* 43(2):231–242.

[14] Eiselt HA, Gendreau M, Laporte G (1995) Arc routing problems, part II: the rural postman problem. *Operations Research* 43(3):399–414.

[15] Elmaghraby W, Keskinocak P (2003) Combinatorial auctions in procurement. In: Billington C, Harrison T, Lee H, Neale J (eds) *The practice of supply chain management*. Kluwer Academic Publishers:245–258.

[16] Epstein R, Henríquez L, Catalán J, Weinstraub GY, Martínez C (2002) A combinatorial auction improves school meals in Chile. *Interfaces* 32(6):1–14.

[17] Ergun O, Kuyzu G, Savelsberg M (2007) Shipper collaboration. *Computers & Operations Research* 34(6):1551–1560.

[18] Figliozzi M, Mahmassani HS, Jaillet P (2002) Modeling online transportation market performance: carrier competition under asymmetric fleet assignment technology. In: Proceedings of the 5th international conference on electronic commerce research (ICECR-5), CD-ROM. Centre de Recerce sur le Transports, Montréal, Canada.

[19] Gavish B (2003) Combinatorial auctions mathematical formulations and open issues. *International Journal of Information Technology & Decision Making* 2(1):5–27.

[20] Golden BL, Wong RT (1981) Capacitated arc routing problems. *Networks* 11(3):305–315.

[21] Jara-Diaz SR (1988) Multioutput analysis of trucking operations using spatially disaggregated flows. *Transportation Research Part B* 22(3):159–171.

[22] Kalagnanam J, Parkes D (2003) Auctions, bidding and exchange design. In: Simchi-Levi D, Wu SD, Shen Z-J (eds) *Handbook of quantitative supply chain analysis: modeling in the e-business era*. Kluwer Academic Publishers:143–212.

[23] Klemperer P (1999) Auction theory: a guide to the literature. *Journal of Economic Surveys* 13(3):227–286.

[24] Kwon RH, Lee C-G, Ma Z (2007) A carrier's optimal bid generation problem in combinatorial auctions for transportation procurement. *Transportation Research Part E* 43(2):173–191.

[25] Ledyard JO, Olson M, Porter D, Swanson JA, Torma DP (2002) The first use of a combined-value auction for transportation services. *Interfaces* 32(5):4–12.

[26] Lucking-Reley D (2000) Auctions on the Internet: what's being auctioned, and how? *The Journal of Industrial Economics* 48(3):227–252.

[27] Moore EW, Warmke JM, Gorban LR (1991) The indispensable role of management science in centralizing freight operations at Reynolds Metals company. *Interfaces* 21(1):107–129.

[28] Nisan N (2000) Bidding and allocation in combinatorial auctions. In: Proceedings of the Second ACM Conference on Electronic Commerce. Minneapolis, MN:1–12.

[29] Nisan N (2006) Bidding languages for combinatorial auctions. In: Cramton P, Shoham Y, Steinberg R (eds) *Combinatorial auctions*. MIT Press, Cambridge, MA:215–232.

[30] Parkes DC (2000) Optimal auction design for agents with hard valuation problems. In: Moukas A, Sierra C, Ygge F (eds) *Agent mediated electronic commerce II: towards next-generation agent-based electronic commerce systems*. Springer, Berlin Heidelberg New York:206–219.

[31] Pekeč A and Rothkopf MH (2003) Combinatorial auction design. *Management Science* 49(11):1485–1503.

[32] Plummer C (2003) Bidder response to combinatorial auctions in truckload procurement. Master Thesis, Massachusetts Institute of Technology, Cambridge, MA.

[33] Regan A, Song J (2003) Combinatorial auctions for transportation service procurement: the carrier perspective. *Transportation Research Record* 1833:40–46.

[34] Regan A, Song J (2005) Approximation algorithms for the bid construction problem in combinatorial auctions for the procurement of freight transportation contracts. *Transportation Research Part B* 39(10):914–933.

[35] Sandholm T (2002) Algorithm for optimal winner determination in combinatorial auctions. *Artificial Intelligence* 135(1):1–54.

[36] Sheffi Y (2004) Combinatorial auctions in the procurement of transportation services. *Interfaces* 34(4):245–252.

[37] Solomon MM (1987) Algorithms for the vehicle routing and scheduling problems with time window constraints. *Operations Research* 35(2):254–265.

[38] Toth P, Vigo D (2002) *The Vehicle Routing Problem*. Volume 9 of SIAM Monographs on Discrete Mathematics and Applications, Philadelphia, PA.

# Applications of the Double Standard Model for Ambulance Location

Gilbert Laporte[1], François V. Louveaux[2],
Frédéric Semet[3,4], and Arnaud Thirion[2]

[1] Canada Research Chair in Distribution Management, HEC, Canada
    gilbert@crt.umontreal.ca
[2] Facultés Universitaires Notre-Dame de la Paix, Belgium
    francois.louveaux@fundp.ac.be, thirionarnaud@hotmail.com
[3] LAMIH-ROI, Université de Valenciennes et du Hainaut-Cambrésis, France
    frederic.semet@univ-valenciennes.fr
[4] GRIS, Département d'administration de la santé, Université de Montréal, Canada

**Summary.** This paper describes classical and advanced ambulance location models developed over the past 35 years. One of these models, called the *Double Standard Model* (DSM) maximizes double demand coverage with a fixed number of ambulances. A dynamic version of DSM was developed and tested on data from the Island of Montreal. The static version was successfully applied to data from Montreal, Austria and Wallonia.

**Key words:** Ambulance location and relocation, Double coverage, Dynamic model, Probabilistic model

## 1 Introduction

The design and provision of efficient and cost effective ambulance services is a problem faced by all major municipal and regional authorities throughout the world. This is an area in which operations researchers have traditionally had an important impact. Their role is likely to increase in coming years with the development of powerful metaheuristics and their interaction with geographic information systems and advanced telecommunication technologies. A central problem arising in ambulance fleet management is to decide where to locate ambulances in order to provide adequate population coverage. Since the early 1970s, there has been a steady evolution in the models and algorithms proposed for this type of problem. The aim of this paper is to report on some recent developments in the area of ambulance location.

L. Bertazzi et al. (eds.), *Innovations in Distribution Logistics*, Lecture Notes
in Economics and Mathematical Systems 619, DOI: 10.1007/978-3-540-92944-4,

Formally, ambulance location models are defined on a graph $G = (V \cup W, A)$ where $V$ is the a node set representing aggregated demand points, $W$ is a set of potential ambulance location site, and $A = \{(i, j) \in V \times W, i \neq j\}$ is an arc set. With each arc $(i, j)$ is associated a travel time $t_{ij}$. A demand point $i \in V$ is covered by site $j \in W$ if and only if $d_i \leq r$, where $r$ is a preset coverage standard. Let $W_i = \{j \in W : t_{ij} \leq r\}$ be the set of location site covering demand point $i$. Several objectives and constraints are possible. For example, one could determine the minimum number of ambulances needed to cover all demand, or maximize the demand covered with a given number of ambulances. More involved models aimed at covering some demand points several times have also been proposed. In addition, dynamic and probabilistic models have been put forward.

In the Location Set Covering Model (LCSM) (Toregas *et al.* [23]) the aim is to minimize the number of ambulances needed to cover all demand points. The model uses binary variable $x_j$ equal to 1 if and only if an ambulance is located at $j$:

$$\text{(LSCM) Minimize} \sum_{j \in W} x_j \tag{1}$$

$$\text{subject to} \sum_{j \in W_i} x_j \geq 1 \quad (i \in V), \tag{2}$$

$$x_j \in \{0, 1\} \ (j \in W). \tag{3}$$

In the Maximum Covering Location Problem (MCLP) (Church and ReVelle [4]) $p$ ambulances are given and the aim is to cover the largest possible demand $z(p)$. Denote by $d_i$ the demand at node $i \in V$ and let $y_i$ be a binary variable equal to 1 if and only if $i$ is covered by at least one ambulance:

$$\text{(MCLP) Maximize} \ z(p) = \sum_{i \in V} d_i y_i \tag{4}$$

$$\text{subject to} \sum_{j \in W_i} x_j \geq y_i \quad (i \in V), \tag{5}$$

$$\sum_{j \in W} x_j = p \tag{6}$$

$$x_j \in \{0, 1\} \ (j \in W), \tag{7}$$

$$y_i \in \{0, 1\} \quad (i \in V). \tag{8}$$

In practice, one can repeatedly solve MCLP with increasing values of $p$ and select a solution offering a good compromise between $p$ and $z(p)$.

The main drawback of LCSM and MCLP is that once an ambulance has been dispatched to a site its coverage becomes lost. To remedy this situation, several classes of more sophisticated models have been proposed. In the first class, multiple coverage of each demand point is ensured. In the second class, the probability that an ambulance will be available at any moment is explicitly taken into account. A third class of models allow for ambulance relocation in real-time. In other words, ambulance location models have evolved from static to probabilistic to dynamic settings. For recent surveys of ambulance location models, see Marianov and ReVelle [17], Brotcorne et al. [3], Goldberg [14], and Cordeau et al. [5].

This study focuses on the Double Standard Model (DSM) of Gendreau et al. [10] which belongs to the first class and has also been extended to handle dynamic ambulance relocation (Gendreau et al. [11]). Applications to Canadian data are provided by these authors. This model has recently been applied to the Austrian and Belgian contexts, by Doerner et al. [8] and Thirion [22], respectively.

The remainder of this paper is organized as follows. Section 2 presents an overview of some advanced ambulance location models, with an emphasis on DSM. In Sect. 3, we summarize the applications of the DSM to the Canadian, Austrian and Belgian cases. Conclusions follow in Sect. 4.

## 2 Advanced Ambulance Location Models

Our review of advanced ambulance location models covers the three classes of extensions presented in the introduction but, for logical reasons, the probabilistic case is described immediately after the dynamic case.

### 2.1 Multi-Coverage Static Models

One of the first extensions of MCLP model has been the use of several vehicle types and the introduction of the requirement that each demand point should be covered with one vehicle of each type (Schilling *et al.* [20]). Daskin and Stern [6] have later proposed a hierarchical objective to first maximize the demand covered more than once, and then the demand covered exactly once. Hogan and ReVelle [15] consider only one vehicle type but have developed two Backup Coverage Problems called BACOP1 and BACOP2. In BACOP1 the number $p$ of ambulances is sufficient to cover the demand at least once, and the objective is to maximize the total demand covered at least twice. In BACOP2 the objective is a convex linear combination of the demand covered once or at least twice.

The Double Standard Model (DSM) of Gendreau et al. [10] works with two coverage standards $r_1$ and $r_2$, with $r_1 < r_2$, as specified by the United States

Emergency Medical Services Act of 1973. A proportion $\alpha$ of the demand must be covered within $r_1$, while the entire demand must be covered within $r_2$. In the DSM, the objective is to maximize the demand covered twice within $r_1$ using $p$ ambulances, at most $p_j$ of which are located at $j \in W$, subject to the double coverage constraints. Let $W_i^1 = \{j \in W : t_{ij} \leq r_1\}$ and $W_i^2 = \{j \in W : t_{ij} \leq r_2\}$. The integer variable $x_j$ denotes the number of ambulances located at $j \in W$ and the binary variable $y_j^k$ is equal to 1 if and only if the demand at node $i \in V$ is covered $k$ times ($k = 1$ or 2) within $r_1$. The formulation is then:

$$\text{(DSM) Maximize} \quad \sum_{i \in V} d_i y_i^2 \qquad (9)$$

$$\text{subject to} \quad \sum_{j \in W_i^2} x_j \geq 1 \quad (i \in V), \qquad (10)$$

$$\sum_{i \in V} d_i y_i^1 \geq \alpha \sum_{i \in V} d_i \qquad (11)$$

$$\sum_{j \in W_i^1} x_j \geq y_i^1 + y_i^2 \quad (i \in V), \qquad (12)$$

$$y_i^2 \geq y_i^1 \quad (i \in V), \qquad (13)$$

$$\sum_{j \in W} x_j = p \qquad (14)$$

$$x_j \leq p_j \quad (j \in W), \qquad (15)$$

$$y_i^1 \in \{0, 1\} \quad (i \in V), \qquad (16)$$

$$y_i^2 \in \{0, 1\} \quad (i \in V), \qquad (17)$$

$$x_j \geq 0 \text{ and integer} \quad (j \in W). \qquad (18)$$

In this model, the objective function computes the demand covered twice within $r_1$ time units. Constraints (10) mean that all demand is covered within $r_2$. The left-hand side of (12) represents the number of ambulances covering node $i$ within $r_1$ units, while the right-hand side is equal to 1 if $i$ is covered once within $r_1$ units, and equal to 2 if it is covered at least twice within $r_1$ units. The combination of constraints (11) and (12) ensures that a proportion $\alpha$ of the demand is covered within $r_1$. Constraints (13) state that node $i$ cannot be covered at least twice if it is not covered at least once.

## 2.2 Multi-Coverage Static Models

The only known dynamic ambulance relocation model is due to Gendreau et al. [11]). Their *Dynamic Double Standard Model* (DDSM) can be used to redeploy ambulances in real-time whenever a call is made. In addition to the DSM constraints, the DDSM includes the following requirements: 1) vehicles moved in successive redeployments cannot always be the same; 2) repeated round trips between the same two location sites must be avoided; 3) long trips between the initial and final location sites must be avoided.

The dynamic aspect of the redeployment model is captured by time-dependent constants $M_{jl}^t$ equal to the cost of repositioning, at time $t$, ambulance $l$ from its current site to site $j \in W$. This includes the case where site $j$ coincides with the current location of the ambulance, i.e., $M_{jl}^t = 0$. The constant $M_{jl}^t$ captures some of the history of ambulance $l$. If it has been moved frequently prior to time $t$, then $M_{jl}^t$ will be larger. If moving ambulance $l$ to site $j$ violates any of above constraints, then the move is simply disallowed. Binary variables $x_{jl}$ are equal to 1 if and only if ambulance $l$ is moved to site $j$. The constraints of the DDSM are similar to those of the DSM, with the extra constraint that at most one deployment per ambulance is allowed. The objective at time $t$ is:

$$(\text{DDSM}^t) \text{ Maximize } \sum_{i \in V} d_i y_i^2 - \sum_{j \in W} \sum_{l=1}^{p} M_{jl}^t x_{jl} \tag{19}$$

## 2.3 Probabilistic Models

Probabilistic models take into account the fact that ambulances are not always available to answer a call. Each ambulance has a probability $q$, called *busy fraction*, of being unavailable. It is computed as the ratio of the total time spent by all ambulances on all calls to the total ambulance time available. If $i \in V$ is covered by $k$ ambulances, then the expected demand covered at that node is $E_{i,k} = d_i(1 - q^k)$ and the marginal contribution of the $k^{th}$ ambulance is : $E_{i,k} - E_{i,k-1} = d_i(1 - q)q^{k-1}$.

In the Maximum Expected Covering Location Model (MEXCLP) (Daskin [7]), up to $p$ ambulances may be located in total, and more than one vehicle may be located at the same node. Let $y_{ik}$ be a binary variable equal to 1 if and only if node $i \in V$ is covered by at least $k$ ambulances. The model is as follows :

$$\text{Maximize } \sum_{i \in V} d_i(1 - q)q^{k-1}y_{ik} \tag{20}$$

$$\text{subject to } \sum_{j \in W_i} x_j \geq \sum_{k=1}^{p} y_{ik} \qquad (i \in V), \tag{21}$$

$$\sum_{j \in W} x_j \leq p \tag{22}$$

$$x_j \geq 0 \text{ and integer} \qquad (j \in W), \tag{23}$$

$$y_{ik} \in \{0, 1\} \ (i \in V, k = 1, \ldots, p). \tag{24}$$

The validity of this model stems from the fact that the marginal contribution of the $k^{th}$ ambulance is concave in $k$. Therefore, if $y_{ik} = 1$, then $y_{ih} = 1$ for $h \leq k$. Since the objective is to be maximized, both (21) and (22) will be satisfied as equalities. It follows that the two sides of (21) will be equal to the number of ambulances covering node $i \in V$.

A dynamic implementation of MEXCLP, called TIMEXCLP, in which travel speeds change dynamically over time has been developed by Repede and Bernardo [18], while Goldberg et al. [13] have worked with stochastic travel times. These authors have also developed a formula to compute the probability that a site will be covered at any time. A related model used to locate physicians' cars in the Montreal region was recently developed by Gendreau, Laporte and Semet [12].

ReVelle and Hogan [19] have developed two versions of the Maximal Availability Location Problem, called MALP I and MALP II, in which each demand point is covered with the probability $\alpha$, given that an ambulance is unavailable with probability $q$. Further studies on the estimate of the busy fraction have been conducted by Batta et al. [2] and by Marianov and ReVelle [16]. Finally, Ball and Lin [1] have developed an extension of LSCM which contains a linear constraint on the number of ambulances required to achieve a given reliability level.

# 3 Applications of the Double Standard Model

The Double Standard Model has been applied by three groups of researchers. The data used in these studies originate from Montreal (Gendreau, Laporte and Semet [10], [11]), from the eight rural provinces of Austria (Doerner et al. [8]), and from part of Wallonia (Thirion [22]).

## 3.1 Application to Montreal

The first application of the Double Standard Model was made to the Island of Montreal data, using the population distribution of 1986 (Statistics Canada [21]). The demand points are defined by the centroids of the $|V| = 2'521$ census tracts which range from two to 7'000 inhabitants. The total population is 1'758'600. For this study, four sets W of potential location sites, with $|W| = 40, 50, 60$ and $70$ were used, and the number $p$ of ambulances was 25, 30, 35 or 40.

Because of its large scale, this instance was solved heuristically. An upper bound $\bar{z}$ on the objective function value of DSM was first computed by solving the linear relaxation of the integer program at the root of the search tree by means of CPLEX. A first heuristic called CPLEX2 was obtained by solving the integer linear program and stopping 1) at the optimum, or 2) after 100'000 branch-and-bound nodes, or 3) as soon as a feasible solution of value $\underline{z} \leq 0.99 \, \bar{z}$ was reached.

A second heuristic, called TABU, consisted of applying tabu search starting from a solution derived from the linear relaxation. Consider the values $\bar{x}_j$ taken by the $x_j$ variables in the continuous solution. If these values are integer, then the solution is feasible and optimal. Otherwise, $\lfloor \bar{x}_j \rfloor$ ambulances are allocated to each site $j$ and $p - \sum_{j \in W} \lfloor \bar{x}_j \rfloor$ additional ambulances are allocated to the sites $j$ for which $x_j > 0$. This initial solution may be feasible or infeasible. The solution space is then explored according to the usual tabu search rules. The basic move consists of relocating an ambulance to one of its five closest neighbour sites where it is still possible to locate an ambulance. However, a simple application of this move would often yield uncovered areas and for this reason the neighbour solution is obtained by moving not only a single ambulance, but a sequence of ambulances in order to maximize the objective function at each step of the sequence. More specifically, a move can be described as a set of $r$ pairs $(j_t, j_t')$ $(t = 1, \ldots, r)$, where an ambulance is moved from $j_t$ to $j_t'$.

The objective function used during the search is not the original objective function $z$ defined by (9), but a modified hierarchical objective:

$$z' = z + M_1 z_1 + M_2 z_2$$

where $M_1$ and $M_2$ are two weights satisfying $M_1 > M_2 > 1$. The functions $z_1, z_2$ are defined as follows:

$$z_1 = |\{i \in V : \sum_{j \in W_i^2} x_j \geq 1\}| \tag{25}$$

and

$$z_2 = \min \{\alpha, \sum_{i \in V} d_i y_i^1 / \sum_{i \in V} d_i\}. \tag{26}$$

The tabu search algorithm stops after 1000 iterations without improvement, or whenever it has identified a feasible solution within 1% of $\bar{z}$.

Table 1 compares lower bounds values obtained by TABU and CPLEX2 to $\bar{z}$ and gives the computation times in seconds needed to run these two heuristics on a Sun Sparcstation 1000. Both algorithms consistently yield near optimal solution. It can be seen that CPLEX is slightly better than TABU, but has more unstable computation times. The solution with 40 potential sites and 25 ambulances is depicted in Fig. 1.

The tabu search algorithm developed for the static double standard location problem was extended by Gendreau, Laporte and Semet [11] to the dynamic case. The same algorithm is applied at each instant $t$ at which a call is made, with the modified objective function (19), resulting in a redeployment plan. Because it takes about around 3 min to run the algorithm, it is preferable to precompute solutions during the time elapsed between two consecutive calls. More, specifically, for each site $j$ at which an ambulance is currently positioned, one can determine a relocation plan under the hypothesis the ambulance located at $j$ would be the one dispatched to answer the next call. In order to speed up the solution process, several scenarios can be considered simultaneously by using parallel computing. When the next call

**Table 1.** DSM results for the Island of Montreal

| $m$ | $p$ | TABU $\sqrt{z}$ | CPLEX2 $\sqrt{z}$ | TABU time (s) | CPLEX2 time (s) |
|---|---|---|---|---|---|
| 40 | 25 | 0.995 | 0.995 | 282 | 349 |
|    | 30 | 0.997 | 0.998 | 227 | 205 |
|    | 35 | 0.992 | 0.995 | 259 | 330 |
|    | 40 | 0.999 | 1.000 | 188 | 157 |
| 50 | 25 | 0.992 | 0.993 | 692 | 2999 |
|    | 30 | 0.994 | 0.995 | 403 | 396 |
|    | 35 | 0.995 | 0.994 | 186 | 227 |
|    | 40 | 0.999 | 0.999 | 175 | 185 |
| 60 | 25 | 0.995 | 0.996 | 354 | 491 |
|    | 30 | 0.992 | 0.993 | 353 | 316 |
|    | 35 | 0.996 | 0.998 | 238 | 234 |
|    | 40 | 0.996 | 0.999 | 194 | 207 |
| 70 | 25 | 0.991 | 0.993 | 698 | 1855 |
|    | 30 | 0.994 | 0.996 | 333 | 331 |
|    | 35 | 0.995 | 0.996 | 270 | 225 |
|    | 40 | 0.998 | 0.999 | 201 | 221 |

occurs, the corresponding redeployment plan is implemented and new solutions are recomputed. There always exists a risk that a scenario cannot be computed in time for the next call, in which case no redeployment is implemented.

In order to test the feasibility of this approach, the tabu search algorithm was run on a network of 16 Sun Ultra-1/140 workstations (spec int95:5.87; spec fp95:8.38; ram: 64M; Solaris 7 exploitation system) and PVM was used for the parallel implementation. Six simulated data sets derived Urgences Santé data in Montreal were used to run the tests. Each set corresponds to a 7 h period and contains an average of 130 calls served by an average of 54 ambulances. These calls were distributed into four categories: urgent calls requiring one ambulance : 80%; urgent calls requiring two or three ambulances : 3%; less urgent calls: 10%; pending calls : 7%. The covering radiuses were $r_1 = 7$ min and $r_2 = 15$ min. With the proposed system, all calls were covered within with $r_2$, and 98% of urgent calls were covered within $r_1$ with an average of 3.5 min (the desired response time set by Urgences Santé for urgent calls is 7 min 90% of the time). Less urgent calls were served within an average of 9 min. The algorithm was capable of precomputing in time a redeployment plan in 95% of all cases. The only case where this was not possible is when two calls arrived within less than 32 seconds of each other. Out of all calls, 62% required no relocation and 99.59% of all relocation involved at most five ambulances with an average of 2.08. Thirty-three scenarios extracted from the

**Fig. 1.** DSM solution for the Island of Montreal. Number of ambulances located at a site (1 or 2)

simulation data were solved exactly by CPLEX. On these scenarios the tabu search heuristic produces solutions within 2% of the optima.

### 3.2 Application to Austria

Doerner *et al.* [8] have applied a model derived from DSM to locate ambulances in Austria. In their model, constraint (10) and (11) are treated as soft constraints, and the fraction of the demand $w_i$ covered per ambulance for each demand point $i$ is bounded above by a constant $w_0$. This requirement is also treated as a soft constraint. The objective function to be maximized is therefore:

$$z' = z - M_1 z_1 - M_2 z_2 - M_3 z_3,$$

where $z$ is defined by (9) and $M_1$, $M_2$, $M_3$ are positive constants. The functions $z_1$, $z_2$, $z_3$ are defined as follows:

$$z_1 = |\{i \in V : \sum_{j \in W_i^2} x_j = 0\}| \tag{27}$$

$$z_2 = \alpha - \min \; \{\alpha, \sum_{i \in V} d_i y_i^1 / \sum_{i \in V} d_i\} \qquad (28)$$

and

$$z_3 = \sum_{i \in V} \max \; \{0, d_i / \sum_{j \in W_i^2} x_j - w_0\} \qquad (29)$$

The model was solved by the tabu search algorithm of Gendreau, Laporte and Semet [10], and also by an ant colony optimization (ACO) heuristic inspired from that of Doerner *et al.* [9]. Basically, several location plans are determined, and ambulances are located at the selected sites. If fewer than $p$ ambulances have been assigned though this process, the remaining ambulances are randomly assigned to the remaining sites. A measure of the attractiveness of the solution is computed and is used to assign a "pheromone value" to the solution. More specifically, the pheromone values $\tau_{js}$ are numbers associated to the decision of locating $s$ ambulances at site $j$. The pheromone values are periodically updated in order to influence the design of ulterior solutions. The higher the value of $\tau_{js}$ is, the more likely $s$ ambulances will be located at site $j$ in future iterations. The authors have also applied a local search mechanism, similar to that of the tabu search algorithm of Gendreau, Laporte and Semet [10] in order to improve the successive solutions constructed by their ACO algorithm. The ACO algorithm generated solutions of similar quality to those produced by tabu search but required significantly more computation time.

Both algorithms were applied to locate 1'952 ambulances over 460 bases in all of Austria, except Vienna. In two cases where it was possible to compute an optimal solution with an exact algorithm, the optimum was found. Figure 2 depicts the best solution identified for the Province of Salzburg.

Doerner *et al.* [8] have clearly demonstrated the practicality of the Gendreau, Laporte and Semet tabu search algorithm [10] over a different data set and with a slightly modified model. In our opinion, the main advantage of this model lies in the assignment of a maximum demand to any ambulance. This feature seems to be absent from previous ambulance location models.

### 3.3 Application to Wallonia

Finally, Thirion [22] has recently applied and compared several ambulance location models to a zone comprising 247 communes in the provinces of Namur and Brabant Walloon in the French speaking part of Belgium. This zone is currently served by 46 ambulances, 25 of which are located on nine sites within the zone, and 21 are located on 12 sites outside the zone but can still answer calls inside the zone. The set $W$ is made up of 259 sites comprising the 247 communes of the zone, and the 12 current location sites outside the zone which are considered to be fixed. Currently,

**Fig. 2.** DSM solution for the Province of Salzburg. Number of ambulances located at a site (1 or 2)

126 communes (51.01%) are covered by an ambulance within 8 min, while five cannot be reached within 15 min. In 2005, 77.47% of the 19'197 calls made to the ambulance services could be reached within 8 min.

A straight application of the LCSM shows that 26 ambulances would be necessary to cover all demand points within 8 min, while successive applications of the MCLP show that only eight sites are necessary to cover 82.33% of all calls within 8 mins and 15 sites yield a 95.34% coverage.

The MEXCLP was also applied to these data. The busy fraction $q$ was estimated as 17.79% in 2005. It was found with this model that 12 ambulances would be needed to cover 81.26% of all demand, taking into account the temporary unavailability of some ambulances. To obtain a 95% coverage, 28 ambulances would be needed. The data should be contrasted with those of the MCLP which ensures a zero busy fraction.

Finally, the DSM was also applied to the same data. It was found that 20 ambulances are necessary to cover 82.38% of all calls twice within eight min, and 30 ambulances are necessary to bring the proportion up to 95.22%. Optimally relocating the 25 ambulances currently located inside the zone while leaving the location of the 21 outside ambulances unchanged would yield a 91.31% double coverage of all calls, as opposed to the current proportion of 75.88%.

These results clearly show that the application of models such as the MEXCLP and the DSM, which are based on different principles, can yield a significant improvement in simple or double coverage by simply reallocating the available ambulances to different sites. Figure 3 shows the locations of the 25 ambulances of the zone in the current situation, and as suggested by the MEXCLP and DSM solutions.

## 4 Conclusions

We have described some classical and advanced ambulance location models developed over the past 35 years. One of these models, called the *Double Standard Model* (DSM), maximizes double coverage with a fixed number of ambulances. A dynamic version of the DSM was developed by Gendreau, Laporte and Semet [11]. The static and dynamic versions of the DSM were successfully tested on data from the Island of Montreal. A slightly modified version of the DSM was later applied to Austrian data. Recently, the application of DSM to Walloon data has shown that significant coverage improvements could be reached without increasing the number of ambulances now in use.

## Acknowledgments

This work was partly supported by the Canadian Natural Sciences and Engineering Council under grants 39682-05, by the FEDER program of the European Community, and by the Région Nord Pas de Calais, France. This support is gratefully acknowledged.

## References

[1] Ball MO, Lin FL (1993) A reliability model applied to emergency service vehicle location. *Operations Research* 41:18-36.

[2] Batta R, Dolan JM, Krishnamorthy NN (1989) The maximal expected covering location model revisited. *Transportation Science* 23:277-287.

[3] Brotcorne L, Laporte G, Semet F (2003) Ambulance location and relocation models. *European Journal of Operational Research* 147:451-463.

[4] Church RL, ReVelle CS (1974) The maximal covering location problem. Papers of the Regional Science Association 32:101-118.

[5] Cordeau JF, Laporte G, Potvin J-Y, Salvesbergh MWP (2007) Transportation on demand. In : Barnhart C, Laporte G (eds) *Transportation*, Handbooks

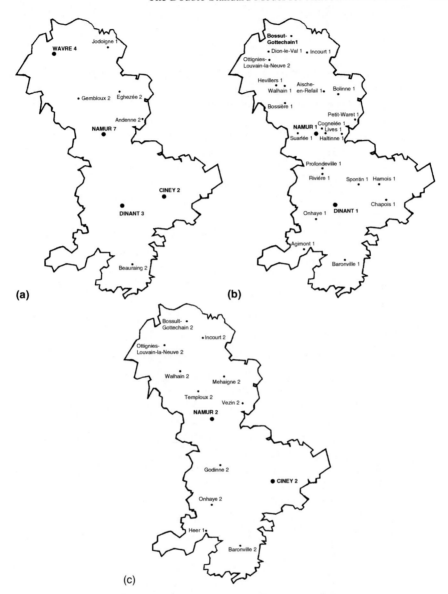

**Fig. 3.** Location of 25 ambulances in the Namur and Brabant Walloon provinces : (a) current situation; (b) MEXCLP solution ; (c) DSM solution

in Operations Research and Management Science, Volume 14. Elsevier, Amsterdam.

[6] Daskin MS, Stern EH (1981) A hierarchical objective set covering model for emergency medical service vehicle deployment. *Transportation Science* 15:137-152.

[7] Daskin MS (1983) The maximal expected covering location model: formulation, properties, and heuristic solution. *Transportation Science* 17:48-70.

[8] Doerner KF, Gutjahr WJ, Hartl RF, Karall M, Reimann M (2005) Heuristic solution of an extended double-coverage ambulance location problem for Austria. *Central European Journal of Operations Research* 13:325-340.

[9] Doerner KF, Gutjahr WJ, Hartl RF, Strauss C, Stummer C (2006) Nature inspired metaheuristics in multiobjective activity crashing. Forthcoming in *Omega*.

[10] Gendreau M, Laporte G, Semet F (1997) Solving an ambulance location model by tabu search. *Location Science* 5:75-88.

[11] Gendreau M, Laporte G, Semet F (2001) A dynamic model and parallel tabu search heuristic for real-time ambulance relocation. *Parallel Computing* 27:1641-1653.

[12] Gendreau M, Laporte G, Semet F (2006) The maximal expected demand coverage relocation problem for emergency vehicles. *Journal of the Operational Research Society* 57:22-28.

[13] Goldberg JB, Dietrich R, Chen JM, Mitwasi MG, Valenzuela T, Criss E (1990) Validating and Applying a Model for Locating Emergency Medical Vehicles in Tucson, AZ. *European Journal of Operational Research* 49:308-324.

[14] Goldberg JB (2004) Operations research models for the deployment of emergency services vehicles, *EMS Management Journal* 1:20-39.

[15] Hogan K, ReVelle CS (1986) Concept and applications of backup coverage. *Management Science* 32:1434-1444.

[16] Marianov V, ReVelle CS (1994) The queuing probabilistic location set covering problem and some extensions. *Socio-Economic Planning Sciences* 28:167-178.

[17] Marianov V, ReVelle CS (1995) Siting emergency services. In: Drezner Z. (ed) *Facility Location: A Survey of Applications and Methods*. Springer-Verlag, New York.

[18] Repede JF, Bernardo JJ (1994) Developing and validating a decision support system for locating emergency medical vehicles in Louisville, Kentucky. *European Journal of Operational Research* 75:567-581.

[19] ReVelle CS, Hogan K (1989) The maximum availability location problem. *Transportation Science* 23:192-200.

[20] Schilling DA, Elzinga DJ, Cohon J, Chuch RL, ReVelle CS (1979) The TEAM/FLEET models for simultaneous facility and equipment sitting. *Transportation Science* 13:163-175.

[21] Statistics Canada (1991) GEO: fichier de conversion des codes postaux.

[22] Thirion A (2006) Modèles de localisation et de réallocation d'ambulances. Application aux communes en provinces de Namur et Brabant Wallon. M.Sc. Dissertation, Facultés Universitaires Notre-Dame de la Paix, Namur, Belgium.

[23] Toregas C, Swain R, ReVelle C.S., Bergman L. (1971) The location of emergency service facilities. *Operations Research* 19:1363–1373.

# A Pricing Algorithm for the Vehicle Routing Problem with Soft Time Windows

Federico Liberatore[1], Giovanni Righini[2], and Matteo Salani[3]

[1] Dipartimento di Tecnologie dell'Informazione, Università degli Studi di Milano, Italy
   fliberatore@crema.unimi.it
[2] Dipartimento di Tecnologie dell'Informazione, Università degli Studi di Milano, Italy
   righini@dti.unimi.it
[3] TRANSP-OR, École Politechnique Fédérale de Lausanne, Switzerland
   matteo.salani@epfl.ch

**Summary.** The Vehicle Routing Problem with Soft Time Windows consists in computing a minimum cost set of routes for a fleet of vehicles of limited capacity that must visit a given set of customers with known demand, with the additional feature that each customer expresses a preference about the time at which the visit should occur. If a vehicle serves the customer out of its specified time window, an additional cost is incurred. Here we consider the case with penalties linearly depending on the time windows violation. We present an exact optimization algorithm for the pricing problem which arises when the vehicle routing problem with soft time windows is solved by column generation. The algorithm exploits bi-directional and bounded dynamic programming with decremental state space relaxation.

**Key words:** Combinatorial optimization, Vehicle routing, Column generation, Dynamic programming, Time windows

## 1 Introduction

In distribution logistics it is common that customers impose constraints on the arrival and departure time of the vehicles visiting them for pick-up or delivery operations. Therefore any planning algorithm for optimally routing and scheduling a fleet of vehicles must comply with these restrictions. The scientific literature in logistics optimization is rich of references to the vehicle routing problem with time windows (VRPTW): for a recent survey the reader is referred to Cordeau et al. [5] and Kallehauge et al. [12].

Soft time windows do not represent constraints but rather preferences about the time at which visits should occur at customers' locations: if a customer is visited

L. Bertazzi et al. (eds.), *Innovations in Distribution Logistics*, Lecture Notes
in Economics and Mathematical Systems 619, DOI: 10.1007/978-3-540-92944-4,
© 2009 Springer-Verlag Berlin Heidelberg

out of his preferred time window, a penalty is incurred in terms of additional costs that are charged to the distributor/collector. The main advantage of routing with soft time windows is that a feasible plan may include more visits than in the case in which the same time windows are imposed as "hard" constraints; solutions implying a small violation of one or more time window constraints and that therefore would have been discarded as infeasible for the VRPTW, may be discovered in this way and hence more profitable plans can be produced. However the optimization algorithm must take into account the additional cost terms coming from the penalties; as illustrated in the remainder this implies a number of algorithmic problems that must be suitably addressed to make use of the mathematical optimization techniques developed so far for the VRPTW.

Several heuristic algorithms have been proposed for the vehicle routing problem with soft time windows (VRPSTW), starting with the early work by Sexton and Choi [14], which concerned the pick-up and delivery version of the problem; Koskosidis et al. [13] developed an optimization-based heuristic; Balakrishnan [1] developed several constructive heuristics, while Taillard et al. [15] and Chiang and Russell [4] presented tabu search heuristics; more recently Ibaraki et al. [9] studied acceleration techniques for local search algorithms in the case of multiple soft time windows.

In this paper we present an algorithm for the resource constraint elementary shortest path problem with soft time windows, which forms the basis to develop a branch-and-price algorithm for the exact optimization of the VRPSTW. Recent examples of branch-and-price algorithms for the VRPTW are those of Desaulniers et al. [6] and Jepsen et al. [11].

## 2 Problem Formulation

The VRPSTW is defined as follows: a graph $G(\mathcal{V}, \mathcal{A})$ is given, where the vertex set $\mathcal{V}$ is made of a set $N$ of $N$ customers and two vertices, numbered 0 and $N + 1$ representing the depot, where $V$ vehicles are located. Non-negative weights $t_{ij}$ and $c_{ij}$ are associated with each arc $(i, j) \in \mathcal{A}$; they represent respectively the traveling time and the transportation cost on each arc $(i, j) \in \mathcal{A}$; traveling times are given by shortest path lengths and therefore they satisfy the triangle inequality.

A positive integer demand $q_i$ is associated with each vertex $i \in N$ and a capacity $Q$ is associated with each vehicle. A non-negative integer service time $\theta_i$ and a time window $[a_i, b_i]$, defined by two non-negative integers, are also associated with each vertex $i \in N$; if the service at vertex $i$ starts inside its time window $[a_i, b_i]$ no penalty is incurred; if a vehicle starts servicing the customer at vertex $i$ before time $a_i$ or after time $b_i$, then a *linear* penalty has to be paid, which is proportional to the anticipation or delay through non-negative coefficients $\alpha_i$ and $\beta_i$ respectively. Indicating with $T_i$ the starting time of service at vertex $i$ the penalty term $\pi_i(T_i)$ is defined as follows:

$$\pi_i(T_i) = \begin{cases} \alpha_i(a_i - T_i) & \text{if } T_i \leq a_i \\ 0 & \text{if } a_i \leq T_i \leq b_i \\ \beta_i(T_i - b_i) & \text{if } T_i \geq b_i. \end{cases}$$

However the vehicles are allowed to wait at no cost at any time along their routes.

Column generation algorithms for routing problems rely upon a set covering reformulation, as follows.

$$\text{minimize} \sum_{f \in \mathcal{F}} w_f z_f$$

$$\text{subject to} \sum_{f \in \mathcal{F}} x_{if} z_f \geq 1 \qquad \forall i \in \mathcal{N} \qquad (1)$$

$$-\sum_{f \in \mathcal{F}} z_f \geq -V \qquad (2)$$

$$z_f \in \{0, 1\} \qquad \forall f \in \mathcal{F} \qquad (3)$$

where $\mathcal{F}$ is the set of feasible vehicle routes, $w_f$ is the cost of route $f \in \mathcal{F}$, that is the sum of the costs of the arcs in the route plus the penalties due to the soft time windows violations, and $x_{if}$ is the number of times route $f \in \mathcal{F}$ visits customer $i \in \mathcal{N}$. The linear relaxation of this set covering reformulation usually yields very tight lower bounds (see for instance Bramel and Simchi-Levi [3] and the references therein). However since in general $\mathcal{F}$ contains an exponential number of columns, only a subset $\mathcal{F}'$ is kept in a restricted linear master problem (RLMP) and further feasible routes are generated by the iterated solution of a pricing problem. The pricing problem consists in finding routes with negative reduced cost or proving that none exists. The reduced cost of route $f \in \mathcal{F}$ is:

$$\overline{w}_f = w_f - \sum_{i \in \mathcal{N}} x_{if} \lambda_i + \lambda_0$$

where $(\lambda, \lambda_0)$ is the vector of non-negative dual variables corresponding to constraints (1) and (2) in the linear restricted master problem.

The capacity constraints as well as the penalties are taken into account in the pricing subproblem: in particular the former ones restrict the set $\mathcal{F}$ of feasible routes, while the latter ones contribute to determine the cost of each route.

Hence the pricing subproblem turns out to be a resource constrained shortest path problem with soft time windows. A vehicle must go from vertex 0 to vertex $N + 1$, visiting a subset of the other vertices; no cycles are allowed. Because there are non-negative prizes $\lambda$ associated with the vertices, negative cost cycles can occur. Therefore the requisite that the path must be elementary does not come for free from cost minimization but it must be explicitly enforced. The capacity constraint is taken into account as a resource constraint: each vehicle leaving the depot has $Q$ units of available resource and every time it visits a customer with demand $q_i$ it consumes $q_i$ units of resource. The objective is to minimize the cost, given by the sum of the costs of the arcs traversed plus the sum of the penalties for anticipation and delay, minus the sum of the prizes collected at the vertices visited.

The main difficulty in dealing with soft time windows in column generation algorithms is that the possibility of trading time vs. cost generates an infinite number of possible solutions of the pricing problem that do not dominate one another. In algorithmic terms this means that the pricing algorithm, usually a dynamic programming algorithm, must take into account an infinite number of Pareto-optimal states. The pricing subproblem turns out to be an elementary shortest path problem with continuous resources on a graph with negative-cost arcs. The resource constrained elementary shortest path problem (RCESPP) on graphs with negative cost cycles is strongly NP-hard (see Dror, [7]) and it has been recently investigated by Righini and Salani who proposed bi-directional bounded dynamic programming algorithms [17] and decremental state space relaxation [16]. Another recent contribution on the subject is due to Irnich [10]. Hereafter we extend this research stream and we show how it can be adapted to the case with soft time windows implying continuous resources. In particular we present experimental evaluations of the bi-directional search technique [17], coupled with decremental state space relaxation [16] [18], when dynamic programming labels represent in a compact way an infinite number of non-dominated states.

## 3 The Algorithm

The resource constrained elementary shortest path problem with soft time windows is solved to optimality by a bi-directional dynamic programming algorithm.

### 3.1 States, Labels and Extension

A *state* associated with vertex $i \in N$ represents a path from the depot 0 to $i$. Different states associated with the same vertex correspond to different feasible paths reaching that vertex.

When a vehicle reaches a vertex it can start the service immediately or it can wait and start the service at a later time in order to reduce costs in case of early arrival. Therefore from each feasible state an infinite number of feasible states can be generated. For this reason our dynamic programming algorithm must take into account an infinite number of non-dominated states and this is done by grouping them into *labels*. Each label corresponds to an infinite number of states associated with the same path.

A label associated with vertex $i \in N$ is a tuple $L_i = (S, i, r, C(T_i))$, where $S$ is a binary vector indicating the vertices visited along the path, $i$ is the last reached vertex, $r$ is the amount of capacity consumed up to $i$, $C$ is the cost of the path, $T_i$ is the time at which the service at vertex $i$ begins. In each label the function $C(T_i)$ describes the trade-off between cost and time. This function is piecewise linear and convex, because it is the sum of piecewise linear and convex functions, like the one shown in Fig. 1, one for each visited vertex. Its domain ranges from earliest possible arrival time to infinity.

In bi-directional dynamic programming these states are called *forward states* and in the same way we define *backward states*, corresponding to paths from vertex $i$ to the final depot $N + 1$ represented by labels $(S, i, r, C(T_i))$, where $T_i$ is the time at

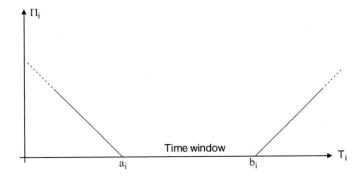

**Fig. 1.** A soft time window at a generic vertex $i \in \mathcal{N}$: a linear penalty $\pi_i$ is incurred depending on the service starting time $T_i$

which the service at vertex $i$ begins.

*Remark.* To keep a perfect symmetry in the description of the algorithm, one could equivalently define $T_i$ in backward labels as the time at which the service at vertex $i$ terminates and consider backward time windows $[a_i + \theta_i, b_i + \theta_i]$, specifying the range in which the service at vertex $i$ should preferably terminate. In the exposition we preferred not to introduce backward time windows, in order not to make the notation unnecessarily complicated, and hence we refer to service starting time and to the original time windows both in forward and in backward labels.

The dynamic programming algorithm iteratively extends all feasible forward and backward labels to generate new forward and backward labels respectively. The extension of a forward label corresponds to appending an additional arc $(i, j)$ to a path from 0 to $i$, obtaining a path from 0 to $j$, while the extension of a backward label corresponds to appending an additional arc $(j, i)$ to a path from $i$ to $N + 1$, obtaining a path from $j$ to $N + 1$.

To avoid negative cost cycles a dummy resource is associated with each vertex $i \in \mathcal{N}$: there is only one unit available for each dummy resource and it is consumed when the corresponding vertex is visited. The binary vector $S$, indicating the vertices already visited, is therefore a resource consumption vector related to the dummy resources. It is initialized at 0 at vertex 0. Note that $S$ does not keep any information about the order in which the vertices are visited. When a forward label $(S, i, r, C(T_i))$ is extended to a vertex $j$, a new forward label $(S', j, r', C'(T_j))$ is generated and the update rule is:

$$S'_k = \begin{cases} S_k + 1 & \text{if } k = j \\ S_k & \text{if } k \neq j. \end{cases}$$

A label $(S, i, r, C(T_i))$ corresponds to an elementary path only if $S_k \leq 1 \ \forall k \in \mathcal{N}$.

The resource consumption $r$ indicates the amount of capacity used. When a vehicle leaves the depot 0 all the resource is available, that is $r = 0$, and the extension rule is:

$$r' = r + q_j.$$

A label $(S, i, r, C(T_i))$ is feasible only if $r \leq Q$.

Finally the cost $C(T_i)$ is initialized with the function $C(T_0) = 0 \ \forall T_0 \geq 0$ at the depot 0. At each extension the cost depends on both traveling time and penalties and it is updated according to the formula:

$$C'(T_j) = C(T_j - (\theta_i + t_{ij})) - \lambda_i/2 + c_{ij} - \lambda_j/2 + \pi_j(T_j),$$

where $\lambda_i = -\lambda_0$ if $i = 0$ and $\lambda_j = -\lambda_0$ if $j = N+1$. In this expression the cost function of the predecessor is evaluated at $T_i = T_j - (\theta_i + t_{ij})$, which is the latest time instant at which the service at vertex $i$ should begin to allow starting the service at vertex $j$ at time $T_j$.

Figure 2 shows an example of forward extension. In graphical terms, the cost function $C(T_i)$ is shifted to the right by the service time at vertex $i$, that is $\theta_i$, plus the traveling time $t_{ij}$ spent to reach vertex $j$, it is shifted up by the traveling cost $c_{ij}$ minus the prizes $\lambda_i/2 + \lambda_j/2$ collected for visiting the vertices. Then it is summed to the penalty term $\pi_j$, which depends on the arrival time $T_j$. If $C(T_i)$ is piecewise linear and convex, then the resulting function $C'(T_j)$ preserves both these properties. The number of segments in these piecewise linear functions is increased by at most two at every extension.

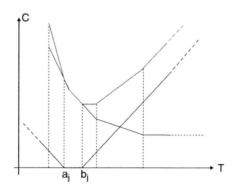

**Fig. 2.** Forward extension of a label of vertex $i$ to vertex $j$. The $C'(T_j)$ function resulting from the extension is the sum of the $C(T_i)$ function of the extended label suitably shifted and the penalty function $\pi_j(T_j)$

The extension rules for backward labels are symmetrical to those above. The cost function at the final depot is initialized as $C(T_{N+1}) = 0 \ \forall T_{N+1} \leq T^{max}$, where $T^{max}$ is defined below.

## 3.2 Dominance Rules

In general the effectiveness of any dynamic programming algorithm heavily depends on the number of states generated. Hence it is essential to fathom feasible states which cannot lead to optimal solutions.

In our case this is done in two ways: first, by the deletion of parts of the linear functions in all labels and, second, by suitable dominance rules.

Consider a generic forward label like that represented in Fig. 3. Since all penalty functions have a positive penalty term for delays, the rightmost part of the function certainly has positive first derivative. Since waiting at no cost is allowed, all states in this part of the function are dominated by the states of the same label with smaller values of both time and cost. Therefore these dominated states can be replaced by states with the same cost of the dominating ones, as shown in Fig. 3. In graphical terms this means that the rightmost part of the piecewise linear function in each label is a horizontal unbounded segment, replacing all the segments with positive first derivative resulting from the last extension. A symmetric argument holds for backward labels, where a horizontal segment replaces the leftmost part of the piecewise linear functions with negative first derivative.

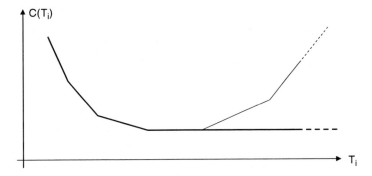

**Fig. 3.** States on the ascending part of the piecewise linear function are dominated: the same value in time can be reached at a smaller cost

The second way to eliminate dominated states is by dominance tests, that are performed each time labels are extended. Let $L' = (S', i, r', C'(T_i'))$ and $L'' = (S'', i, r'', C''(T_i''))$ be two labels associated with vertex $i$; let $(S', i, r', C', T_i')$ and $(S'', i, r'', C'', T_i'')$ be two states belonging to $L'$ and $L''$ respectively, corresponding to two points on the piecewise linear functions $C'(T_i')$ and $C''(T_i'')$. Then the former state dominates the latter only if

$$S' \leq S''$$
$$r' \leq r''$$
$$C' \leq C''$$
$$T_i' \leq T_i''$$

and at least one of the inequalities is strict. When a dominance rule is applied to single states and the test succeeds, the effect is simply to delete one of the two states. In our case, where each label represents an infinite number of states, the effect of the dominance test is to delete some parts of the piecewise linear functions, as shown in Fig. 4. The states surviving the dominance test are those of minimum cost for each feasible value of the service starting time.

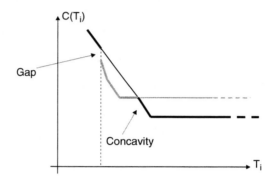

**Fig. 4.** As an effect of the dominance test, some parts of the piecewise linear functions are deleted. Only the states drawn with heavy lines are non-dominated and survive the test

This holds for both forward and backward labels. As a consequence, the resulting piecewise linear functions may have gaps and are not convex in general, as the small example in Fig. 4 shows.

When a new label is generated, it is compared to all the labels currently associated with the same vertex and the dominance test is applied for each comparison. Let $L' = (S', i, r', C'(T_i))$ and $L'' = (S'', i, r'', C''(T_i))$ be the labels of vertex $i$ that are compared. Then if $S' = S''$ and $r' = r''$ then the two labels are merged into one, so that at most one piecewise linear function is stored for each feasible combination of $S$, $i$ and $r$; in this case the non-dominated states form a new piecewise linear function possibly with vertical gaps and concavities as the one formed by the points in heavy lines in Fig. 4. In the other case, for instance (w.l.o.g.) if $S' \leq S''$ and $r' \leq r''$, then the dominated states in $C''(T_i)$ are deleted and hence the resulting piecewise linear function can have also horizontal gaps, as the black heavy lines in Fig. 4.

### 3.3 Profitable Time Windows

The combined effect of soft time windows and dual prizes induces profitable time windows in the pricing problem. Consider a vertex $i$ with dual prize $\lambda_i$, soft time window $[a_i, b_i]$ and penalty coefficients $\alpha_i$ and $\beta_i$. Any visit to the vertex occurring out of the time window $[a_i - \lambda_i/\alpha_i, b_i + \lambda_i/\beta_i]$ is not profitable, because the penalty to be paid would be larger than the prize gained. Such a non-profitable visit can always be skipped from any path $\mathcal{P}$, yielding another path $\mathcal{P}'$ not worse than $\mathcal{P}$, owing to the triangle inequality. We do not explicitly include these profitable time windows in our model as hard constraints, but we use this observation to detect unreachable vertices, as explained in the next paragraph. Owing to this observation we also define a maximum allowed arrival time $T^{max} = \max_{i \in \mathcal{N}}\{b_i + \lambda_i/\beta_i + t_{i\,N+1}\}$, which is the latest possible arrival time at the final depot for an optimal path.

## Unreachable Vertices

A vertex $j$ is unreachable from a label $L = (S, i, r, C(T_i))$ when its demand $q_j$ cannot be fulfilled with the remaining capacity $Q - r$ or when its profitable time window would be violated by any visit starting from label $L$, that is $T_i + \theta_i + t_{ij} > b_j + \lambda_j/\beta_j$ for all feasible values of $T_i$. For the purpose of the dominance test, unreachable vertices can be counted as if they were already visited, exploiting an idea of Feillet et al. [8]. Indicating with $U'$ and $U''$ the characteristic vectors of the sets of unreachable vertices in two labels $L'$ and $L''$, the condition

$$S' \leq S''$$

in the dominance test is replaced by

$$S' + U' \leq S'' + U'',$$

which is a weaker and therefore more effective sufficient condition.

## Search Policy

Labels are extended according to the vertices they are associated with. The vertices are cyclically visited in increasing order of the starting time $a_i$ of their soft time windows; for each vertex the algorithm extends all its labels that have not yet been extended. At each vertex all the labels not yet extended are sorted according to a hierarchical criterion: first they are sorted by increasing overall length of the path; in case of ties they are sorted by increasing amount of capacity consumed; in case of further tie, they are sorted by number of vertices visited.

## Joining Forward and Backward States

In our bi-directional dynamic programming algorithm forward and backward paths must be joined together to produce complete paths from vertex 0 to vertex $N + 1$. Let $L^{fw} = (S^{fw}, i, r^{fw}, C^{fw}(T_i))$ be a forward label and $L^{bw} = (S^{bw}, j, r^{bw}, C^{bw}(T_j))$ be a backward label. The join between $L^{fw}$ and $L^{bw}$ is subject to feasibility conditions on the resources. In particular the feasibility test on dummy resources $S$ imposes that a same vertex can not be visited by both paths:

$$S_k^{fw} + S_k^{bw} \leq 1 \quad \forall k \in \mathcal{N}.$$

Moreover the consumption of capacity in the overall resulting path can not exceed the overall amount of available capacity, that is:

$$r^{fw} + r^{bw} \le Q.$$

If the join is feasible, then the cost of the resulting path can be obtained as the minimum of the function

$$C(T) = C^{fw}(T) - \lambda_i/2 + c_{ij} - \lambda_j/2 + C^{bw}(T + \theta_i + t_{ij}).$$

This function may have several local minima, as shown in Fig. 5. However the detection of the global minimum takes time linear in the number of discontinuity points of the two piecewise linear functions, since it requires a merge operation between two sorted lists.

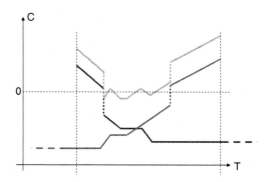

**Fig. 5.** When a forward and a backward labels are joined, the two corresponding piecewise linear functions are summed up. The resulting piecewise linear function may have multiple local minima. All its points below 0 correspond to negative reduced cost routes

### Resource-Based Bounding

Since all forward and backward states generated by the bi-directional search algorithm are tentatively joined, it is crucial to reduce their number as much as possible. To this purpose we select a critical resource, whose consumption is monotone along the paths, and we do not extend states in which at least half of the available amount of that resource has already been consumed. We tried using $r$, $T$ and $|S|$ as critical resources. These choices are compared in the next section.

### Decremental State Space Relaxation

Decremental state space relaxation (DSSR) was independently introduced by Righini and Salani [16] and by Boland et al. [2] to reduce the number of states to be explored by dynamic programming. The basic idea is that the elementary path constraints are not imposed on all the vertices of the graph but rather on a subset of critical vertices. If the optimal solution of the pricing problem is not elementary, one or more vertices visited more than once are inserted into the critical vertex set and the dynamic programming algorithm is executed again. In this way at each iteration a lower bound for the pricing problem is computed. Computational experiments reported in [16] show that the number of vertices that must be defined as critical in order to obtain a feasible solution is usually only a small fraction of $N$ and therefore the pricing problem is solved to proven optimality in a fraction of the computing time required by dynamic programming when multiple visits are forbidden at all vertices.

In our algorithm we map each label $(S, i, r, C(T_i))$ onto a new label $(\hat{S}, i, r, C(T_i))$, where $\hat{S}$ is the set of critical vertices, initially empty. This is done for both forward and backward labels. The dominance rule is then modified, replacing condition $S' \le S''$ by $\hat{S}' \le \hat{S}''$. Also in this case the detection of unreachable critical vertices is useful to enhance the effect of dominance tests. Indicating with $\hat{U}$ the set of unreachable critical vertices, the dominance test is $\hat{S}' + \hat{U}' \le \hat{S}'' + \hat{U}''$.

## DSSR Policies

In Boland et al. [2] several different policies to update the critical set were considered. A computational comparison between them is also reported in Righini and Salani [18]. In the algorithm for the VRPSTW, we used the policy called HMO according to the terminology of [2]: it consists in adding to the critical vertex set one vertex at each iteration, selecting one at random among those visited the largest number of times.

## Initialization of the critical vertex set

Also the initialization of the critical vertex set can be done in different ways. In Righini and Salani [18] several policies were compared. The aim is to identify a subset of vertices that have a high probability to belong to the final critical vertex set. Let us define $f_{ij}$ to be a measure of the "cycling attractiveness" of a vertex $i$ with respect to a vertex $j$ as the ratio of the prize $\lambda_i$ over the duration of the cycle $i$-$j$-$i$:

$$f_{ij} = \lambda_i / (\theta_i + t_{ij} + \theta_j + t_{ji}).$$

Now we define an ordering of the vertices based on the following four criteria:

- Highest Cycling Attractiveness (HCA): order by $\max_{j \in \mathcal{N} \setminus \{i\}} \{f_{ij}\}$.
- Total Cycling Attractiveness (TCA): order by $\sum_{j \in \mathcal{N} \setminus \{i\}} f_{ij}$.
- Weighted Highest Cycling Attractiveness (WHCA): order by $\max_{j \in \mathcal{N} \setminus \{i\}} \{f_{ij}(b_i - a_i)\}$.
- Weighted Total Cycling Attractiveness (WTCA): order by $\sum_{j \in \mathcal{N} \setminus \{i\}} f_{ij}(b_i - a_i)$.

In general these criteria give different results and none of them is reliable to reveal the necessary vertices to be put in the critical vertex set. Let us define $HCA_m$, $TCA_m$, $WHCA_m$ and $WTCA_m$ to be the sets containing the first $m$ vertices according to each of the ordering criteria above. We use as an initial critical vertex set the one obtained from the intersection of these four sets. By a suitable choice of the value of $m$ the initial critical vertex set can be set large with the aim of reducing the number of iterations (high $m$) or small to reduce the probability of inserting unnecessary vertices into it (low $m$). In our computational experiments we set $m = 8$.

## 4 Experimental Results

For our experiments we used instances derived from the well-known Solomon's VRPTW benchmark. In particular we considered the instances in the class *random* with 100 customers. The integer coefficients $\alpha_i$ and $\beta_i$ for each vertex were generated at random with uniform probability distribution in the range $[1, 5]$. The values of the $t_{ij}$ data are the Euclidean distances rounded down to the nearest multiple of 0.1. The values of the $c_{ij}$ data are given by $\theta_i + t_{ij}$. The vehicles capacity was set to 200.

All tests were performed on a PC equipped with an Intel Core Duo T2500 2x1.0 GHz processor, with 1024 MB RAM. The algorithms were coded in C++. We used GLPK as a linear programming solver for the RLMP.

Tables 1, 2 and 3 report on the experimental comparison between the bi-directional bounded dynamic programming algorithm without DSSR and with DSSR, when the critical resource used for stopping the extension in both directions is respectively the amount of capacity used, the time elapsed and the number of vertices visited. For all algorithms we report the number of stored labels and the computing time. For the DSSR algorithms we also report the number of iterations and the number of critical vertices at the end.

A time-out of 15 min was imposed to all tests. A dash in the "Time" column indicates that the time-out was exceeded.

**Table 1.** Comparison between bi-directional bounded dynamic programming algorithms without and with DSSR, using capacity as a critical resource

| Instance | D.P. | | DSSR | | | |
|---|---|---|---|---|---|---|
| | Labels | Time | Labels | Iterations | Critical | Time |
| r101_100 | 3,636 | 0.30 | 1,218 | 1 | 0 | 0.13 |
| r102_100 | 29,190 | 4.24 | 5,882 | 4 | 5 | 5.83 |
| r103_100 | 70,654 | 18.08 | 8,351 | 5 | 7 | 14.19 |
| r104_100 | 141,422 | 54.92 | 10,893 | 6 | 7 | 25.00 |
| r105_100 | 9,248 | 0.94 | 2,357 | 4 | 3 | 1.47 |
| r106_100 | 47,889 | 9.95 | 7,251 | 4 | 5 | 8.00 |
| r107_100 | 93,888 | 26.47 | 8,755 | 4 | 5 | 12.78 |
| r108_100 | 166,226 | 69.97 | 11,446 | 6 | 7 | 26.59 |
| r109_100 | 26,092 | 3.91 | 5,498 | 5 | 4 | 5.98 |
| r110_100 | 64,594 | 16.20 | 7,332 | 4 | 4 | 9.64 |
| r111_100 | 80,498 | 21.22 | 9,431 | 5 | 6 | 19.11 |
| r112_100 | 191,198 | 86.86 | 11,979 | 5 | 5 | 26.44 |

**Table 2.** Comparison between bi-directional bounded dynamic programming algorithms without and with DSSR, using time as a critical resource

| Instance | D.P. | | DSSR | | | |
|---|---|---|---|---|---|---|
| | Labels | Time | Labels | Iterations | Critical | Time |
| r101_100 | 1,935 | 0.09 | 724 | 1 | 0 | 0.06 |
| r102_100 | 37,698 | 11.69 | 9,792 | 4 | 5 | 19.24 |
| r103_100 | 219,389 | 311.25 | 20,219 | 5 | 7 | 83.36 |
| r104_100 | 485,653 | – | 37,272 | 6 | 7 | 225.36 |
| r105_100 | 6,667 | 0.69 | 2,060 | 4 | 3 | 1.34 |
| r106_100 | 82,709 | 58.22 | 12,461 | 4 | 5 | 28.80 |
| r107_100 | 308,385 | 549.99 | 20,847 | 4 | 5 | 68.24 |
| r108_100 | 481,206 | – | 38,076 | 6 | 7 | 230.87 |
| r109_100 | 28,577 | 6.67 | 5,587 | 5 | 4 | 9.84 |
| r110_100 | 136,516 | 162.34 | 12,764 | 4 | 4 | 38.69 |
| r111_100 | 161,772 | 135.19 | 18,773 | 5 | 6 | 74.61 |
| r112_100 | 430,117 | – | 27,943 | 5 | 5 | 206.75 |

**Table 3.** Comparison between bi-directional bounded dynamic programming algorithms without and with DSSR, using the number of vertices visited as a critical resource

| Instance | D.P. | | DSSR | | | |
|---|---|---|---|---|---|---|
| | Labels | Time | Labels | Iterations | Critical | Time |
| r101_100 | 4,518 | 0.30 | 1,367 | 1 | 0 | 0.14 |
| r102_100 | 373,100 | – | 26,470 | 4 | 5 | 101.81 |
| r103_100 | 415,083 | – | 48,253 | 5 | 7 | 566.94 |
| r104_100 | 521,151 | – | 249 | 6 | 7 | – |
| r105_100 | 22,764 | 4.08 | 3,808 | 4 | 3 | 3.38 |
| r106_100 | 468,522 | – | 32,710 | 4 | 5 | 186.64 |
| r107_100 | 448,413 | – | 47,882 | 4 | 5 | 453.78 |
| r108_100 | 558,889 | – | 66,627 | 6 | 7 | – |
| r109_100 | 175,621 | 155.84 | 18,498 | 5 | 5 | 61.78 |
| r110_100 | 304,154 | – | 25,526 | 4 | 4 | 120.25 |
| r111_100 | 420,081 | – | 54,034 | 5 | 6 | 767.61 |
| r112_100 | 521,833 | – | 59,536 | 5 | 5 | – |

A first observation that can be drawn from the three tables reported, is that DSSR yields a remarkable improvement. In the worst cases only 7 vertices among 100 needed to be considered as critical in order to obtain a feasible solution.

A second observation concerns the choice of the critical resource: for the benchmark instances considered here capacity was definitely the most useful critical resource. Bounding on the basis of the number of vertices visited was particularly ineffective since the limit is set to half of the vertices, that is 50, while the average number of vertices in optimal paths is between 6 and 7. It should be remarked that Righini and Salani [16] obtained their best results on the same data-set with hard time windows using the elapsed time as a critical resource. Hence the relaxation of the time windows constraints makes the problem definitely more difficult, owing to an increased number of feasible labels.

A third observation concerns multiple pricing. The time spent by a pricing algorithm in column generation may be excessive if only one column is added to the RLMP at each iteration. An important feature of the algorithm described here is that it allows multiple pricing, which has become a common practice in column generation algorithms. At each join operation multiple columns with negative reduced cost can be found and inserted into the RLMP.

Moreover, by a trivial modification of the dominance test, the algorithm can be easily transformed into a heuristic, to accelerate the pricing phase.

## 5 Conclusions

In this paper we have presented an exact pricing algorithm for the vehicle routing problem with soft time windows, an important optimization problem in distribution logistics, for which no exact optimization algorithm has been published so far at the best of our knowledge.

The pricing algorithm is based on bi-directional and bounded dynamic programming with decremental state space relaxation.

We have described how an infinite number of non-dominated states are represented in a compact way by means of piecewise linear functions, whose description is stored in the labels extended by the dynamic programming algorithm; we have also presented the corresponding dominance rules and other algorithmic details.

The outcome of our computational experiments shows that this technique yields significant improvements compared to classical dynamic programming algorithms.

The next step will be to develop a branch-and-price algorithm for the exact optimization of the VRPSTW, exploiting this effective pricing technique. For this purpose the main developments needed are: (1) accurate and fast heuristics to compute primal bounds; (2) heuristic pricing algorithms to further speed up the search for profitable columns; (3) branching strategies compatible with the structure of the pricing problem.

## Acknowledgments

We thank M.Grazia Speranza and the organizers and participants of the International

Workshop on Distribution Logistics 2006 (Brescia), where a preliminary version of this research was presented. We also acknowledge the support of ACSU – Associazione Cremasca Studi Universitari to the OptLab, where this research was done.

# References

[1] Balakrishnan N (1993) Simple heuristics for the vehicle routeing problem with soft time windows *Journal of the Operational Research Society* 44:279–287.

[2] Boland N, Dethridge J, Dumitrescu I (2006) Accelerated label setting algorithms for the elementary resource constrained shortest path problem *Operations Research Letters* 34:58–68.

[3] Bramel J, Simchi-Levi D (2001) Set-covering-based algorithms for the capacitated VRP. In: Toth P, Vigo D (eds) *The vehicle routing problem*. SIAM Monographs on Discrete Mathematics and Applications, Philadelphia.

[4] Chiang WC, Russell RA (2004) A metaheuristic for the vehicle-routeing problem with soft time windows *Journal of the Operational Research Society* 55:1298–1310.

[5] Cordeau JF, Desaulniers G, Desrosiers J, Solomon MM, Soumis F (2001) The VRP with time windows. In: Toth P, Vigo D (eds) *The vehicle routing problem*. SIAM Monographs on Discrete Mathematics and Applications, Philadelphia.

[6] Desaulniers G, Lessard F, Hadjar A (2006) Tabu search, generalized $k$-path inequalities, and partial elementarity for the vehicle routing problem with time windows. Les Cahiers du GERAD 45, GERAD, Canada.

[7] Dror M (1994) Note on the complexity of the shortest path models for column generation in VRPTW *Operations Research* 42:977–978.

[8] Feillet D, Dejax P, Gendreau M, Gueguen C (2004) An exact algorithm for the elementary shortest path problem with resource constraints: application to some vehicle routing problems *Networks* 44:216–229.

[9] Ibaraki T, Imahori S, Kubo M, Masuda T, Uno T, Yagiura M (2005) Effective local search algorithms for routing and scheduling problems with general time-window constraints *Transportation Science* 39:206–232.

[10] Irnich S (2006) Resource extension functions: properties, inversion, and generalization to segments Technical report, RWTH Aachen University, Germany.

[11] Jepsen M, Petersen B, Spoorendonk S, Pisinger D (2006) A non-robust Branch-and-Cut-and-Price algorithm for the Vehicle Routing Problem with Time Windows. Technical report 06-03, DIKU University of Copenhagen, Denmark.

[12] Kallehauge B, Larsen J, Madsen OBG, Solomon MM (2005) Vehicle routing problem with time windows. In: Desaulniers G, Desrosiers J, Solomon MM (eds) *Column generation*. Springer, New York.

[13] Koskosidis YA, Powell WB, Solomon MM (1992) An optimization based heuristic for vehicle routeing and scheduling with soft time window constraints *Transportation Science* 26:69–85.

[14] Sexton T, Choi Y (1986) Pickup and delivery of partial loads with soft time windows *American Journal of Mathematical and Management Sciences* 6:369–398.

[15] Taillard E, Badeau P, Gendreau M, Guertin F, Potvin JY (1997) A tabu search heuristic for the vehicle routing problem with soft time windows *Transportation Science* 31:170–186.

[16] Righini G, Salani M (2005) New dynamic programming algorithms for the resource-constrained elementary shortest path problem. Note del Polo - Ricerca 69, DTI, University of Milan, Italy. Accepted for publication on Networks.

[17] Righini G, Salani M (2006) Symmetry helps: bounded bi-directional dynamic programming for the elementary shortest path problem with resource constraints *Discrete Optimization* 3:255–273.

[18] Righini G, Salani M (2006) Dynamic programming for the orienteering problem with time windows. Note del Polo - Ricerca 91, DTI, University of Milan, Italy.

# Effective Algorithms for a Bounded Version
# of the Uncapacitated TPP

Renata Mansini[1] and Barbara Tocchella[2]

[1] Department of Electronics for Automation, University of Brescia, Italy
    rmansini@ing.unibs.it
[2] Department of Electronics for Automation, University of Brescia, Italy
    barbara.tocchella@ing.unibs.it

**Summary.** Let us consider a set of markets plus a depot and a set of products. Each product is made available at a given price in a subset of markets. The distance between each couple of markets and between each market and the depot is known. The Uncapacitated Traveling Purchaser Problem with Budget constraint (UTPP-B) looks for a simple cycle starting at and ending to the depot which visits a subset of markets at the minimum traveling cost while purchasing all products at a global cost that does not exceed a defined budget threshold. Although the problem arises in several application domains very few contributions exist in the literature for the UTPP-B. We propose and compare two solution algorithms for the problem, an enhanced local search heuristic and a Variable Neighborhood Search (VNS) approach. UTPP benchmark instances with additional budget constraints are used for computational experiments. Heuristic performances are compared to exact solution values provided in [13] while solving with a single-objective hierarchical approach a bi-objective UTPP.

**Key words:** Traveling purchaser problem, Budget constraint, Local search, Variable neighborhood search

## 1 Introduction

The Traveling Purchaser Problem (TPP), originally proposed by Ramesh [9] is a generalization of the Traveling Salesman Problem and can be stated as follows. Consider a set of markets $M := \{1, \ldots, m\}$ plus a depot (indexed 0) and a set of products $K := \{1, \ldots, n\}$. A traveling cost between each couple of markets and between each market and the depot $c_{ij}$, $\forall i, j \in M \cup \{0\}$, is given. Each product $k, k \in K$, can be purchased in a given cluster of markets $M_k \subseteq M$ at a nonnegative price $f_{ki}$ depending on the market $i, i \in M_k$. The problem objective is to find a cycle starting at the depot

L. Bertazzi et al. (eds.), *Innovations in Distribution Logistics*, Lecture Notes
in Economics and Mathematical Systems 619, DOI: 10.1007/978-3-540-92944-4,
© 2009 Springer-Verlag Berlin Heidelberg

and purchasing each product while minimizing the sum of the traveling costs plus the purchasing costs. This problem is also known as uncapacitated TPP to distinguish it from the capacitated one. The latter is a generalization where for each product $k$ a required amount $d_k$ to be purchased is specified and a defined quantity $q_{ki}$ is made available at market $i$, $i \in M_k$, where $0 < q_{ki} \leq d_k$ and $\sum_{i \in M_k} q_{ki} \geq d_k$ for all $k \in K$ and $i \in M_k$.

In this paper we analyze a bounded version of the uncapacitated TPP where the purchasing costs are removed from the objective function and added to the problem constraints by specifying a threshold $B$ (the budget) as upper bound on the total amount which can be spent to purchase products. The problem looks for a minimum traveling costs cycle such that the demand for all products is satisfied at a global purchasing cost which does not exceed the budget threshold $B$. We identify such problem as the uncapacitated TPP with budget constraint (UTPP-B). The introduction of this bounded version aims at avoiding the sum in the same objective function of two cost measures which may be extremely different in nature. This is indeed the main drawback of the TPP formulation which does not take into account how, in many practical applications, the traveling costs are represented as distances or traveling times whereas the purchasing costs are measured as currency. Moreover, a trade-off frequently characterizes these two objectives since reducing the purchasing cost may imply an increase in the distance traveled.

Whereas many contributions can be found on the UTPP in the literature, its bounded version UTPP-B, although its recognized practical relevance, has received a limited attention. The uncapacitated version of the TPP was originally introduced by Burstall [1] and Ramesh [9] and has found several applications in scheduling and routing contexts. The problem is known to be *NP-hard* in the strong sense, reducing to the Traveling Salesman Problem (TSP) when each market offers a product which is not provided by the remaining ones. It also contains the uncapacitated facility location problem and the set cover problem as special cases. Due to its computational hardness several heuristic procedures have been proposed to solve both the capacitated and the uncapacitated version of the problem (see, for instance, Golden et al. [3], Ong [7], Pearn and Chien [8], Voss [15], Renaud et al. [11] and Mansini et al. [6]). Branch and Bound exact algorithms have been studied by Singh and van Oudheusden [14] while recently Laporte et al. [5] have introduced a branch-and-cut procedure for the capacitated version of the problem which is able to solve problems with up to 200 markets and 150 products.

On the contrary, the bounded version UTPP-B is cited and analyzed in very few works. In Riera-Ledesma and Salazar-González [13] the authors have introduced the bi-objective TPP, i.e. a bi-criteria version of the TPP where minimizing the purchasing cost and the traveling distances are two separate objectives. The bi-objective is a generalization of the TPP whose solution provide insight into the trade-off between the two costs. The authors tackle the problem by generating the set of all supported and non-supported efficient points in the objective space. For each efficient point in the objective space a Pareto optimal solution in the decision space is computed by solving a single-objective problem while bounding the remaining cost function. In the case the single objective is represented by traveling costs while purchasing costs are bounded in the constraints the problem solved is the UTPP-B. The authors provide exact solutions by means of a Branch and Cut approach for instances with up to 100 markets and 200 different products. The uncapacitated TPP with budget-constraint is also analyzed in Ravi and Salman [10] as an application of telecommunication network design. In this paper the authors introduce an algorithm for the bi-criteria version of the UTPP with metric distances based on the rounding of an LP relaxation solution whose worst-case ratio is poly-logarithmic. They also show that for a special case of the UTPP which models the ring-star network design problem with proportional costs a constant factor approximation algorithm exists.

In the present paper, we propose two solution algorithms for the UTPP-B. To the best of our knowledge these procedures represent the first heuristics proposed in the literature to solve this bounded version of the TPP. In particular, the first procedure is based on an enhanced local search scheme where a neighbor solution is obtained from the current one by removing $l$-consecutive markets and inserting as many markets as required to restore solution feasibility. The heuristic uses a neighborhood similar to that proposed in [12] with the addition of a simple tabu structure introduced to avoid cycling. We will refer to such algorithm as enhanced local search since with respect to a pure local search scheme it varies the neighborhood during the search. The procedure is very efficient and, in those instances where budget constraint is not too tight (i.e. not too close to the minimum purchasing cost), it is also able to provide very effective solutions. The second algorithm is based on a Variable Neighborhood Search scheme (see Hansen and Mladenović [4] and references therein) where a sequence of neighborhoods $N_q(\cdot)$ is introduced each one characterized by a different parameter $q$ representing the number of markets randomly inserted in a current solution. This procedure has the merit to show how a simple and straightforward application of a VNS structure along with an effective local search can be enough to produce high quality solutions.

The paper is organized as follows. In Sect. 2, the mathematical formulation of the UTPP with budget constraint is described. The solution algorithms are introduced in Sect. 3, while Sect. 4 is devoted to computational results. We have tested the proposed algorithms on benchmark instances of the uncapacitated TPP to which a budget constraint has been added. More precisely, algorithms performance has been compared to the optimal solution value provided in [13], where the authors deal with the UTPP with budget constraint to compute Pareto optimal solutions of a bi-objective TPP. Finally, conclusions and future developments are drawn in Sect. 5.

## 2 Problem Formulation

The *uncapacitated TPP-B* can be formally defined on an undirected complete graph $G = (V, E)$, where $V := \{0\} \cup M$ is the vertex set with vertex 0 representing the depot and $E := \{(i, j) : i, j \in V, i < j\}$ is the edge set. The problem looks for a simple cycle in $G$, starting at and ending to vertex 0, which visits a subset of vertices at a minimum traveling cost while purchasing all products at a global purchasing cost which does not exceed the budget threshold $B$. To avoid explicit consideration of trivial cases we have assumed that none of the markets can provide each of the $n$ products at the cheapest cost.

To formulate the uncapacitated version of the problem we have introduced the following three sets of decision variables:

$$x_{ij} := \begin{cases} 1 & \text{if edge } (i, j) \text{ belongs to the optimal cycle} \\ 0 & \text{otherwise;} \end{cases} \quad \forall (i, j) \in E,$$

$$y_i := \begin{cases} 1 & \text{if market } i \text{ belongs to the optimal cycle} \\ 0 & \text{otherwise;} \end{cases} \quad \forall i \in V,$$

$$z_{ki} = \begin{cases} 1 & \text{if product } k \text{ is purchased at market } i \\ 0 & \text{otherwise;} \end{cases} \quad \forall k \in K, \forall i \in M_k.$$

For any $S \subset V$, we define as $E(S) := \{(i, j) \in E : i, j \in S, i < j\}$ and as $\delta(S) := \{(i, j) \in E : i \in S, j \in V \setminus S\}$. Moreover, we indicate as $M^*$, the set $M^* := \{0\} \cup \{i \in M : \exists k \in K \text{ such that } i \in M_k \text{ and } |M_k| = 1\}$, representing the markets which necessarily have to make part of any feasible solution.

The uncapacitated TPP with budget-constraint can be formulated as follows:

$$(UTPP\text{-}B) \quad \min \quad v := \sum_{(i,j)\in E} c_{ij}x_{ij} \tag{1}$$

$$\sum_{(i,j)\in\delta(\{i\})} x_{ij} = 2y_i \quad \forall i \in V \tag{2}$$

$$\sum_{(i,j)\in\delta(S)} x_{ij} \geq 2y_t \quad \forall S \subseteq M \text{ and } \forall t \in S \tag{3}$$

$$\sum_{k\in K}\sum_{i\in M_k} f_{ki}z_{ki} \leq B \tag{4}$$

$$\sum_{i\in M_k} z_{ki} = 1 \quad \forall k \in K \tag{5}$$

$$z_{ki} \leq y_i \quad \forall k \in K \text{ and } \forall i \in M_k \tag{6}$$

$$y_i = 1 \quad \forall i \in M^* \tag{7}$$

$$x_{ij} \in \{0, 1\} \quad \forall (i, j) \in E \tag{8}$$

$$y_i \in \{0, 1\} \quad \forall i \in V \setminus M^* \tag{9}$$

$$z_{ki} \in \{0, 1\} \quad \forall k \in K \text{ and } \forall i \in M_k \tag{10}$$

Objective function (1) establishes the minimization of the routing costs. The set of constraints (2) ensures that the degree of each market $i$ in the solution ($y_i = 1$) has to be equal to two, i.e. each selected market is visited only once. The inequalities (3) ensure connectivity. Constraint (4) imposes a budget threshold equal to $B$ on the total purchasing costs. Equalities (5) guarantee that each product $k$ is purchased, whereas for inequalities (6) it is not possible to purchase a product $k$ at an unvisited market. Equalities (7) impose that all the markets belonging to set $M^*$ are necessarily selected in any feasible solution. Constraints (8)–(10) are binary conditions on the variables.

# 3 Solution Algorithms

In this section we describe the two algorithms proposed to solve the UTPP-B. Given a current solution $s$, we call external the markets which do not belong to it and define as *Tabu* the set of markets in $s$ which are forbidden to exit the solution in the current iteration.

The first proposed algorithm is called *EJEMO* (*EJEct* and *MOve*). Its detailed description is provided in Fig. 1. *EJEMO* receives as input a feasible solution $s^I$ and provides as output a new, possibly improved, feasible solution $s^E$. The algorithm consists in an enhanced local search procedure which moves from a current solution $s$ to its neighbor $s'$ by removing a chain of $l$ markets and inserting as many markets as required to restore feasibility (procedure ChainEjection$(s, l)$). Inserted markets (set $S$) become tabu for a maximum of $r_{max}$ iterations during which they cannot be removed from the current solution. If the new solution $s'$ has a better value with respect to $s$, the parameter $l$ (initially set to its maximum value $l_{max}$) remains unchanged (internal while loop) and the algorithm moves from $s$ to $s'$. Otherwise the parameter $l$ is reduced and the search is restarted from $s$. When $l$ reaches its minimum value $l_{min}$ (for loop), the algorithm diversifies the search generating a new solution $s''$ by using the procedure Shaking$(s, h)$. This procedure randomly inserts in the current solution $s$ as many markets as necessary to increase its traveling costs by at least a predefined percentage $h$ (initially set to its maximum value $h_{max}$). The procedure aims at possibly escaping from a local minimum by leading the search towards unexplored regions of the solution space. After the insertion of the new markets, the heuristic starts again with $l$ set to a new initial value $l_{max}$ computed as a fixed percentage of the markets making part of the current solution, whereas the diversification parameter $h$ is reduced by a constant percentage $h_0$. The algorithm's stopping rule is given by $h = h_{min}$ (main while loop).

Procedure ChainEjection$(s, l)$ is a two-step routine. The first step tries to reduce the length of the current cycle $s$ by removing the chain of $l$ consecutive markets that yields the largest reduction in the traveling costs. At this aim the procedure finds the set $P(s)$ of all possible paths with $l + 1$ edges belonging to the current solution $s$ and which do not contain markets belonging to $M^*$ or markets which are tabu (since they have been inserted in one of the previous iterations). If $P$ is void an aspiration rule is applied which allows to consider tabu markets for paths construction. For each path $p \in P(s)$, the procedure calculates the routing cost reduction produced by its removal and the path yielding the highest reduction is removed by directly joining its extreme vertices. The second step controls if the new cycle is feasible and,

**Fig. 1.** Pseudo-code of *EJEMO* algorithm

---

INPUT:

initial solution $s^I$.

OUTPUT:

final solution $s^E$.

MAIN LOOP:

$h := h_{max}, r := r_{max};$

$s^E := s^I, v(s^E) := v(s^I);$

$s := s^E, v(s) := v(s^E);$

**while** $(h > h_{min})$ {

    **for** $(\, l := l_{max}; l > l_{min}; l := l - 1\,);$ {

        $s' := \mathbf{ChainEjection}(s, l);$

        Let $S$ be the set of markets inserted to restore feasibility;

        **if** $r = 0$ **then** $\{\, r := r_{max}\, \};$

        Add markets in $S$ to $Tabu$ with tabu tenure $r$;

        $r := r - 1;$

        **while** $(\, v(s') < v(s)\,)$ {

            $s := s', v(s) := v(s');$

            $s' := \mathbf{ChainEjection}(s, l);$

            let $S$ be the set of markets inserted to restore feasibility;

            **if** $r = 0$ **then** $\{\, r := r_{max}\, \};$

            Add markets in $S$ to $Tabu$ with tabu tenure $r$;

            $r := r - 1;$

        } **end while**

        **if** $v(s) < v(s^E)$ **then** $\{\, s^E := s, v(s^E) := v(s)\, \};$

    } **end for**

    $s'' := \mathbf{Shaking}(s, h);$

    $s := s'', v(s) := v(s'')\,;$

    $h := h - h_0;$

} **end while**

---

if not, it restores feasibility by consecutively applying, if necessary, two feasibility procedures (`GetAll(s)` and `BudgetRestore(s)`).

Procedure `GetAll(s)` receives as input a solution $s$ which is infeasible with respect to products purchasing. It iteratively inserts external markets one after the other until all products result to be bought. Since the procedure aims at satisfying products demand while minimizing routing costs, the markets to enter the solution are selected and inserted according to cheapest insertion rule and without considering purchasing costs. All inserted markets become *tabu* with a tabu tenure value $r$ depending on the iteration. Every $r_{max}$ iterations the tabu tenure $r$ reduces to zero and the set *Tabu* becomes void. This means that when entering a solution a market will remain tabu for a number of iterations equal to the current value of parameter $r$ where $r$ is less than or equal to $r_{max}$.

The solution finally provided by GetAll($s$) may not be feasible with respect to the budget threshold constraint. In such case the procedure BudgetRestore($s$) is applied. As a first step, such procedure finds all the subsets of markets external to solution $s$ with cardinality one and two such that, if singularly inserted in the current solution, guarantee the budget to be satisfied. Then, the subset which allows to get feasibility with the smallest increase in the traveling costs, is added to the current solution. Markets insertion is made by *cheapest insertion* rule. If no subset of cardinality less than or equal to two is enough to restore budget feasibility, the procedure iteratively inserts one by one the external markets, previously sorted according to a saving cost rule, until the budget constraint is satisfied. More precisely, when inserting one or two markets in a solution, the corresponding saving in the purchasing costs is exactly computed as the difference between the total purchasing costs before and after markets insertion. The markets inserted become *tabu* for a number of iterations equal to $r$ as described above.

*EJEMO* is very quick and has the main advantage of depending on only two main parameters: the one establishing the length of the chain to be removed (parameter $l$) and that controlling diversification and imposing a maximum number of restarts (parameter $h$). For parameter $l$ we have also tested the variant of the algorithm in which $l$ is initially set to its minimum value $l_{min}$ and then increased up to $l_{max}$. Since this version has provided a worse performance we have abandoned it.

Finally, the idea of removing chains of markets is also used in the algorithm proposed in [12]. With respect to such local search heuristic in our case markets inserted into a solution become tabu with a predefined tabu tenure. The length of the tabu list is not constant but reduces at a constant rate from $r_{max}$ to zero. As soon as tabu tenure is zero it is reset to its maximum value $r_{max}$.

The second proposed algorithm uses a modification of the previous heuristic as a local search to be applied within a Variable Neighborhood Search (VNS) scheme. Figure 2 shows the pseudo-code of the implemented algorithm.

In a Variable Neighborhood Search scheme a sequence of neighborhoods $N_q(\cdot)$ is introduced each one characterized by a different parameter $q$. If intensification of the search is required, the initial value of $q$ is set to a minimum value so that the closest neighborhood is selected first. The parameter will be increased only if no solution improvement can be obtained. If diversification of the search is required, the initial value for $q$ is set to a maximum value and then decreased.

In our case the algorithm explores increasing neighborhoods of a current solution (the VNS parameter $q$ is initialized to its minimum value $q_{min}$) and moves from this solution to a new one restarting the whole procedure if and only if an objective function improvement has been obtained. The algorithm receives as input an initial

**Fig. 2.** Pseudo-code of *Variable Neighborhood Search* algorithm

---

INPUT:

initial solution $s^I$.

OUTPUT:

final solution $s^{VNS}$.

MAIN LOOP:

$s^{VNS} := s^I; v(s^{VNS}) := v(s^I);$

**for** $(q := q_{min}; q <= q_{max}; q := q + q_{step})$ {

    $s' := \mathbf{RandomIns}(s^{VNS}, q);$

    **for** $(l := l_{max}; l > l_{min}; l := l - 1)$ {

        $s'' := \mathbf{Improvement}(s', l);$

        **if** $( v(s'') < v(s') )$ {

            $s' := s'';$

            $v(s') := v(s'');$

            $l := l_{max};$

        } **end if**

    } **end for**

    **if** $( v(s') < v(s^{VNS}) )$ {

        $s^{VNS} := s', v(s^{VNS}) := v(s');$

        $q := q_{min};$

    } **end if**

} **end for**

---

solution $s^I$ and provides as output the possibly improved solution $s^{VNS}$ which is initialized to $s^I$.

Given $q$ and the solution $s^{VNS}$, a new solution $s'$ belonging to the neighborhood $N_q(s^{VNS})$ is generated from $s^{VNS}$ by inserting $q$ external random markets (RandomIns($s^{VNS}, q$) procedure). Insertion is made according to the cheapest insertion rule. Then a local search procedure is applied to $s'$. The local search routine starts by setting the parameter $l$ to its maximum value $l_{max}$ (computed as a predefined percentage of the markets making part of the current solution). Then procedure Improvement() is called. As procedure ChainEjection() in *EJEMO*, Improvement($s', l$) routine is based on the ejection of $l$ consecutive markets from a solution $s'$. However, in this case, the chain to be ejected is not selected a priori as the one which yields the highest travel costs reduction, but the method iteratively evaluates each feasible solution which can be obtained by removing $l$ consecutive markets and by inserting as many external markets as necessary to restore the feasibility (through GetAll() and BudgetRestore() procedures). As far as an improving solution is obtained the evaluation of $l$-markets chains ejection is interrupted (first improvement rule), the algorithm moves to the new solution, while the parameter $l$ is reset to a new maximum value $l_{max}$ (depending on the number of markets in the current solution) and the local search is restarted. If no improvement is achieved, after evaluating all feasible solutions obtainable by the elimination of $l$ consecutive markets, the parameter $l$ is decreased by one unit and the procedure is repeated. When $l = l_{min}$ the local search stops providing the best solution $s'$ found. If this solution is better than the incumbent one $s^{VNS}$, the algorithm moves to this new solution and the VNS procedure restarts with $q := q_{min}$ and $s^{VNS} := s'$, otherwise the parameter $q$ is increased by a value $q_{step}$. Parameter $q$ controls search intensification and diversification. When no better solutions can be found $q$ is increased allowing the search to explore farthest and possibly more promising regions of the solution space. The algorithm stops when $q$ takes its maximum value $q_{max}$. The proposed VNS is not as efficient as EJEMO algorithm but, on average, provides more effective solutions.

Both proposed solution algorithms receive as input an initial feasible solution $s^I$ consisting of two steps. In the first one a construction heuristic is applied: the markets are sorted according to a predefined rule and then added one by one to a cycle until all products have been purchased. In the second step, the previously obtained solution is improved by possibly reducing its traveling costs with a TSP-improving heuristic. If, at step one, the constructed solution results to be infeasible with respect to the budget threshold $B$, the BudgetRestore() routine is called. In the construction heuristic we have used the nearest neighborhood rule to select the market which has to be inserted next and we have started insertion with markets belonging to set $M^*$. As TSP-improving heuristic we have applied GENIUS algorithm (see Gendrau et al. [2]).

## 4 Computational Results

Proposed algorithms have been coded in C and run on a PC Pentium with 3.5 GHz and 2 GB of RAM.

Since no instances for the TPP with budget constraint are available in the literature, we have used the Euclidean instances proposed by Laporte et al. [5] for the uncapacitated TPP (Class 3 instances). We have set the budget constraints equal to the optimal purchasing costs provided by Riera Ledesma and Salazar Gonzalez in [13] when solving the bi-objective formulation of the Uncapacitated TPP on the same set of instances. In this paper, the authors have developed an iterative procedure to determine the solutions efficient set of their bi-criteria problem in the objective space. For each efficient point in the objective space, a Pareto optimal solution in the decision space is computed. The first step of this general method determines, for each instance, two initial efficient points through a two phases procedure which optimizes hierarchically the two objective functions represented by the purchasing and the routing costs. More precisely, an efficient point is computed by first minimizing the purchasing costs over all markets and then by setting this value as a constraint in a problem which minimizes the routing costs. The other one is obtained by first solving the problem that looks for the cycle at minimum cost while purchasing all products and then by finding the minimum purchasing costs over the subset of selected markets.

For each efficient point obtained by solving a given instance of the bi-objective problem a corresponding instance for the UTTP-B is generated by setting its budget threshold $B$ equal to the optimal purchasing costs of the efficient point. This means that for each instance of Class 3 two different instances for the UTPP-B are created. Notice that while the instance corresponding to the first efficient point will have a tight budget threshold (the optimal purchasing costs correspond to the minimum purchasing costs), this is not the case for that corresponding to the second one. It is immediate to see that the optimal solutions for these UTPP-B instances correspond to the optimal traveling costs of the corresponding efficient points. We will identify the instances corresponding to the first efficient point and characterized by a budget equal to the minimum purchasing costs as *tight-budget instances*, while we will refer to those instances associated to the second efficient point as *untight-budget instances*.

We recall that instances in Class 3 are characterized by integer coordinate vertices generated in a $[0,1,000] \times [0,1,000]$ square according to a uniform distribution and Euclidean distances. Each product $k$ is associated with $|M_k|$ randomly selected markets, where $|M_k|$ is randomly generated in $[1, |V| - 1]$. Product prices are generated in the interval $[1,500]$ according to a discrete uniform distribution. In [13] the authors provide optimal solutions (initial efficient points) for instances with $m = 50,100,150,200$ markets and $n = 50,100$ products. For each combination of $m$ and $n$ they solve five different instances made available on the web page *http://webpages.ull.es/users/jjsalaza*. This means that, for our problem, we have generated and solved 80 instances altogether, 40 of which are tight-budget instances and 40 untight-budget ones. Moreover, given the random nature of the proposed solution algorithms, each of them has been run 5 times (5 trials) over each instance. The following two tables provide the average, min and max percentage errors with respect to the optimal solution value out of these 5 trials for the 40 tight-budget instances (Table 1) and for the 40 untight-budget ones (Table 2), respectively.

Each table is divided into two parts, the first one provides the computational results for heuristic *EJEMO* whereas the second one those for the VNS procedure. The column headings in each table have the following meaning: symbol # identifies, given $m$ and $n$, the instance solved; $B$ provides the budget threshold (the optimal purchasing costs of the corresponding efficient point); *opt* gives the optimal solution value (optimal traveling costs); *aver.*, *min* and *max* refers to the average, the minimum and the maximum percentage error of the heuristic solution value from the optimal one out of the 5 trials; finally, *sec* provides the average computational time in seconds out of the 5 trials.

For heuristic *EJEMO* we have set the parameters $h_{max} = 35\%$ and $h_{min}$ and $h_0$ equal to 1%. The decision has been taken after trying different values for the maximum percentage $h_{max}$ equal to 20%, 25%, 40%, and 50%, respectively. The maximum value for the tabu tenure $r_{max}$ has been set to 4, while parameter $l_{max}$ and

**Table 1.** Computational results: tight-budget instances

| n | m | # | B | opt | EJEMO aver. | min | max | sec | VNS aver. | min | max | sec |
|---|---|---|---|---|---|---|---|---|---|---|---|---|
| 50 | 50 | 1 | 76 | 3629 | 3.25 | 2.65 | 3.69 | 1.99 | 0.71 | 0.61 | 0.74 | 4.12 |
| | | 2 | 66 | 3511 | 5.74 | 3.59 | 9.83 | 1.88 | 1.12 | 0.00 | 3.25 | 4.23 |
| | | 3 | 70 | 3164 | 1.99 | 1.04 | 3.60 | 1.77 | 0.05 | 0.00 | 0.25 | 3.95 |
| | | 4 | 80 | 3439 | 1.19 | 0.55 | 1.48 | 1.81 | 0.00 | 0.00 | 0.00 | 1.87 |
| | | 5 | 72 | 4116 | 5.52 | 4.64 | 6.75 | 2.04 | 0.47 | 0.00 | 2.36 | 3.28 |
| | 100 | 1 | 151 | 4050 | 2.20 | 0.49 | 3.04 | 2.72 | 0.00 | 0.00 | 0.00 | 11.18 |
| | | 2 | 137 | 4442 | 6.66 | 5.11 | 7.99 | 2.75 | 0.10 | 0.00 | 0.18 | 9.74 |
| | | 3 | 134 | 4174 | 17.48 | 8.53 | 22.76 | 2.76 | 0.68 | 0.00 | 1.92 | 10.97 |
| | | 4 | 159 | 4117 | 2.79 | 1.51 | 3.84 | 2.82 | 0.00 | 0.00 | 0.00 | 6.46 |
| | | 5 | 145 | 4110 | 5.51 | 4.70 | 7.20 | 2.77 | 0.00 | 0.00 | 0.00 | 7.30 |
| | 150 | 1 | 224 | 4851 | 9.35 | 7.87 | 11.44 | 6.31 | 2.80 | 1.46 | 4.16 | 13.53 |
| | | 2 | 199 | 4434 | 8.21 | 4.53 | 13.01 | 5.46 | 1.16 | 0.59 | 1.65 | 14.73 |
| | | 3 | 193 | 4609 | 2.26 | 1.48 | 3.62 | 5.76 | 0.00 | 0.00 | 0.02 | 17.49 |
| | | 4 | 222 | 4692 | 2.22 | 2.22 | 2.22 | 5.93 | 0.00 | 0.00 | 0.00 | 19.67 |
| | | 5 | 215 | 5018 | 8.41 | 6.30 | 12.71 | 6.70 | 0.00 | 0.00 | 0.00 | 20.49 |
| | 200 | 1 | 299 | 4816 | 4.94 | 3.09 | 7.56 | 10.51 | 0.13 | 0.00 | 0.25 | 29.56 |
| | | 2 | 258 | 4564 | 5.67 | 3.22 | 9.14 | 9.21 | 0.52 | 0.00 | 2.61 | 29.91 |
| | | 3 | 262 | 4259 | 8.32 | 7.54 | 10.24 | 9.52 | 0.01 | 0.00 | 0.02 | 31.96 |
| | | 4 | 291 | 4684 | 1.96 | 1.96 | 1.96 | 10.48 | 0.00 | 0.00 | 0.00 | 22.88 |
| | | 5 | 282 | 5184 | 6.25 | 4.44 | 7.66 | 9.60 | 1.01 | 0.00 | 1.81 | 22.52 |
| 100 | 50 | 1 | 58 | 2930 | 5.24 | 3.38 | 6.83 | 15.33 | 0.00 | 0.00 | 0.00 | 32.47 |
| | | 2 | 53 | 3091 | 1.99 | 0.49 | 3.49 | 12.85 | 0.46 | 0.03 | 1.59 | 23.84 |
| | | 3 | 71 | 2977 | 2.57 | 0.97 | 6.18 | 13.93 | 0.19 | 0.00 | 0.37 | 33.75 |
| | | 4 | 62 | 3311 | 7.70 | 2.66 | 11.87 | 13.67 | 0.38 | 0.00 | 1.00 | 37.85 |
| | | 5 | 67 | 2890 | 1.42 | 1.31 | 1.63 | 13.27 | 0.00 | 0.00 | 0.00 | 21.59 |
| | 100 | 1 | 123 | 3842 | 5.23 | 0.96 | 9.24 | 4.09 | 0.03 | 0.03 | 0.03 | 137.36 |
| | | 2 | 121 | 3767 | 9.25 | 5.89 | 12.37 | 3.69 | 1.28 | 0.00 | 6.24 | 197.63 |
| | | 3 | 137 | 3725 | 7.21 | 4.62 | 9.96 | 4.22 | 0.23 | 0.00 | 0.62 | 142.59 |
| | | 4 | 121 | 3749 | 6.57 | 3.55 | 9.98 | 3.99 | 1.06 | 0.00 | 5.25 | 167.48 |
| | | 5 | 129 | 3893 | 2.17 | 1.64 | 2.80 | 3.30 | 0.11 | 0.00 | 0.28 | 141.94 |
| | 150 | 1 | 177 | 4312 | 10.63 | 4.17 | 13.01 | 4.28 | 0.93 | 0.00 | 4.66 | 324.90 |
| | | 2 | 188 | 4584 | 8.89 | 3.77 | 13.46 | 4.71 | 1.34 | 0.00 | 6.65 | 365.70 |
| | | 3 | 191 | 4724 | 8.82 | 5.86 | 12.47 | 4.67 | 1.36 | 0.28 | 2.46 | 251.93 |
| | | 4 | 189 | 4649 | 10.90 | 7.14 | 13.85 | 5.40 | 1.01 | 0.00 | 2.41 | 234.38 |
| | | 5 | 192 | 4502 | 7.01 | 2.95 | 10.35 | 4.77 | 0.75 | 0.33 | 0.71 | 264.40 |
| | 200 | 1 | 230 | 4124 | 7.49 | 4.92 | 10.14 | 2.79 | 0.00 | 0.00 | 0.00 | 253.17 |
| | | 2 | 261 | 5120 | 9.79 | 7.01 | 11.19 | 3.16 | 1.20 | 0.29 | 2.09 | 271.57 |
| | | 3 | 246 | 4660 | 8.51 | 4.59 | 10.52 | 2.86 | 1.06 | 0.00 | 1.59 | 272.74 |
| | | 4 | 245 | 4673 | 7.01 | 4.86 | 8.88 | 3.06 | 0.03 | 0.00 | 0.00 | 251.32 |
| | | 5 | 258 | 4699 | 6.01 | 2.43 | 9.45 | 3.13 | 0.44 | 0.00 | 0.87 | 336.05 |
| Average | | | | | 6.11 | 3.72 | 8.44 | 5.75 | 0.52 | 0.09 | 1.40 | 101.21 |

**Table 2.** Computational results: untight-budget instances

| n | m | # | B | opt | EJEMO aver. | min | max | sec | VNS aver. | min | max | sec |
|---|---|---|---|-----|-------------|-----|-----|-----|-----------|-----|-----|-----|
| 50 | 50 | 1 | 176 | 1684 | 0.00 | 0.00 | 0.00 | 1.99 | 0.00 | 0.00 | 0.00 | 1.00 |
| | | 2 | 143 | 935 | 0.00 | 0.00 | 0.00 | 1.88 | 0.00 | 0.00 | 0.00 | 0.66 |
| | | 3 | 158 | 1422 | 0.00 | 0.00 | 0.00 | 1.77 | 0.04 | 0.00 | 0.00 | 0.60 |
| | | 4 | 242 | 1228 | 0.00 | 0.00 | 0.00 | 1.81 | 0.00 | 0.00 | 0.00 | 0.53 |
| | | 5 | 269 | 1317 | 0.00 | 0.00 | 0.00 | 2.04 | 0.00 | 0.00 | 0.00 | 1.00 |
| | 100 | 1 | 277 | 2120 | 2.03 | 1.84 | 2.08 | 2.72 | 0.96 | 0.00 | 1.60 | 0.87 |
| | | 2 | 309 | 1854 | 0.00 | 0.00 | 0.00 | 2.75 | 0.00 | 0.00 | 0.00 | 0.88 |
| | | 3 | 238 | 1614 | 0.00 | 0.00 | 0.00 | 2.76 | 0.00 | 0.00 | 0.00 | 1.03 |
| | | 4 | 287 | 2822 | 0.35 | 0.00 | 0.89 | 2.82 | 0.00 | 0.00 | 0.00 | 1.47 |
| | | 5 | 263 | 2340 | 0.29 | 0.00 | 1.24 | 7.82 | 0.00 | 0.00 | 0.00 | 1.04 |
| | 150 | 1 | 394 | 2409 | 0.00 | 0.00 | 0.00 | 29.20 | 0.14 | 0.00 | 0.71 | 1.27 |
| | | 2 | 343 | 1815 | 0.00 | 0.00 | 0.00 | 5.22 | 0.00 | 0.00 | 0.00 | 1.52 |
| | | 3 | 443 | 1917 | 0.05 | 0.05 | 0.05 | 21.09 | 0.00 | 0.00 | 0.00 | 1.04 |
| | | 4 | 411 | 2143 | 0.00 | 0.00 | 0.00 | 1.71 | 0.00 | 0.00 | 0.00 | 0.92 |
| | | 5 | 362 | 2800 | 0.50 | 0.50 | 0.50 | 12.36 | 0.30 | 0.00 | 0.50 | 1.95 |
| | 200 | 1 | 446 | 2933 | 0.14 | 0.00 | 0.38 | 2.50 | 0.27 | 0.00 | 0.68 | 1.94 |
| | | 2 | 447 | 1989 | 0.03 | 0.00 | 0.05 | 23.44 | 0.17 | 0.00 | 0.85 | 2.04 |
| | | 3 | 466 | 1860 | 0.00 | 0.00 | 0.00 | 1.98 | 0.00 | 0.00 | 0.00 | 1.36 |
| | | 4 | 451 | 2414 | 0.00 | 0.00 | 0.00 | 16.87 | 0.00 | 0.00 | 0.00 | 4.75 |
| | | 5 | 478 | 3148 | 0.10 | 0.10 | 0.10 | 27.36 | 0.00 | 0.00 | 0.00 | 1.99 |
| 100 | 50 | 1 | 137 | 1335 | 0.18 | 0.00 | 0.30 | 13.17 | 0.00 | 0.00 | 0.00 | 2.63 |
| | | 2 | 186 | 832 | 0.12 | 0.12 | 0.12 | 11.39 | 0.00 | 0.00 | 0.00 | 0.77 |
| | | 3 | 156 | 1475 | 0.00 | 0.00 | 0.00 | 8.23 | 0.00 | 0.00 | 0.00 | 1.47 |
| | | 4 | 146 | 1586 | 1.56 | 1.07 | 2.40 | 6.07 | 0.34 | 0.00 | 0.57 | 4.62 |
| | | 5 | 111 | 2391 | 0.00 | 0.00 | 0.00 | 2.57 | 0.00 | 0.00 | 0.00 | 10.94 |
| | 100 | 1 | 211 | 1919 | 0.60 | 0.31 | 0.68 | 1.58 | 0.18 | 0.05 | 0.68 | 22.94 |
| | | 2 | 266 | 1662 | 1.13 | 0.60 | 1.26 | 3.20 | 0.18 | 0.18 | 0.18 | 7.14 |
| | | 3 | 239 | 1597 | 0.00 | 0.00 | 0.00 | 1.47 | 0.00 | 0.00 | 0.00 | 23.51 |
| | | 4 | 248 | 1406 | 2.39 | 0.21 | 3.84 | 1.51 | 0.00 | 0.00 | 0.00 | 2.60 |
| | | 5 | 249 | 2709 | 0.04 | 0.04 | 0.04 | 3.82 | 0.04 | 0.04 | 0.07 | 12.93 |
| | 150 | 1 | 380 | 1863 | 0.65 | 0.00 | 2.20 | 1.93 | 0.02 | 0.00 | 0.05 | 3.07 |
| | | 2 | 321 | 2505 | 0.01 | 0.00 | 0.04 | 2.17 | 0.00 | 0.00 | 0.00 | 21.88 |
| | | 3 | 421 | 1898 | 0.00 | 0.00 | 0.00 | 1.99 | 0.00 | 0.00 | 0.00 | 1.33 |
| | | 4 | 337 | 2292 | 0.02 | 0.00 | 0.09 | 3.82 | 0.00 | 0.00 | 0.00 | 29.81 |
| | | 5 | 339 | 2848 | 0.01 | 0.00 | 0.04 | 1.90 | 0.00 | 0.00 | 0.00 | 18.23 |
| | 200 | 1 | 408 | 1483 | 3.84 | 3.84 | 3.84 | 3.09 | 1.94 | 0.00 | 3.24 | 37.70 |
| | | 2 | 438 | 2683 | 0.63 | 0.48 | 0.78 | 1.22 | 0.23 | 0.11 | 0.26 | 29.58 |
| | | 3 | 485 | 2364 | 0.00 | 0.00 | 0.00 | 1.61 | 0.00 | 0.00 | 0.00 | 8.24 |
| | | 4 | 429 | 3024 | 0.11 | 0.03 | 0.20 | 2.25 | 0.01 | 0.00 | 0.03 | 14.27 |
| | | 5 | 470 | 2308 | 0.00 | 0.00 | 0.00 | 7.82 | 0.00 | 0.00 | 0.00 | 1.53 |
| Average | | | | | 0.37 | 0.23 | 0.53 | 6.27 | 0.12 | 0.01 | 0.24 | 7.08 |

$l_{min}$ are equal to 10% and 1% of the number of the markets making part of the current solution, respectively.

For algorithm VNS we have easily set $q_{max} = 20$ and $q_{min} = q_{step} = 1$ as the best parameter values, whereas some testing has been required in order to choose the initial value for parameter $l$ which is resulted to be more crucial. At this aim we have solved several instances with different values of parameter $l_{max}$. In Fig. 3 we report the average percentage errors with respect to the optimal solution value found by the VNS approach out of the 5 tight-budget instances (tight-budget line in the graph) and the 5 untight-budget instances (untight-budget line in the graph) with 100 markets and 100 products for five different values of the parameter $l_{max}$ equal to 5, 10, 15, 20, and 25%, respectively. For this testing we have considered only one algorithm trial. We can notice how, in both the cases, the better algorithm performance can be obtained by setting $l_{max}$ equal to 5% of the number of markets in the current solution. This result has been confirmed by other experiments carried out on instances with a different number of markets and products.

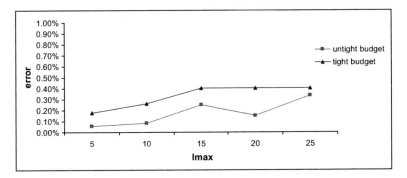

**Fig. 3.** Tuning of the parameter $l_{max}$ over instances with 100 markets and 100 products

By comparing the two tables it is evident how tight-budget instances are the most difficult ones. In such instances *EJEMO* algorithm behaves rather poorly providing an average error equal to 6.11% out of the 40 solved instances (given the five trials). Moreover, it is never able to find the optimal solution value (cfr. column *min* in Table 1). Such bad performances are, on average, due to ChainEjection() procedure which fails to eject the right markets. Indeed, the simple selection rule used by such procedure to choose the chain of markets to be removed behaves acceptably good only when the number of good quality solutions is large enough as for untight-budget instances. On the contrary the Variable Neighborhood Search procedure yields an average error equal to 0.52% and finds the optimal solution value in 32 out of the 40 instances solved (considering the 5 trials). From the efficiency point of view, the computational time required by *EJEMO* is almost constant for all instances (its average value out of all instances is equal to 5.75 s), whereas the average computational time for the VNS algorithm grows quickly with the problem size reaching an average value of 276.97 s for instances with 100 markets and 200 products.

Both heuristics performance improves when untight-budget instances are taken into account. Table 2 shows that $EJEMO$ provides an average percentage error out of all instances equal to 0.37% with a maximum error never larger than 3.84%. Moreover, given the five trials, it is able to find the optimal solution values in 27 out of 40 solved instances. The results obtained by the VNS approach are even better: this procedure yields an average percentage error equal to 0.12% out of the 40 solved instances and gets the optimal solution value (given the five trials) in all but three instances. The average computational times are close to 7 s for both the heuristics.

To conclude, if we consider tight-budget instances the ranking of the proposed algorithms in terms of solution values suggests VNS as a first choice followed by $EJEMO$. Nevertheless, if running time is of high relevance $EJEMO$ may be used even if at a price of a lower performance. On the contrary for untight-budget instances procedure VNS is the dominant one providing solutions which are more effective and as efficient as those yielded by $EJEMO$.

## 5 Conclusions

In this paper we have analyzed a variant of the uncapacitated Traveling Purchasing Problem based on the minimization of the traveling costs while bounding the purchasing costs in the constraints. For this problem, which has found very little attention in the literature, we have introduced two different solution algorithms. The first one is an enhanced local search procedure which generates neighbor solution by ejecting chain of consecutive vertices from the current solution and is based on a simple tabu structure to avoid cycling ($EJEMO$ algorithm). The second one is a straightforward implementation of a Variable Neighborhood Search scheme which uses a modification of the previous algorithm as local search routine.

Computational results on benchmark instances created for the problem have shown how $EJEMO$ is an efficient and effective procedure when the solved instances are characterized by a large budget threshold (untight-budget instances). On the other side, the procedure may produce bad results when the number of good quality solutions is small as for the tight-budget instances.

On the contrary the VNS approach always provides effective solutions and in the case of untight-budget instances it is also efficient showing an average computational time equivalent to that of $EJEMO$.

As future developments we will study a dynamic version of the problem for the capacitated case. As a first step in this direction we will analyze the performance of the proposed heuristic algorithms as well as that of a new multi-start variant of the VNS approach when applied to the capacitated version of the problem. The best resulting approach will be used as solution algorithm for the dynamic routing problem.

# References

[1] Burstall M.R. (1966) A heuristic method for a job sequencing problem. *Operational Research Quarterly* 17: 291–304.

[2] Gendreau M., Hertz A., Laporte G. (1992) New insertion and postoptimization procedures for the Traveling Salesman Problem. *Operations Research* 40(6): 1086–1094.

[3] Golden B. L., Levy L., Dahl R. (1981) Two generalizations of the Traveling Salesman Problem. *Omega* 9: 439–445.

[4] Hansen, P., Mladenović N. (2001) Variable Neighborhood Search: principles and applications. *European Journal of Operational Research* 130: 449–467.

[5] Laporte G., Riera-Ledesma J., Salazar-Gonzàlez J.J. (2003) A branch-and-cut algorithm for the undirected Traveling Purchaser Problem. *Operations Research* 51(6): 940–951.

[6] Mansini R., Pelizzari M., Saccomandi R. (2005) An effective tabu search algorithm for the capacitated Traveling Purchaser Problem. Technical Report 2005-10-49, Dipartimento di Elettronica per l'Automazione, University of Brescia.

[7] Ong H.L. (1982) Approximate algorithms for the Traveling Purchaser Problem. *Operations Research Letters* 1: 201–205.

[8] Pearn W. L., Chien R. C. (1998) Improved solutions for the Traveling Purchaser Problem. *Computers and Operations Research* 25: 879–885.

[9] Ramesh T. (1981) Traveling purchaser problem. *Operations Research* 18: 78–91.

[10] Ravi R., Salman S. (1999) Approximation algorithms for the Traveling Purchaser Problem and its variants in network design. In: Proceedings of the 7th Annual European Symposium on Algorithms, Lecture Notes in Computer Science vol. 1643. Spriger, Berlin: 29–40.

[11] Renaud J., Boctor F. F., Laporte G. (2003) Heuristics for the traveling purchaser problem. *Computers and Operations Research* 30(4): 491–504.

[12] Riera-Ledesma J., Salazar-González J. J. (2005) A heuristic approach for the Traveling Purchaser Problem. *European Journal of Operational Research* 162(1): 142–152.

[13] Riera-Ledesma J., Salazar-González J. J. (2005) The Biobjective Traveling Purchaser Problem. *European Journal of Operational Research* 160: 599–613.

[14] Singh K. N., Van Oudheusden D. L. (1997) A branch and bound algorithm for the traveling purchaser problem. *European Journal of Operational Research* 97: 571–579.

[15] Voß S. (1996) Dynamic tabu search strategies for the traveling purchaser problem. *Annals of Operations Research* 63: 253–275.